Women in Engineering and Science

Series Editor
Jill S. Tietjen, Greenwood Village, CO, USA

The Springer Women in Engineering and Science series highlights women's accomplishments in these critical fields. The foundational volume in the series provides a broad overview of women's multi-faceted contributions to engineering over the last century. Each subsequent volume is dedicated to illuminating women's research and achievements in key, targeted areas of contemporary engineering and science endeavors. The goal for the series is to raise awareness of the pivotal work women are undertaking in areas of keen importance to our global community.

Margaret Bailey • Laura Shackelford
Editors

Women in Mechanical Engineering

Energy and the Environment

Springer

Editors
Margaret Bailey
Department of Mechanical Engineering
Rochester Institute of Technology
Rochester, NY, USA

Laura Shackelford
English Department
Rochester Institute of Technology
Rochester, NY, USA

ISSN 2509-6427 ISSN 2509-6435 (electronic)
Women in Engineering and Science
ISBN 978-3-030-91545-2 ISBN 978-3-030-91546-9 (eBook)
https://doi.org/10.1007/978-3-030-91546-9

This Springer imprint is published by the registered company Springer Nature Switzerland AG
The registered company address is: Gewerbestrasse 11, 6330 Cham, Switzerland

Preface

One of the most challenging and promising questions that has grown out of the wide-reaching upheavals of 2020 and 2021 amidst the painful realities of the COVID-19 pandemic is the question of what we as individuals, communities, scholars, and societies would like to preserve and transform in the months and years to come, knowing that these events will have lasting impacts on so many dimensions of our lives that were previously taken for granted.

Two faculty members from the Rochester Institute of Technology (RIT)—Margaret Bailey, Ph.D., P.E., Professor of Mechanical Engineering, and Laura Shackelford, Ph.D., Professor of English—edited this unique and timely collection, *Women in Mechanical Engineering: Energy and the Environment*, to jointly reflect on these questions. The book is one volume in the impressive Springer Publishing Women in Engineering Series with Jill Tietjen, P.E. (Society of Women Engineers Fellow and Past-President) as editor-in-chief (read more at https://www.springer.com/gp/campaigns/women-in-engineering). In this edited collection, we gather new visions, feature transformational work, and address pressing challenges in the areas of energy and the environment in Mechanical Engineering. The book highlights the contributions of diverse women engineers who are at the forefront of these changes and addressing them daily, while also navigating research and professional domains that are historically gendered to the disadvantage of women.

Exploring scholarly research, professional trajectories, disciplinary shifts, proven methods, personal insights and attitudes, or a combination of these, the book's chapters tell an important story about the field of Mechanical Engineering in the areas of energy and the environment, as seen from the perspective of these remarkable contemporary women engineers (with degrees in Mechanical Engineering and other engineering disciplines—civil, chemical, nuclear, electrical, environmental, material science, systems, etc.) and from within academe, industry, and government. Individual chapters highlight areas such as renewable energy, batteries and energy storage, power generation and distribution, sustainability, engineering and public policy, combustion and emissions, and engineering education. Chapter authors are members of the National Academies, winners of major awards and recognitions that include National Science Foundation Awards and Fellowships, as well as SWE, ASME, ASEE, and IEEE Award winners and Fellows.

Gathering Stories

The editorial vision and impetus for this collection grew out of just such experiences of seeking out and finding role models and learning from other Mechanical Engineers, both from the past and the present, who had accomplished great things and actively worked to improve women's opportunities and visibility in these fields, as well. Co-editor, Margaret Bailey, found her inspiration from several women—past and present. Below is a first-person account of her inspiration journey which led to the creation of this volume:

I'm from a small, rural, farming community in the western part of the Pocono Mountains in Pennsylvania and I didn't know any engineers when I entered Penn State to study architectural engineering after high school. On my first day at college I arrived early for my chemistry class – it was a large room with about 200 seats. I sat down and got settled in, pulling out my book and pencil. When I looked up and began to watch students filing in, I realized that there were no women among them. After straining my eyes and neck I finally saw a few women in seats scattered here and there in that large space. In my high school the brightest students were at least half women. That was my paradigm which quickly shifted. I spent five years studying engineering and never had one women professor teach me a technical course. And yet, most of the best students in my classes were women. I often wondered why that was and thought that one day maybe I would be an engineering professor. Just like when I was a freshmen in chemistry, I've needed to deliberately look for women to serve as possible engineering role models in my career. About 20 years ago, I found one in a woman named Kate Gleason.

Kate Gleason (1865–1933) was an American engineer and businesswoman known both for being an accomplished woman in the predominantly male field of engineering and for her philanthropy. She is featured in the opening "Trailblazers" historic chapter within this volume and she serves as an example of an amazing woman whose story was forgotten, only to be rediscovered and accurately told 75 years after her death. I came to know of Kate Gleason when I accepted the Kate Gleason endowed chair position at RIT in 2003 and had the opportunity to learn about her from Jan Gleason who was married to Kate's great-nephew. Kate was famous among people like Henry Ford and mentored by Susan B. Anthony and yet her story was erased for about 75 years after her death until Jan Gleason became curious, did research, and pieced back together Kate's story. Getting acquainted with Kate Gleason's story helped fuel my strong interest in history and deep appreciation for storycraft.

Kate Gleason has been nominated for entry into the Women's Hall of Fame in Seneca Falls many times and a strong proponent of these nominations is Jill Tietjen, editor of the volume series where this book appears. I met Jill six years ago and our connection was nearly instant due to mutual interests in history, Kate Gleason, and women in engineering. She told me about this book series and she wondered if I might be interested in editing a volume. About that same time, Laura Shackelford an English professor at RIT who I had known and admired for many years was named

Director of the new Center for Engaged Storycraft. Laura focuses her research on narrative and literary practices and the transformations they have gone through in relation to science since World War II, especially as these sciences impact understandings of women, gender, and race. I thought that it would be wonderful to co-edit a volume for Jill's series with Laura. I'm very grateful that Laura accepted my proposition! Together we created a plan for this volume and the response to our call for chapters during the Covid-19 summer of 2020 was stronger than we had anticipated. The response was overwhelming.

Kate Gleason inspired me to work with Jill and Laura to create this volume. Her story was almost lost and that would have been a loss for all of us. It made me wonder how many other women's stories in the early years of engineering were lost. I find Kate Gleason to be an inspiration and the young people, especially the women, enrolled in the Kate Gleason College of Engineering where I teach are also inspired by her story. I wanted to learn from contemporary women engineers about what their journeys have been like. And I thought that others could learn and become inspired through their words and experiences.

Book Aims and Organization

Our volume on *Women in Mechanical Engineering: Energy and the Environment* has a primary aim of showcasing the important technical scholarship and contributions of women currently working in areas of mechanical engineering. The volume contextualizes the present field of women and their technical scholarship in relation to three inspiring and influential predecessors and then proceeds to showcase remarkable technical contributions of women engineers to the areas of energy and the environment today—here the book is divided into three sections which focus on New Perspectives, Research/Technical, and Career Journeys. The chapters focus on scholarly research, professional trajectories, disciplinary shifts, proven methods, personal insights and attitudes, or a combination of these.

The chapters are written in a narrative style in order to help readers to reflect on how the author's work in engineering has impacted her life and how being a woman in engineering has impacted her work. Authors adopted a variety of approaches, stylistically, in telling their stories. Combined, the chapters tell an important story about the field of Mechanical Engineering, and examples of what you will read in this volume include:

- A hybrid micro narrative of the author's personal and professional journeys which answers questions—how/why they found engineering, how did their career unfold, how did one position or assignment build upon and perhaps complement others, who are their role models, what were key points of transition and connections?

- How the author's work in engineering has impacted their life and how being a woman in engineering has impacted her work? How has being a woman informed their work?
- A sharing of how curiosity, challenges, or other attitudes and modes of thinking and working sparked and motivated the author's work while highlighting sources of inspiration, challenges, successes, and moments of resilience.
- Considerations of what might have facilitated the author's self-efficacy in earlier stages of their career and efforts they made to facilitate, mentor, or guide professionals in the early stages of their careers.
- Exploring personal values and how these map to disciplinary values. Considering which norms within the engineering discipline helped the author succeed and which norms may have possibly hindered their progress.
- Addressing shifts in priorities, practices, and research amidst recent transformations as a result of COVID-19.

Intended Audiences

Chapter authors inspire and inform readers by giving first-hand accounts of their experiences currently working in and transforming mechanical engineering in these diverse areas of energy and the environment. Contributors' chapters provide models for people at all stages of their education, career, and life, helping others learn from their critical decisions, their hopes and dreams, and their strategies of resilience in the face of challenges. The volume's authors address a wide array of potential readers, including:

- Young people interested in Mechanical Engineering: Energy and Environment
- Women in industry as well as academia and other professionals interested in the field of Mechanical Engineering and interested in the careers of current engineers as models for their own career trajectories
- Engineering leaders/educators/professionals seeking a current snapshot of the field from the perspective of nearly 30 women engineers
- Women interested in reading and reflecting on others' experiences in these fields and their strategies for advancing in research, professionally and personally in spite of obstacles
- First year college students who are studying narrative or enrolled in an introduction to engineering course
- People engaged in advocacy efforts in support of women in engineering, and other under-represented populations within engineering
- Young professionals moving from technical to managerial positions within engineering
- People interested in what other things besides engineering that you can do with an Engineering degree

Conclusions

Our hope is that the volume will engage, educate, and inspire readers who may be either very familiar or relatively unfamiliar with engineering. We invite you to vicariously experience what it is like working in these fields and be left with a sense of hope and inspiration. We think you will find these women engineers' stories positively moving, as they provide their first-person perspectives and journeys. We also hope that in presenting contemporary, real-world models and methods of working in Mechanical Engineering in these pivotal areas of energy and environment, these stories can be literally moving, in revealing models or methods that might support your own professional development, resilience, empowerment, and purpose.

As editors, we applaud the authors in creating chapters that raise and answer questions of how being a woman informs their work and embrace the changes that they've documented and promoted through their life stories, thus far.

Rochester, NY, USA

Margaret Bailey
Laura Shackelford

Contents

Section I
Introduction

Energetic Trailblazers: Kate Gleason, Edith Clarke, and Mária Telkes

Jill S. Tietjen and Margaret Bailey

Introduction

The three women featured in this chapter – Kate Gleason (1865–1933), Edith Clarke (1883–1959), and Mária Telkes (1900–1995) – were truly "energetic trailblazers" during their lifetimes and their legacies today are very much alive. They each continue to have a transformative impact well-beyond their individual lives and careers. The chapter is presented in chronological order, describing each of these women's work in related fields and begins during the early decades of the 1900s.

After more than 50 years of action led in large part by Susan B. Anthony of Rochester, New York, the years 1900–1920 were very active years for women's suffrage. At the end of those two decades, during which Anthony died and many others took up the mantle, the USA ratified the Nineteenth Amendment to the US Constitution which granted full voting rights to American women (Equality Day – August 26, 1920).

Kate Gleason was a close friend and mentee of Susan B. Anthony and also lived in Rochester, New York. This fact helps set the stage for a wider view of this moment of women's suffrage activism and the variety of fronts on which women were forging ahead and breaking new ground for women in their own ways. Although Gleason is not as widely recognized as being a part of a larger community of change, she served as a significant pioneer and trailblazer.

While creating and editing this chapter, we wondered if these three women would have known each other. It's likely that Edith Clarke and Mária Telkes knew each other or knew of each other. Both were awarded the Society of Women

J. S. Tietjen (✉)
Technically Speaking, Inc., Greenwood Village, Colorado, USA

M. Bailey
Department of Mechanical Engineering, Rochester Institute of Technology, Rochester, NY, USA
e-mail: Margaret.Bailey@rit.edu

M. Bailey, L. Shackelford (eds.), *Women in Mechanical Engineering*, Women in Engineering and Science, https://doi.org/10.1007/978-3-030-91546-9_1

Engineers' (SWE) Achievement Award within 2 years of each other. SWE was founded in 1949–1950 and, in 1952, Mária Telkes was the first recipient of the new society's Achievement Award. Two years later, in 1954, Edith Clarke received the award. According to SWE's history:

> *"A woman who receives an Achievement Award is often described as a fast tracker, a doer, an achiever, highly motivated. A person who gets things done* [1]". Until 1966, the Achievement Award was the only award SWE presented and *"it was one of the few avenues of honor and acknowledgment women engineers were likely to have received. It remains the Society's highest honor"* [1].

Kate Gleason died in 1933 and therefore never knew of the Society of Women Engineers. However, she was well-known during her lifetime and in 1930 she attended the Second World Power Conference in Berlin, Germany, as the American Society of Mechanical Engineers' representative, along with 3900 other engineers and scientists from at least 38 countries around the world. Professor Albert Einstein of Germany and Dr. H. Foster Bain, Secretary of the American Institute of Mining and Metallurgical Engineers of America [2], gave two of the seven main plenary addresses at the conference [3, 4]. Clarke and Telkes likely knew of this large-scale international conference and may have heard of the woman representing American Society of Mechanical Engineers (ASME).

There are several similarities between the three women and overlapping themes linking their lives and work. All three were alive during World War I. The careers of Gleason and Clarke were impacted by the war primarily due to the exodus of men called into service. "By an irony of fate, war, always bitterly denounced by women, has advanced them in the engineering profession" [5]. Born in 1900, Telkes grew up in Budapest, Hungary, and would have personally felt the war's impact as a teenager growing up in that city which experienced high inflation and food shortages during the war. Other similarities among the three include their early initial interests in science and engineering, the recognition that each achieved during their lifetimes, and that each engaged in numerous professional activities over their careers.

As you read this chapter, we offer some points for consideration:

- How these women strategically navigated professional and academic terrain that was not designed for them, likely quite inhospitable in many ways, and found ways to pursue their work and their passion and stand out among their male counterparts
- How they were all able to trust their knowledge and insight and act decisively on these, especially in contexts where they were all but certainly going to be presumed to be lacking in knowledge, skills, or simply wrong
- How they were able to seek out opportunities that others hadn't been able to see and even to see opportunities precisely because they were presumed to be outsiders in these fields
- How they demonstrated determination that others would not stand in their way
- How they all had an impact that went well beyond their work, career, or professional life at any given moment and influential legacies that continue well beyond their own lives

We have several questions that remain about the three subjects of this chapter. Did these women know each other? Ever meet? What role did MIT play in enabling Clarke and Telkes to enter professional and academic life? What was it like for these women to be working in male disciplines and environments?

We also pose the following thought for consideration in light of the truly remarkable achievements of the three women featured. There is a risk in telling these stories of repeating the myth of the lone, genius, self-made person (usually masculine), and so it might be worth thinking about how these women excelled and were geniuses, certainly, yet also saw themselves as part of a larger community of change – whether for women's rights, affordable housing, education, solar solutions to energy shortages, and so forth.

Kate Gleason (1865–1933)

Kate Gleason's march into history began when she was 11 years old and her older stepbrother Tom died. Tom had worked with their father in his new business, and his death left the family devastated and the business short-handed. Kate decided that being a girl wasn't going to stop her and soon made herself indispensable around the office. Although she studied mechanical engineering in college for a time, business called her back. She became the first woman to sell machine tools and equipment internationally, an early woman bank president, the first woman to take a company out of bankruptcy, and later became a developer of large-scale low-cost housing. An ardent supporter of women's suffrage, Kate was also the first female member of the American Society of Mechanical Engineers (ASME). "By 1918, her work had so impressed the American Society of Mechanical Engineers that she was unanimously elected to membership as its first woman member" [6]. This trailblazing woman's legacy lives on through the Rochester Institute of Technology where the Kate Gleason College of Engineering stands as the first engineering college to be named after a woman. As Gleason said, "I wanted one thing – to demonstrate that a business woman can work as well as a man" and she adopted her motto "*Possum volo*" – "I can if I will" [7, 8].

The oldest child of William Gleason and his second wife, Ellen McDermott Gleason, Catherine Anselm Gleason (Kate) had an older half-brother Tom and three young siblings [9]. Having been mechanically inclined as a child as well as intelligent, Gleason began reading books about engineering and machines by the time she was nine [10]. She was also an ardent tomboy:

> My girlish ambitions were fiercely personal. I felt keenly that girls in this world were accorded second place, and I resented being second. My grandmother always wanted my brothers to take her about; she thought little of me. Friends and neighbors used to watch me, and shake their heads and remark:
>
> "She should have been a boy." They were justified, for I was trying my best to be as nearly a boy as I could. I wore my hair short and straight in a day when girls wore long curls or braids. I played with the boys. They didn't want me, but I earned my right.

> If we were jumping from the shed roofs I chose the highest spot; if we vaulted fences I picked the tallest. I was husky and able, and to this I added a bit of recklessness that carried me through. It took just that added bit of daring to outdo the rest. I carried that lesson into business. A bold front, determination, and the willingness to risk more than the crowd, plus some common sense, and hard work, wins out [8].

Kate's half-brother, Tom, helped her father out in the business – a machine tool shop in Rochester, New York – The Gleason Works [6, 7]. William Gleason had invented the bevel gear planar, which eliminated hand cutting, in 1874 [11]. Shortly after Tom died of typhoid, when Kate was 11, she was terribly saddened to hear her father exclaim, "Oh, if Kate had only been a boy!" The very next Saturday, she "walked down to the shop, mounted a stool and demanded work." Although such behavior was completely outside of the norm for girls of her era, her father might not have objected because suffragist Susan B. Anthony, a resident of Rochester, New York, was a friend of her mother's [8, 9]! Her father gave her some bills to make out and she worked regularly from that point on. When she was 14, she decided that she could do the bookkeeping [8].

At age 19, in 1884, Kate became the first woman engineering student to enroll at Cornell University in the Mechanical Arts (now mechanical engineering) program. But her father's business fell upon hard times and Kate was called home to work [9, 10].

> That was my first big sorrow and my heart broke utterly. I took Father's letter out on the campus and sat under a tree where I thought no one would find me, and wept and wept. I had planned to finish the engineering course. I was the only woman in it, and it meant so much. . .
>
> Since I was nine I had been reading books on machines and engineering; my one year had given me the essentials of the profession. The rest I could do, and did do, for myself. My fierce determination to equal the young men I left at college served as a spur, and I worked with every bit of energy I possessed [8].

Gleason was able to briefly return to Cornell later, but health issues required her to return home [10]. She was never able to complete her degree although she did take additional courses at the Sibley College of Engraving and Mechanics Institute (today the Rochester Institute of Technology) and learned the rest of what she needed to be called an engineer through on-the-job training and self-education [9, 10].

From 1890 to 1913, Kate Gleason (Fig. 1) was the chief sales representative of Gleason Works as well as its Secretary, Treasurer. Her first sales call was to Ohio in 1888 to sell "machines," explaining the complex technology and its benefits to potential customers [6, 11, 12]. She learned two lessons from her first sales call: "There is no sense in being scared of anyone; he may be more scared of you than you of him," and "It pays to be first in any field, if you can" [8]. On subsequent business calls where every client was a man, her reputation often preceded her. Gleason took to heart the advice she had received from one of her mentors, Susan B. Anthony: "Any advertising is good. Get praise if possible, blame if you have to. But never stop being talked about" [8].

Fig. 1 Kate Gleason. (Courtesy of Rochester Institute of Technology)

Tool sales in the USA became significantly depressed due to the Panic of 1893, but the European economy was still strong. Thus, Gleason embarked on a long trip to Europe (there and back took 2 months and she travelled on a cattle steamer, not an ocean liner) and returned to the USA with orders from England, Scotland, France, and Germany [10]. Thus, she became one of the first international sales people in her industry and laid the foundation for a significant portion of the company's business, as is still the case today [6, 9]. Interestingly, due to ill health, her doctor had recommended that she sojourn to Atlantic City, New Jersey, but she chose Europe as she had no customers in Atlantic City [6]! Due to the business downturn, she also recommended to her father that Gleason Works transition to gears and gear planing machines. Five years after this recommendation, Gleason Works was no longer in the tool making business [8].

She became a successful businesswoman in spite of her gender. Gleason was sometimes credited with Gleason Works' inventions that she often vehemently denied creating while highlighting the rightful inventors within the company. In addition, she was not at all interested in marriage as she was more interested in her career. Her increasing fame and her untoward behavior (the proper place of women at the time was in the home and education was still thought to endanger a woman's health) led to a rift with her brothers, and she left the family business in 1913 [6, 10].

In 1914, Gleason became the first woman in New York State to be appointed receiver by a bankruptcy court [12]. The company was the Ingle Machine Company in East Rochester, New York. At the time of her appointment, the company's stock was worthless and it was significantly in debt. Under her leadership, a year and a half later, the debt was paid off. By 1917, the company had earned one million dollars [6].

In 1916, Gleason was one of the first women elected to the Rochester Chamber of Commerce and the first woman elected to the Rochester Engineering Society [6].

Fig. 2 Home and Garden in Concrest (1930s). (Courtesy of the Village of East Rochester)

Sometime during the 1913–1919 time frame, Gleason became the first woman believed to have been elected to membership in Verein Deutscher Ingenieure, the German Engineering Society [12].[1] She was also active in the suffrage movement.

In 1917, she became the first woman with no family ties to become the president of a national bank [6, 9]. She was elected President of the First National Bank of East Rochester when the President was called to service during World War I. She had been a Director of the bank prior to this election as well as a business partner of the former president. During her tenure, she helped to start eight new businesses in East Rochester. One of the projects for which the bank had made a loan was to a local developer who had failed to complete his project. This project, the Concrest and Marigold Gardens, sparked Gleason's interest in low-cost, standardized, concrete homes. The community was a subdivision of 100 six-room houses as well as a country club, golf course, and park. The homes included such amenities as mixing faucets, built-in bookcases, and a mirror in the kitchen [6, 12, 13]. After nearly a century, the homes in Concrest and Marigold Gardens subdivision still exist today with minimal repairs [14] (see Fig. 2, 3, and 4).

Gleason became very interested in providing low-cost housing options due to her experiences with workers in manufacturing and financial companies. She developed new designs for such housing that were made of concrete and used a new pouring method that she developed. Her homes were described in the magazine, *Concrete*, in 1921. The article was titled "How a Woman Builds Houses to Sell at a Profit for $4,000." "Concrete Kate," as she was nicknamed, also became the first woman

[1] Many sources say 1913 – many records of the organization were destroyed in World War II and the date cannot be confirmed.

Fig. 3 Homes in Concrest (1930s). (Courtesy of the Village of East Rochester)

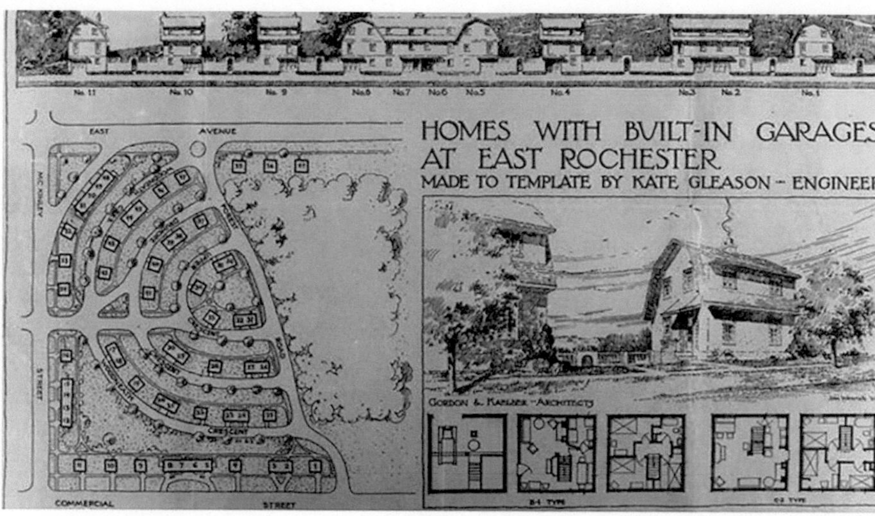

Fig. 4 Map of Concrest. (Courtesy of the City of East Rochester)

member of the American Concrete Institute in 1919 [4, 6, 9, 11]. Gleason applied techniques she had observed while studying operations at a car manufacturing facility:

> My inspiration for mass production methods came from visits to the Cadillac factory, where Mr. Leland showed me the assembly of eight-cylinder engines. [4]

In 1924, Gleason was called to Berkeley, California, to advise the city on rebuilding after a fire. Newspaper articles of the time reference her as a builder and architect.

She purchased property in Sausalito, California, intending to replicate her low-cost concrete home designs and built a few homes. The property, however, was needed for the approach to the Golden Gate Bridge, so she gave most of her holdings to a boy's orphanage [6, 9].

Gleason also purchased an estate in Septmonts, France, in 1924 and restored that structure while providing assistance to the village's recovery in the aftermath of World War I. The property included a castle tower and battlements from the twelfth century. She built a library and a motion picture theatre there as a memorial to the First Division of the American Expeditionary Force [6, 8, 10, 12].

In 1927, Gleason bought land in South Carolina that included 12 miles of beaches and the entirety of Dataw Island. She built homes and a hotel in Beaufort, South Carolina. At the time of her death, she was building an artists' and writers' colony on Lady's Island, which was completed by her sister. The Beaufort Memorial Hospital occupies land that she gave to the town. The Kate Gleason Memorial Park is on land next to the hospital [6, 9, 11, 12].

In 1930, Kate Gleason traveled to Berlin, Germany, where she represented the ASME at the World Power Conference [4].

Today, Gleason Corporation is one of the world's top purveyors of gear-making equipment, including machines, tooling, and technologies, for vehicles to airplanes, power tools to wind turbines [9]. And, Kate Gleason's legacy lives on in other ways as well. A charitable fund established from her wealth upon her death later became part of the Gleason Foundation, one of the largest private foundations in Rochester. At the Rochester Institute of Technology (RIT), the Kate Gleason College of Engineering (KGCOE) became the first engineering school in the USA to be named for a woman. There is a Kate Gleason Hall (dormitory) at RIT. A Kate Gleason endowed chair was established there in 2003 for a faculty member who builds "upon the tradition of Kate Gleason as a role model for women in engineering." This chapter's co-author, Margaret Bailey, was the first professor to hold this position (2003–2009), and currently there are two women faculty who hold Kate Gleason endowed chair positions within KGCOE. There is also a Kate Gleason Scholarship, established in 1996, that provides full tuition for female students. The ASME, of which she became the first full woman member in 1918,[2] established a Kate Gleason Award in her memory in 2011. This Award is bestowed upon distinguished women leaders from the engineering profession [9, 11, 13].

Maybe Susan B. Anthony sums it up best in a book inscription to Kate Gleason:

> Kate Gleason, the ideal business woman of whom I dreamed fifty years ago – A worthy daughter of a noble father. May there be many such in the years to come is the wish of. Yours affectionately, Susan B. Anthony, Rochester, NY, December 2, 1903. [10]

[2] Many sources say she became a member of ASME in 1914.

Edith Clarke (1883–1959)

A woman engineer with many firsts to her name, Edith Clarke (Fig. 5) grew up in Maryland without any intentions of even going to college. Orphaned by the time she was 12, Clarke attended boarding school, reached the age of 18 (the age of majority) and then decided to go to college so that she could find interesting work; work that replicated the interest she had discovered while playing duplicate whist (a card game). She spent the principal from her inheritance, against the advice of many family members and friends, to obtain an education because of a remembrance of a conversation she had with her mother years earlier in which her mother indicated her approval of a young man's decision to spend his inheritance on a college education – and who thereafter became a brilliant lawyer [5].

After graduating from Vassar with an A.B. in mathematics and astronomy in 1908 (Phi Beta Kappa), Clarke taught math and science for 3 years in San Francisco and West Virginia. But teaching was not holding her interest, and she decided to pursue becoming an engineer instead. She enrolled as a civil engineering undergraduate student at the University of Wisconsin and remained there for a year [15–17]. Then, she went to work for American Telephone & Telegraph Company (AT&T) as a computing assistant. She intended to return to the University of Wisconsin to complete her engineering studies but found the work so interesting at AT&T that she stayed for 6 years [5]. Clarke is an example of an early "computer" – woman with advanced training in mathematics who performed calculations for engineers (men) [17].

During World War I, Clarke supervised the women at AT&T who did computations for research engineers in the Transmission Department and studied radio at Hunter College and electrical engineering at Columbia University at night.

Fig. 5 Edith Clarke. (Courtesy Walter P. Reuther Library, Wayne State University)

Eventually, she enrolled at the Massachusetts Institute of Technology (MIT) and received her master's degree in electrical engineering in 1919, the first woman awarded that degree from MIT [17]. Upon graduation, she wanted to work for either General Electric (GE) or Westinghouse. But even with her stellar credentials, no one would hire her as an engineer because of her gender – they had no openings for a woman engineer! In 1920, after a long job search, GE offered Clarke a computing job, directing women computers who were calculating the mechanical stresses in turbines for the turbine engineering department at GE [5, 17].

But Clarke wanted to be an electrical engineer! Since that was not the job she was offered and since she wanted to travel the world, she left GE in 1921 to teach physics at the Constantinople Women's College (now Istanbul American College) in Turkey. She was able to also visit France, Switzerland, Italy, Egypt, Austria, Germany, Holland, and England during her time abroad. A year later, GE did offer her a job as an electrical engineer in the central station engineering department. At last, she had found work as interesting as a duplicate whist game [5, 17]! With this offer, she became the first professionally employed female electrical engineer in the USA [18].

Clarke's area of specialty was electric power systems and problems related to its operation. She made innovations in long-distance power transmission and the development of the theory of symmetrical components and circuit analysis [17]. Symmetrical components are a mathematical means by which engineers can study and solve problems of power system losses and performance of electrical equipment. Clarke literally wrote the landmark textbook in this area, *Circuit Analysis of A-C Power Systems, Symmetrical and Related Components* (1943) (Fig. 6) and a second volume in 1950. This textbook, in its two volumes, was used to educate all power system engineers for many years [5, 17, 18].

Clarke published 18 technical papers during her employment at GE reflecting her status as an authority on the topics of hyperbolic functions, equivalent circuits, and graphical analysis within electric power systems. The papers were published in organs including *The General Electric Review*, the American Institute of Electrical Engineers (AIEE)'s *Transactions*, *Electrical Engineering*, and the *AIEE Journal*. Her first paper, titled "Transmission Line Calculations," was published in the *General Electric Review* in June 1923 and describes her mechanical calculator. "Simplified Transmission Line Calculations," which appeared in the *General Electric Review* in May 1926, provided charts for transmission line calculations. She was also involved in the design of hydroelectric dams in Western USA [5, 18].

Clarke received a patent in 1925 (U.S. Patent No. 1,552,113 – Fig. 7) for her "graphical calculator" – a method of considering the impacts of capacity and inductance on long electrical transmission lines. It greatly simplified the calculations that needed to be done. In 1926, she was the first woman to address what is today the Institute of Electrical and Electronics Engineers (IEEE) – at the time it was the AIEE [17]. Her topic was "Steady-State Stability in Transmission Systems" [5]. In 1932, Clarke became the first woman to present a paper before the AIEE; her paper, "Three-Phase Multiple-Conductor Circuits," was named the best paper of the year in the northeastern district. This paper examined the use of multiple conductor

TECHNICAL INFORMATION CENTER
STONE & WEBSTER ENGINEERING CORP.
P. O. BOX 5406
DENVER, COLORADO 80217

CIRCUIT ANALYSIS
OF A-C POWER
SYSTEMS

VOLUME I
Symmetrical and Related Components

By
EDITH CLARKE
CENTRAL STATION ENGINEERING DEPARTMENT
GENERAL ELECTRIC COMPANY

Edith Clarke

One of a Series Written in the Interest of
the General Electric Advanced
Engineering Program

LIBRARY COPY

NEW YORK • JOHN WILEY & SONS, INC.
LONDON • CHAPMAN & HALL, LIMITED

Fig. 6 Title page – Edith Clarke's *Circuit Analysis of A-C Power Systems, Symmetrical and Related Components* (Signed). (Courtesy of Jill S. Tietjen, P.E.)

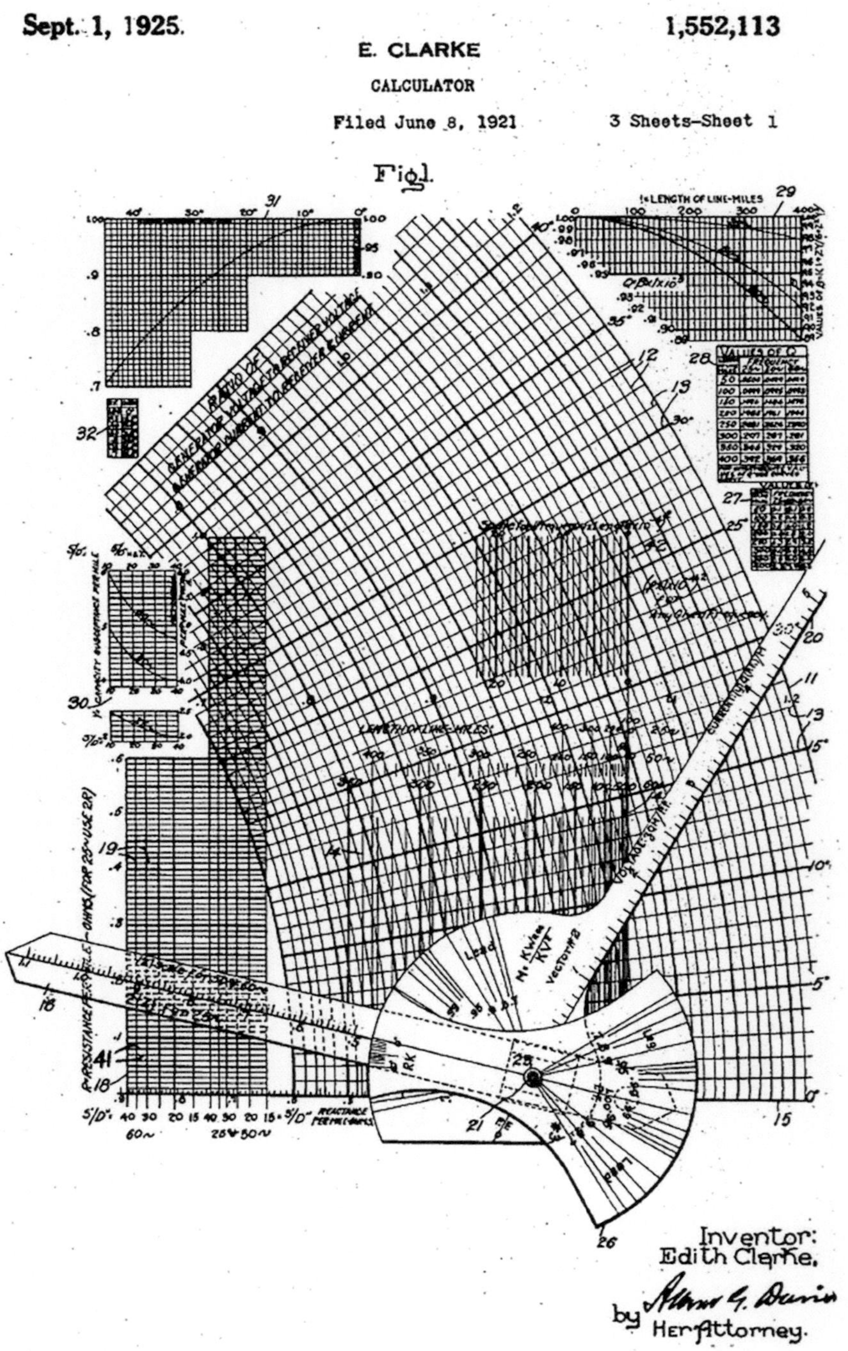

Fig. 7 Page 1 of Edith Clarke's Graphical Calculator Patent – Number 1,552,113

transmission lines with the aim of increasing the capacity of the power lines. In 1948, Clarke was named one of the first three women fellows of IEEE [17]. She had previously become the first female full voting member of IEEE [18]. Clarke was also the recipient of the Woman's Badge from Tau Beta Pi (at a time before women were admitted to membership) [15]. She was one of the few women who were licensed professional engineers in New York State [5].

A year after her retirement from GE in 1945, Clarke became an associate professor of electrical engineering at the University of Texas. In 1947, she rose to full professorship becoming the first woman professor of electrical engineering in the USA [17, 18]. She served on numerous committees and provided special assistance to graduate students through her position as graduate student advisor [17].

In 1954, Clarke received the Society of Women Engineers' Achievement Award "in recognition of her many original contributions to stability theory and circuit analysis." In 2015, she was posthumously inducted into the National Inventors Hall of Fame for her invention of the graphical calculator [18].

Mária Telkes (1900–1995)

A celebrated innovator in the field of solar energy, one of the first people to research practical ways for humans to use solar energy, and the so-called Sun Queen, Mária Telkes was born in Budapest, Hungary, in 1900 [19–21]. She built her first chemistry laboratory when she was 10 years old [22]. Educated at Budapest University as a physical chemist (BA in 1920 and PhD in 1924), she became interested in solar energy as early as her freshman year in college when she read a book titled *Energy Sources of the Future* by Kornel Zelowitch, which described experiments with solar energy that were taking place, primarily in the USA [22].

Telkes served as an instructor at Budapest University after receiving her PhD Her life changed significantly, however, when she traveled to Cleveland, Ohio, to visit her uncle, who was the Hungarian consul. During her visit, she was offered a position as a biophysicist at the Cleveland Clinic Foundation in 1925, working with American surgeon George Washington Crile. During the time (1925–1937) that she was at the Cleveland Clinic Foundation, she worked to create a photoelectric device to record brain waves [16, 21, 23]. She and Crile collaborated in writing a book titled *Phenomenon of Life* to report their findings. Her other work at the Foundation included looking at the source of the energy in brain waves, what happened to that energy when a cell dies, and the changes that occur when a normal cell becomes a cancer cell [20]. She spent her entire professional career in the USA [16].

In 1937, the same year she became a naturalized citizen, Telkes moved to Westinghouse Electric where for 2 years she developed and patented instruments for converting heat energy into electrical energy, so-called thermoelectric devices [16, 20, 21, 23]. In 1939, she began her work with solar energy as part of the Solar Energy Conversion Project at MIT. Initially, her role was working on thermoelectric devices that were powered by sunlight. During World War II, Telkes served as a civilian advisor to the U.S. Office of Scientific Research and Development (OSRD) where she was asked to figure out how to develop a device to convert salt water into drinking water [20, 23].

This assignment resulted in one of her most important inventions, a solar distiller that vaporized seawater and then recondensed it into drinkable water. Its significant advancement used solar energy (sunlight) to heat the seawater so that the salt was separated from the water [20, 23]. This distillation device (also referred to as a solar still) was included in the military's emergency medical kits on life rafts and saved the lives of both downed airmen and torpedoed sailors. It could provide one quart of freshwater daily through the use of a clear plastic film and the heat of the sun and was perfect for use in warm, humid, tropical environments [16, 19, 18, 21, 24, 25]. Later, the distillation device was scaled up and used to supplement the water demands of the Virgin Islands [23]. For her work, Telkes received the OSRD Certificate of Merit in 1945 [20]. The first page of her patent for a solar still is shown in Fig. 8.

Telkes was named an associate research professor in metallurgy at MIT in 1945 [23]. During her years at MIT, she created a new type of solar heating system – one that converted the solar energy to chemical energy through the crystallization of a sodium sulfate solution (Glauber's salt). Previous systems had stored the solar energy in the form of hot water or heated rocks [16, 20]. In 1948, Telkes and architect Eleanor Raymond developed a prototype five-room home built in Dover, Massachusetts [16, 22]. Called the Dover Sun House, this was the world's first modern residence heated with solar energy and it used Telkes's solar heating system [20, 23, 24]. The system was both efficient and cost-effective. It effectively heated the house during cold Massachusetts winters and cooled the house during the summer months. Solar collectors captured sunlight and warmed the air between double layers of glass and a black sheet of metal. That warmed air was then piped into the walls of the house, where it transferred the heat to the sodium sulfate to be stored and used at a later needed time. Thus, the walls of the house became the home's heating system [20, 23].

She next spent 5 years (1953–1958) at New York University (NYU) as a solar energy researcher. At NYU, Telkes established a laboratory dedicated to solar energy research and continued working on solar stills, heating systems, and solar ovens [16, 19]. Her solar ovens proved to be cheap to make, simple, and easy to build and could be used by villagers worldwide. Children could use them and the ovens could be used for any type of cuisine. Tests of the oven showed that it reached 350 °F even when the temperatures outside were in the 60s. This meant the oven could bake bread or cook a roast. Her work also led her to the discovery of a faster way to dry crops. In 1954, she received a $45,000 grant from the Ford Foundation to further develop her solar ovens [11, 19, 25].

After NYU, she worked for Curtis-Wright Company as director of research for their solar energy laboratory (1958–1961). Here, she worked on solar dryers as well as the possible use of solar thermoelectric systems in outer space. She also designed the heating and energy storage systems for a laboratory building constructed by her employer in Princeton, New Jersey. This building included solar-heated rooms, a

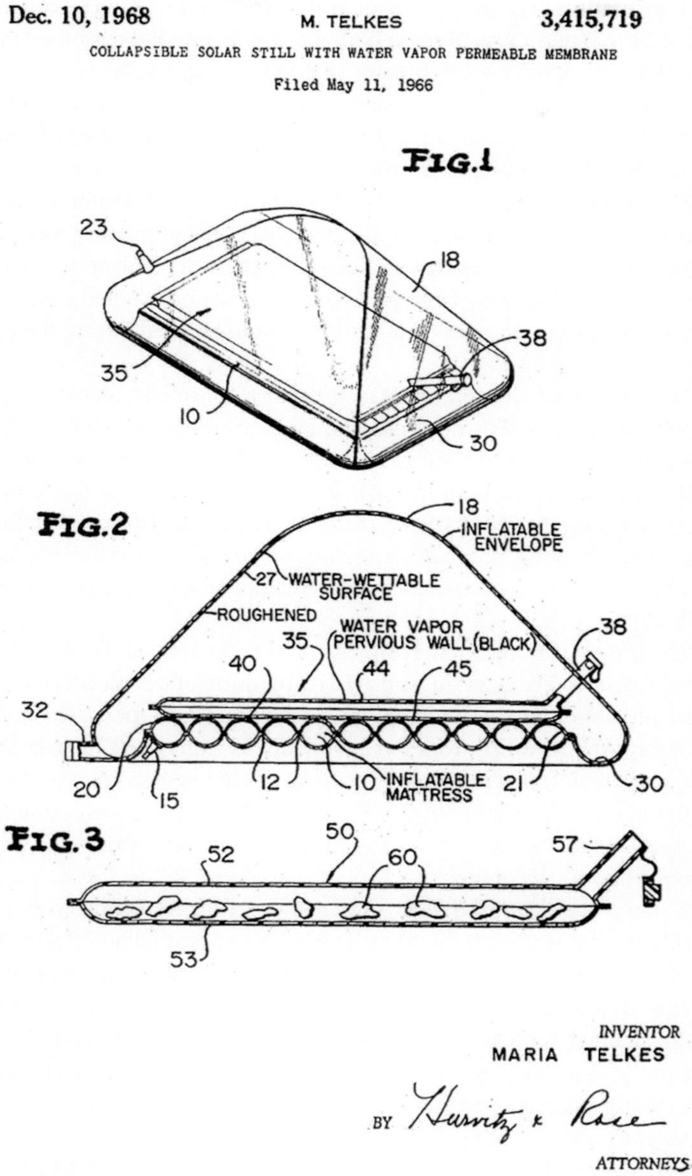

Fig. 8 Patent 3,415,719 Page 1 – Mária Telkes Patent for Collapsible solar still with water vapor permeable membrane

swimming pool, laboratories, solar water heaters, dryers for fruits and vegetables, and solar cooking stoves [20, 22].

In 1961, she moved to Cryo-Therm where she spent 2 years as a researcher working on space-proof and sea-proof materials for use in protecting sensitive equipment from the temperature extremes that would be experienced in those environments. Her work at Cryo-Therm was used on both the Apollo and Polaris projects [16, 19]. Subsequently, she served as the director of Melpar, Inc.'s solar energy laboratory looking at obtaining freshwater from seawater (1963–1969) before returning to academia at the University of Delaware. At the University of Delaware, Telkes served as a professor and research director for the Institute of Energy Conversion (1969–1977) and emerita professor from 1978. Here she worked on materials used to store solar energy as well as heat exchangers that could efficiently transfer energy. The experimental solar-heated building constructed at the University of Delaware, known as Solar One, used her methods. In addition, she researched air-conditioning systems that could store coolness during the night to be used during the heat of the following day [16, 19–21].

After her retirement, she continued to serve as a consultant on solar energy matters. In 1980, after the 1970s oil crisis and a renewed interest nationwide in solar energy, Telkes was involved with a second experimental solar-heated house, the Carlisle House, which was built in Carlisle, Massachusetts. The home had a solar photovoltaic array on the roof to produce electricity, extensive passive solar features to provide space heating, thermal collectors to provide domestic hot water, and many energy conservation measures to reduce electrical and thermal energy requirements [19, 26].

In 1952, Telkes was the first recipient of the Society of Women Engineers' Achievement Award. The citation reads "In recognition of her meritorious contributions to the utilization of solar energy" [22] (see Fig. 9). In 1977, she received the Charles Greely Abbot Award from the American Section of the International Solar Energy Society, which was in recognition of her being one of the world's foremost pioneers in the field of solar energy [19]. In that same year, she was honored by the National Academy of Sciences Building Research Advisory Board for her work in solar-heated building technology. She was a member of the Society of Women Engineers, the American Chemical Society, the Electrochemistry Society, and Sigma Xi (Scientific Research Society). The holder of more than 20 patents (shown in Table 1), in 2012, Telkes was inducted into the National Inventors Hall of Fame [22, 24]. In addition to her patents, Telkes also had many publications on the topics of using the sunlight for heating, thermoelectric/solar generators and distillers, and the electrical conductivity properties of solid electrolytes [24]. She believed so strongly in using solar energy that she said, "Sunlight will be used as a source of energy sooner or later . . . Why wait?" [25].

The three women featured in this chapter – Kate Gleason, Edith Clarke, and Mária Telkes – were each remarkable and energetic trailblazers during their lifetimes. Their legacies today are very much alive through the ideas and devices that they created and through the inspiration that they invoke. They each continue to have a transformative impact well beyond their individual lives and careers.

Fig. 9 Mária Telkes (third from left) receives the first Society of Women Engineers Achievement Award during the 1952 American Society of Civil Engineers *Centennial of Engineering* in Chicago, Illinois. *(l to r)* Rodney Chipp, Beatrice Hicks, Mária Telkes, unknown, Dot Merrill, unknown. (Courtesy Walter P. Reuther Library, Wayne State University)

Table 1 Mária Telkes Patents – Issued by US Patent and Trademark Office

US patent number	Date	Title of patent
2,229,481	January 21, 1941	Thermoelectric couple
2,229,482	January 21, 1941	Thermoelectric couple
2,246,329	June 17, 1941	Heat absorber
2,289,152	July 7, 1942	Method of assembling thermo-electric generators
2,366,881	January 9, 1945	Thermoelectric alloys
2,595,905	May 6, 1952	Radiant energy heat transfer device
2,677,243	May 4, 1954	Method and apparatus for the storage of heat
2,677,367	May 4, 1954	Heat storage unit
2,677,664	May 4, 1954	Composition of matter for the storage of heat
2,808,494	October 1, 1957	Apparatus for storing and releasing heat
2,856,506	October 14, 1958	Method for storing and releasing heat

(continued)

Table 1 (continued)

US patent number	Date	Title of patent
2,915,397	December 1, 1959	Cooking device and method
2,936,741	Mary 17, 1960	Temperature stabilized fluid heater and a composition of matter for the storage of heat therefor
2,989,856	June 27, 1961	Temperature stabilized container and materials therefor
3,206,892	September 21, 1965	Collapsible cold frame
3,248,464	April 26, 1966	Method and apparatus for making large celled material
3,270,515	September 6, 1966	Dew collecting method and apparatus
3,415,719	December 10, 1968	Collapsible solar still with water vapor permeable membrane
3,440,130	April 22, 1969	Large celled material
3,695,903	October 3, 1972	Time/temperature indicators
3,986,969	October 19, 1976	Thixotropic mixture and making of same
4,010,620	March 8, 1977	Cooling system
4,011,190	March 8, 1977	Selective black for absorption of solar energy
4,034,736	July 12, 1977	Solar heating method and apparatus
4,187,189	February 5, 1980	Phase change thermal storage materials with crust forming stabilizers
4,250,866	February 17, 1981	Thermal energy storage to increase furnace efficiency
4,954,278	September 4, 1990	Eutectic composition for coolness storage

References

1. M.L. Bragg. "SWE's highest awards: A look back", *2019 Conference Issue, SWE Magazine*, Nov. 4, 2019. Accessed: Feb. 1, 2021. [Online]. Available: https://alltogether.swe.org/2019/11/swes-highest-awards-a-look-back/,.
2. Association of American State Geologists. "In remembrance – H. Foster Bain". https://www.stategeologists.org/remembrance_by_name, (accessed Feb. 2, 2021).
3. "World power conference at Berlin". *J. Chem. Technol. Biotechnol.*, vol. 49, pp. 531–533, Jun. 1930, doi: https://doi.org/10.1002/jctb.5000492603.
4. R.G. Weingardt, "Engineering legends: Kate Gleason and Roy Peratrovich, Jr," *Leadership and Management in Engineering*, pp. 55–58, Oct. 2002.
5. A.C. Goff, *Women Can Be Engineers*, Ann Arbor, Michigan, U.S.: Edwards Brothers, Inc., 1946.
6. N. Bartels, "The first lady of gearing," *Journal of Gear Manufacturing*, Gear Technology, Sept./Oct. 1997.
7. J. Tierney, "What's in a name? The story of the Kate Gleason College of Engineering," presented at the 12th International Conference of Women Engineers and Scientists (ICWES12), Ottawa, Ontario, Canada, Jul. 28, 2002.
8. H.C. Bennett, "Kate Gleason's adventures in a man's job," *The American Magazine*, pp. 43, 168–175, Oct. 1928.

9. N. Giges, "Kate Gleason," https://www.asme.org/engineering-topics/articles/construction-and-building/kate-gleason (accessed Feb. 2, 2021).
10. M.B. Bailey, "Kate Gleason: The ideal business woman," *Rochester Engineer Magazine*, Jan., 2008.
11. L.M. Sheppard, "Kate Gleason: Leaves her legacy on RIT's College of Engineering," *SWE: Magazine of the Society of Women Engineers*, pp. 34–40, Sept./Oct. 1999.
12. Timeline and other materials prepared by Janis Gleason, Jan. 11, 1999.
13. G. Kass-Simon and P. Farnes, Eds. *Women of Science: Righting the Record*. Bloomington, Indiana, USA: Indiana University Press, 1990.
14. M.K. Hurd, "Concrete houses alive and well after more than 80 years," *Concrete International*, vol. 30, no. 5, pp. 34–38. May, 2008. Accessed Feb. 1, 2021. [Online]. Available: https://www.concrete.org/publications/internationalconcreteabstractsportal/m/details/id/19781.
15. M. Ingels, "Petticoats and slide rules," *Western Society of Engineers*, Sept. 4, 1952.
16. M. Ogilvie and J. Harvey, Eds. *The Biographical Dictionary of Women in Science: Pioneering Lives from Ancient Times to the Mid-20th Century*. Routledge, New York, New York, USA, 2000.
17. P. Proffitt, Ed. *Notable Women Scientists*. Farmington Hills, Michigan, USA: Gale Group, 1999.
18. "E. Clarke." National Inventors Hall of Fame, https://www.invent.org/inductees/edith-clarke (accessed Nov. 30, 2018).
19. "Maria Telkes: The Telkes solar cooker." lemelson@mit.edu, https://lemelson.mit.edu/resources/maria-telkes (accessed Dec. 1, 2018).
20. "Maria Telkes." Encyclopedia.com, https://www.encyclopedia.com/history/encyclopedias-almanacs-transcripts-and-maps/telkes-maria (accessed Dec. 1, 2018).
21. M. Telkes, "Preliminary inventory of the Maria Telkes Papers 1893–2000 (Bulk 1950s-1980s)." Arizona Archives Online, http://www.azarchivesonline.org/xtf/view?docId=ead/asu/telkes_acc.xml (accessed Dec. 1, 2018).
22. "Maria Telkes." Society of Women Engineers – Philadelphia Section, http://philadelphia.swe.org/hall-of-fame-m%2D%2D-z.html (accessed Nov. 30, 2018).
23. J.P. Rafferty, "Maria Telkes: American physical chemist and biophysicist." Encyclopedia Britannica, Inc. https://www.britannica.com/biography/Maria-Telkes (accessed Nov. 30, 2018).
24. "Maria Telkes.", National Inventors Hall of Fame, https://www.invent.org/inductees/maria-telkes (accessed Nov. 30, 2018).
25. A. Boyd, "Engines of Our Ingenuity: No. 2608, Maria Telkes." University of Houston College of Engineering, https://www.uh.edu/engines/epi2608.htm (accessed Dec. 1, 2018).
26. B.E. Nichols and S.J. Strong, *The Carlisle House: An All-Solar Electric Residence*," Lexington, MA, USA: Massachusetts Institute of Technology, Lincoln Laboratory, DOE/ET/20279-133, Jan. 1, 1981.

Jill S. Tietjen, P.E., entered the University of Virginia in the fall of 1972 (the third year that women were admitted as undergraduates – under court order) intending to be a mathematics major. However, midway through her first semester, she found engineering and made all of the arrangements necessary to transfer. In 1976, she graduated with a B.S. in Applied Mathematics (minor in Electrical Engineering) (Tau Beta Pi, Virginia Alpha) and went to work in the electric utility industry.

Galvanized by the fact that no one, not even her Ph.D. engineer father, had encouraged her to pursue an engineering education and that only after her graduation did she discover that her degree was not ABET-accredited, she joined SWE and for more than 40 years has worked to encourage young women to pursue STEM careers. In 1982, she became licensed as a professional engineer in Colorado.

Tietjen started working jigsaw puzzles at age two and has always loved to solve problems. She derives tremendous satisfaction seeing the result of her work – the electricity product that is so reliable that most Americans just take its provision for granted. Flying at night and seeing the lights below, she knows that she had a hand in this infrastructure miracle. An expert witness, she works to plan new power plants.

Her efforts to nominate women for awards began in SWE and have progressed to her acknowledgment as one of the top nominators of women in the country. Her nominees have received the National Medal of Technology and the Kate Gleason Medal; they have been inducted into the National Women's Hall of Fame and state Halls including Colorado, Maryland, and Delaware and have received university and professional society recognition. Tietjen believes that it is imperative to nominate women for awards – for the role modeling and knowledge of women's accomplishments that it provides for the youth of our country.

Tietjen received her MBA from the University of North Carolina at Charlotte. She has been the recipient of many awards including the Distinguished Service Award from SWE (of which she has been named a Fellow and is a National Past President), the Distinguished Alumna Award from the University of Virginia and the Belk College at the University of North Carolina at Charlotte, and she has been inducted into the Colorado Women's Hall of Fame and the Colorado Author's Hall of Fame. Tietjen sits on the board of directors of Georgia Transmission Corporation and served for 11 years on the board of Merrick & Company. Her publications include the bestselling and award-winning books *Her Story: A Timeline of the Women Who Changed America* and *Hollywood: Her Story, An Illustrated History of Women and the Movies.*

Dr. Margaret Bailey, P.E., began her journey as an engineer upon entering the Pennsylvania State University in 1983. After graduating with a B.S. in Architectural Engineering (1988), she worked as a building HVAC design engineer before successfully passing the professional engineers exam in 1993. That same year Bailey began studies toward a Ph.D. and graduated from the University of Colorado in Boulder (1998). Her dissertation involved experimental and analytical research, which gave her the opportunity to address an important question that she had been contemplating, based on her time as a practicing engineer. Could an artificial intelligence-based fault detection system observe the behavior of building mechanical systems when no one was present and alert the building engineers when a fault was likely occurring? Bailey worked under Professor Jan Krieder's advisement while developing this artificial neural network-based system that successfully detected faults within mechanical equipment.

In 1998, she joined the faculty in the Department of Civil and Mechanical Engineering (ME) at the United States Military Academy (USMA) at West Point where she taught courses in thermodynamics and created an active research program. Due to her passion focused on increasing the representation of women within engineering, she was actively involved in the USMA Margaret Corbin Forum, which is dedicated to improving diversity-related issues at the USMA, and she ensured the creation of the SWE USMA Student Section and served as its first advisor. Although she left USMA in 2003 for an ME faculty position at Rochester Institute of Technology (RIT), she has served since 2009 as a member of the USMA Mechanical Engineering Program External Advisory Board. Bailey is now a Professor of ME within the Kate Gleason College of Engineering, RIT. Past appointments include the Founding Director of AdvanceRIT Program (2012–2020), the Co-Chair of the President's Commission on Women (2007–2020), the inaugural Sr. Faculty Associate to the Provost for Women Faculty (2010 –2018), the Founding Director of WE@RIT (2004–2011), and the inaugural Kate Gleason Endowed Chair (2003–2009). She teaches courses and conducts research related to thermodynamics, engineering and public policy, engineering education, and gender in engineering and science. Bailey is the co-author of a widely used engineering textbook, *Fundamentals of Engineering Thermodynamics*, and she has published over 90 refereed journal and conference proceedings papers.

Bailey served as the Principal Investigator (PI) or co-PI on several National Science Foundation funded efforts, including the RIT ADVANCE Institutional Transformation (IT) grant (2012–2019, PI), the ADVANCE IT-Catalyst grant (2008–2011, PI), the ADVANCE Partnership grant focused on faculty salary equity practices and study (2021–2026, co-PI), and a PFE:RIEF grant focused on "Understanding the Relationships Between Gender, Self-Efficacy, and Extracurricular Activities in the Professional Formation of Engineers" (2021–2023, co-PI). The projects that she has led have resulted in impactful new programs, practices, and policies as well as a dedicated ADVANCE unit within the Office of the Provost and the women in engineering center, WE@RIT, within the Kate Gleason College of Engineering. Both of these units have received national recognition. In August 2021, RIT was one of ten universities awarded the National Institute of Health (NIH) Prize for Enhancing Faculty Gender Diversity in Biomedical and Behavioral Science, which is in large part due to the impact of the RIT ADVANCE program. Dr. Bailey is a recipient of the national Maria Mitchell Women in Science Award, the RIT Edwina Award for Gender Diversity and Inclusiveness, the RIT Isaac L. Jordan Sr. Faculty Pluralism Award, and the WEPAN 2008 Women in Engineering Program Award in recognition of the WE@RIT program's offerings and impact for diversifying engineering.

Mechanical Engineering Micronarratives and/as Changing Stories of Women in STEM

Laura Shackelford

This collection gathers a provocative mix of multi-disciplinary research, teaching, and professional experience from contemporary women mechanical engineers transforming distinct areas of energy and environment in the USA and globally. Individually and jointly, these chapters raise and reckon with the question – *How to tell the stories of contemporary women in Mechanical Engineering and related areas of Science, Technology, Engineering, and Math (STEM)*? Engaging with this question, these chapters raise the more challenging and seemingly tautological question – How to narrate and disseminate stories of women in Mechanical Engineering (and related STEM fields) that can *effectively change* the overarching, all-too-familiar story (and experiences) of Women in STEM, which has informed and shadowed women in STEM for much of the twentieth, and now, the start of the twenty-first century?

As you read the stories of this diverse group of women mechanical engineers, each writing at and about distinct stages of their educational, research, career, and life journeys, as these interweave through energy and environment work in academia, industry, public policy, and K-12 education, at times co-habiting several of these areas or bridging multiple countries, at once, it's entirely possible and, I hope, may be quite pleasurable to lose track of the shared methods and aims that traverse these authors' stories. You'll be taken in, and away by specific moments, metaphors like "practicing your game face," educational experiences, cultural references to *Robocop*, or *Star Trek*, familiar pathways, or surprise turns, like an engineering student turned "fashionista" that resonate with you in some impactful, aspirational, cautionary, or another way, perhaps. With this in mind, this chapter attempts to highlight some important, underlying commonalities to the authors' approaches to sharing these personal stories about their educational pathways, remarkable

L. Shackelford (✉)
English Department, Rochester Institute of Technology, Rochester, NY, USA
e-mail: lxsgla@rit.edu

M. Bailey, L. Shackelford (eds.), *Women in Mechanical Engineering*, Women in Engineering and Science, https://doi.org/10.1007/978-3-030-91546-9_2

accomplishments, groundbreaking research trajectories, and everyday experiences working as women mechanical engineers.

Drawing on my own research in narrative theory and storytelling practices of many kinds, my abiding interest in the history and philosophy of contemporary post-World War II information and bioinformatic sciences, as well as in feminist studies of science, I will suggest how and why these authors' methods of telling, gathering, and sharing *micronarratives* of this kind might play a more important role in changing stories of Women in STEM than we, as readers, typically or initially, realize.

To start, by describing the chapters in this collection as *micronarratives,* I want to draw attention to their status as specific, situated, personal, and everyday stories whose focus and delimited scope is, in fact, key to their appeal and impact. The scope and purpose of these stories is largely distinct from grand, overarching *meta-narratives* – the official stories and underlying narrative explanations a society nonconsciously assumes to be universal, common sense accounts of the world, as this concept is described by François Lyotard [1]. *Petits récits* or micronarratives, to the contrary, "affirm their own particularity and partiality; they are about local activities of individuals and communities" [2, p. 79]. The personal stories and experiential accounts of educational experience, career life, and areas of innovative research in mechanical engineering in the late twentieth and early twenty-first century included here, I suggest, recognize and reveal the value of "situated knowledges," or knowledge that is understood in light of its local, interpersonal, technological, material, and context-dependent modes of production, not in spite of the social, cultural, and technical practices that enable and inform it, to borrow feminist historian and philosopher of science Donna Haraway's influential term [3].

Reflecting on their own local, initial conditions, the authors featured here share their memorable micronarratives. At the same time, their focus and specificity frequently open onto larger patterns of behavior, belief, and the gendered, intersectional positionality structuring mechanical engineering professions and industries, and our social systems, more widely. In this regard, they also speak implicitly to the macro-scale of grand narratives of Mechanical Engineering and Women in STEM that we might otherwise take for granted, as absolute and unchanging. The motion of these authors' "butterfly wings" – whether that involves tinkering with the small engine in a family lawnmower or a deep dive into structures of silicon crystals – also resonates at a macro-level and scale that is well-attuned to shifts in the larger field and foci of mechanical engineering to offer an analogy between their micronarratives and the infamous "butterfly effect."

Their stories, in concretizing their experience in all its local complexity and specificity (as authors are both narrators and the subject(s) of their narratives), reveal influential patterns of knowledge-building, experience, resilience, or impediment, which recur across the individual chapters and narratives, and even bridge various national and transnational contexts. I propose, in this respect, that their micronarratives might be read as pieces of a larger, dynamic, shifting, uneven, incomplete puzzle with missing pieces, alongside well-connected ones.[1]

Recounting these educational/professional histories and research trajectories with such nuance and care, in full awareness of their lived complexity, these stories also let us, as readers and fellow (or aspiring) mechanical engineers or co-travelers, appreciate the "structures of feeling" – the emerging (and residual) social and cultural attitudes and orientations we are inclined to share at any given historical moment – as the literary theorist Raymond Williams brilliantly describes shifting cultural dynamics that are both omnipresent and intangible, at once [5]. Their micronarratives literally let us in on these latter dimensions of mechanical engineering and related areas of STEM, which more often fall out of field descriptions when presented in the abstract, as a statistic, or recounted from everywhere and nowhere, at once.

These are just a few of the narrative attributes and affordances at work in these micronarratives that address and open onto changing stories of Women in STEM today. Once appreciated, they might motivate us as readers to revise and rethink the larger, oft-repeated, more official, and, by necessity, generalized story of Women in STEM in light of some of the experiences and accomplishments chronicled here. Approaching these problematics from their local, contextually rich, nuanced (inter) personal perspectives will *not* completely change the story we tell about women in mechanical engineering and related STEM fields of energy and environment as understood over the twentieth and early twenty-first century, nor does it replace macro-level, institutional, governmental, data-driven, or statistical methods and the programmatic action they support. Nonetheless, the gathering, sharing, reflecting, and reconnecting catalyzed by such stories, I suggest, encourages and enables us as academics, practitioners, colleagues, professionals, parents, students, and national and global citizens to read for *both* unrecognized transformations apparent in and across these stories, as well as for the continued patterns and familiar meta-narratives that continue to impact women mechanical engineers, the field, and its knowledge-base in recognizably detrimental, counterproductive ways.

The gathering, sharing, and collaborative (re)telling of stories here (and elsewhere recently) enlist and creatively employ narrative methods of knowledge-building, whose value and import is not always recognized. Thus, in exploring the difference of mechanical engineering micronarratives, as a "difference that makes a difference" – to borrow Gregory Bateson's apt definition of information in terms of its capacity to change a larger system [6] – I will try to underscore what these stories and micronarrative methods (here and elsewhere) contribute to the official story of Women in STEM today. After reading several of the chapters included in this collection, I'm particularly struck by the depth and variety of research areas and new directions these women mechanical engineers are charting and pursuing and the net gain this represents as we face the energy and environmental crises of the present.

[1] The title of Michaela Crawford Reaves' *Filling in the Pieces: Women Tell Their Stories of the Twentieth Century*, a recent collection of contemporary women's stories as oral histories explicitly captures this approach to reading such micronarratives [4].

While not all of the recent changes to the story of Women in STEM are necessarily so positive, or even improvements, it is all the more crucial to recognize these areas of growth and their unfolding potential.

Addressing the Problem of Women in STEM

The overarching meta-narrative, or story of Women in STEM in the USA – one that most academics, engineering professionals, and policymakers are already familiar with – addresses the current role of women in these disciplines and professions as a national problem that requires informed efforts at all levels of education, professional formation, industry, and practice.[2] This narrative catalyzes crucial, ongoing efforts to increase rates of women's inclusion and retention in these areas and, thereby, to lessen the opportunity and knowledge gaps that result when gender is allowed to negatively impact entire fields of knowledge and practice. This well-documented and carefully (re)quantified account of Women in STEM identifies the dire consequences of women's exclusion, diminished roles, and unrecognized contributions in these domains, not only for women but for professionals, industries, academics, and students of all genders, as well as society-writ-large. The story of Women in STEM recognizes women's absences, diminished roles, and underrepresentation in STEM disciplines and professions, quantifying their notable costs to nationally strategic knowledge-production, industry, and labor forces. The official narrative and related discourses surrounding Women in STEM actively confront and contest the belief systems and social biases that impede women's full participation and advancement. As importantly, they increase awareness of the ideologies and blind spots that systematically undermine individual scholars, professionals, and their communities, alike.

With the aim of improving women's participation and the overall strength of STEM fields, national policymakers and organizations, universities, professional groups, progressive industry leaders, and countless STEM professionals and scholars continue to devote significant resources and time to identifying, addressing, and mitigating, if not eradicating, precisely those gender-specific obstacles and gendered value-systems, ideologies, and practices that impede women's full entry, lasting participation, and/or professional and personal success in these influential domains of knowledge and practice. As a result, this is a very familiar story to Mechanical Engineers and other STEM professionals and scholars, to those involved in educating the next generation of scholars and professionals, and a story that is

[2] In this chapter, my focus remains on narrative discourses surrounding Women in STEM in the USA, though it is important to recognize parallel conversations in many other countries, as well as to acknowledge differences in their approaches, policies, and success in addressing similar issues. Several chapters in this volume directly reflect on national and cultural differences impacting women in STEM fields such as Mechanical Engineering, while addressing wider policies and discourses of women in STEM in these other countries is beyond their scope and my own here.

introduced to young people of all genders who aspire to enter into these areas, often well before they might encounter or understand its full meaning or consequence. My daughter's Girl Scout troop, for instance, has enthusiastically participated in STEM events every year since the girls were in the third grade, thoroughly enjoying the female mentors and circuit building or lab activities they introduce to the troop.

As familiar and recognizable as this story of Women in STEM may be in the USA today, it is as important to recognize that it is, equally, a changing story. Admittedly, these changes cannot necessarily be plotted along a single, linear, or a strictly progressive path. Key elements of this story of Women in STEM seem to stubbornly persist and/or disappear only to resurface just when you thought they were history. Yet that does not mean they are not changing in more and less appreciable, if uneven ways worth understanding and expanding on.

The multi-leveled, comprehensive efforts to gather relevant data and track STEM fields and professional trajectories in the USA according to gender and intersecting ethnic, racial, and economic demographics have been, and, I'd suggest, continue to be, invaluable. For instance, I first encountered the national *Women in Science & Engineering* (WiSE) program in the late 1990s during graduate school at Indiana University in Bloomington, where I was studying contemporary American literature and its engagements with post–World War II cybernetics, information, and systems sciences at the core of early digital cultures. Interested in contributions of women and feminist studies and philosophies of science to these emerging post–World War II information and systems sciences, I attended a WiSE event where I made my first direct interdisciplinary encounter with women scientists and engineers who were similarly and very differently interested in the role gender plays in science and engineering. To provide some sense of the longevity and scope of the WiSE program – it is worth noting that WiSE began to emerge in the 1970s as an informal faculty effort to support women in STEM at the University of Michigan, becoming established as one of the first programs supporting women in STEM in the world in the 1980s before receiving the Presidential Award for Excellence in STEM Mentoring in 2000 and a National Exemplary Program Award on 2002 [7]. WiSE continues to support women and their STEM colleagues in a variety of activities and programs nationwide. My current research and professional trajectories have been shaped in significant ways by transformative efforts like WiSE, as have so many other professionals, scholars, and their disciplines judging by the continued prominence and reach of this program, and others like it, after more than 40 years.

It remains equally, if disturbingly, striking how many of these familiar, gender-specific obstacles and impediments to women's advancement and success in STEM fields,[3] nonetheless, continue to inform women's career trajectories and experiences well into the twenty-first century. This poses the intractable question – *How to acknowledge both what's changed in this story of Women in STEM, what hasn't (yet) changed, and for who it has or hasn't?* One approach, which I'll draw on here,

[3] This is also the case in humanities and social sciences, though there are differences in the degree and kind of these barriers, glass walls, and bias.

is provided by the editors of *Women, Science, and Technology: A Feminist Science Studies Reader* in their preface to the third edition [8]. First, they acknowledge "several arenas in which significant change has taken place," noting improvements in

> the increasing representation of women as undergraduate and graduate degree earners in science, mathematics, and engineering; the increasing visibility of women's health care in public policy discussions; the increasing recognition that women scientists and engineers bring useful and (perhaps) distinct experiences to the table in the development, implementation, and adaptation of new discoveries; the increasingly institutionalized commitments of colleges and universities to denounce gender bias in education, employment, and training in science and engineering fields; and the decreasing representation of science and scientists as necessarily masculine. [8, p. xiv]

The editors "honor the change that has taken place" while refusing to "*lay out a claim of discovering that all is well with the world*" [8, p. xiv]. Instead, they recommend a both/and approach that, I'd argue, has become even more essential in the wake of the last US president and his administration's national and global policies towards women, others who affirm non-binary genders or gender fluidity, and towards people of color. Writing before these latest setbacks for women, the editors acknowledge that as "*conversations about diversity and inclusion have taken on national prominence in the United States, and the need to be globally competitive drives efforts to recruit talent to STEM careers, it may seem as if issues of inequality in STEM is are passé. How could anyone still think that only white men are fit for STEM careers?*" they ask [8, p. xix]. Nevertheless, the editors underscore that: "***The issues, unfortunately, are more complicated than simply recruiting more women and people of color into STEM fields – a lot more complicated, in substance, scope, detail, definitions, and debates***" [8, p. xix].

Fast forward 7 years, and it becomes even clearer that as familiar and tired, even *passé*, this story of Women in STEM in the USA may seem today – especially in light of notable increases in the number and diversity of women completing degrees and working in these fields as researchers, engineers, and teachers – the story is, instead, just beginning to reveal its complexity and, from my perspective as an educator and scholar interested in and witnessing these changes, the story continues to provoke and require more nuanced understandings of these, and related questions surrounding practice(s) of science and engineering, and their areas of unseen potential and debilitating blind spots.

Whether one values and/or loathes this complexity, or oscillates between these attitudes on any given day as I sometimes do, it is possible (and important, in my view) to extend the editors' *both/and* approach, pursuing these uneven and unstable shifts and local complexities in the experiences and stories of women in STEM. In fact, this complexity may be just the catalyst necessary to expand conversations surrounding Women in STEM to address the complex shifts in knowledge and belief systems, professional practices, and leadership models, as well as the institutional and educational methods required to thoroughly and permanently change the meaning and impact of gender in science and engineering, and American society, at large, in the foreseeable future. No small task, to be sure, but all the more necessary to work towards.

The predominant, official story of Women in STEM has been rooted in the primary, seemingly binary question of women's inclusion or exclusion and a recognition of this differential inclusion of women as a problem. Building on this primary recognition, while also acknowledging some of the changes that have taken place over (at least) the last 40 years, it may now be possible (and likely pressing) to confront the dynamic, multi-layered dimensions to this story that are so apparent of late. To address these layers, and the kinds of differentially available, combined opportunity and impediment they register, it's crucial to move beyond the former either/or, inclusion/exclusion formulation to confront the both/and, i.e., the more complicated dynamics and intersectional dimensions to gender and value-systems as these connect with social power, national discourses, global politics, and changing twenty-first-century practices and priorities of science and engineering.

Hidden Figures and/or Hidden in Plain Sight?

Enter the diverse and nuanced micronarratives of women working as mechanical engineers in areas of energy and environment included in this collection whose stories are, in fact, part of a broader turn to non-fictional narrative methods and modes of knowledge-building as a crucial source of information and point of access into changing practices of STEM, as seen from the everyday, microperspective of individual engineers, mathematicians, researchers, and their communities. As personal narratives that bring together scholarly, research, and professional trajectories in compelling and informative combinations, these women's accounts of mechanical engineering and research developments in the crucial areas of energy and environment help to flesh out, qualify, and diversify predominant narratives and discourses of Women in STEM in the USA. Through their focused micronarratives, the authors provide "thick description," in an auto-ethnographic sense, revealing their proven and regularly recalibrated processes of doing science and engineering, alongside important insights into these scholarly and professional triumphs and challenges in all of their personal, professional, and experiential depth.

As micronarratives, they capture the seemingly minor, everyday moments of women mechanical engineers, as well as the pivotal moments, awesome discoveries, and life-changing decisions they make and continue to take in stride. These stories recapture experiential dimensions of doing science, and some of the geographic, cultural, economic, racialized, geopolitical, national, regional, and subdisciplinary differences in working and teaching and living as a mechanical engineer in energy and environment in this focused way.

Momentarily situating the micronarratives collected here alongside Margot Lee Shetterly's award-winning nonfiction narrative *Hidden Figures: The American Dream and the Untold Story of the Black Women Mathematicians Who Helped Win the Space Race* [9][4] will further clarify the value and shift of emphasis driving what is a more pervasive turn to non-fictional micronarratives as a source of reunderstanding the history and current trajectories of women in STEM. As one of the most famous and impressive examples of the recent, broader turn to non-fictional micronarratives, Shetterly's narrative approach in *Hidden Figures* is instructive and helps to shed light on similar and similarly impactful characteristics of the micronarratives gathered here, in spite of their different areas of emphasis, different historical contexts, and Shetterly's primary (though not exclusive) focus on black women at NASA's Langley campus.

Shetterly's nonfictional narrative account insightfully weaves together the distinct lives and experiences of four remarkable, trailblazing black women – Dorothy Vaughan, Mary Jackson, Katherine Johnson, and Christine Darden – whose "West Computing" group contributed centrally to NASA's space efforts at the height of the Cold War. Shetterly retraces the connections between their individual personal and professional stories, the emergent story of women 'computers', the debilitating infrastructures and discourses supporting racial segregation and sexism at NASA in 1950s Virginia, and her own story of growing up in Hampton, Virginia with these women as impactful family friends and fixtures in her community.

Shetterly's narrative poignantly illustrates the importance of sharing micronarratives such as these in her personal account of these largely forgotten black women scientists and professionals. Bringing these "hidden figures" to a broader, contemporary American audience, Shetterly's interwoven stories make quite clear how the full acknowledgement of their contributions, perspectives, and lived experience can shed transformative light on unseen and disremembered dimensions of America's space race, emerging STEM cultures, gendered and racialized hierarchies in 1950s America, gendered and racialized practices of science, mathematics, and computing, in addition to fully recognizing the contributions Dorothy Vaughan, Mary Jackson, Katherine Johnson, and Christine Darden individually and jointly have made to STEM, their communities, and country.[5] In the opening pages of her narrative, Shetterly stresses:

> Even if the tale had begun and ended with the first five black women who went to work at Langley's segregated west side in May 1943 – the women later known as the 'West Computers' – I still would have committed myself to recording the facts and circumstances of their lives. Just as islands – isolated places with unique, rich biodiversity – have relevance for the ecosystems everywhere, so does studying seemingly isolated or overlooked people and events from the past turn up unexpected connections and insights to modern life. The idea that black women had been recruited to work as mathematicians at the NASA installation in the South during the days of segregation defies our expectations and challenges much of what we think we know about American history. It's a great story, and that alone makes it worth telling. [9, p. xv]

[4] Shetterly's 2016 nonfiction narrative, *Hidden Figures,* was also released as a feature film in the same year.

[5] Notably, Shetterly notes in her acknowledgments that much of "the history that has come together in these pages wasn't so much hidden as unseen – fragments patiently biding their time in footnotes and family anecdotes and musty folders before returning to view" [8, p. 267].

Shetterly's gathering of these women's micronarratives and careful contextualization of their community of black professionals working in Virginia, at this pivotal time in American history is certainly "*a great story,*" especially in that it continues to have a significant impact on the way contemporary readers and audiences across the political spectrum understand and tell the story of black women's research and professional work in mathematics, engineering, aerospace, and computing. She draws crucial attention to the debilitating circumstances and extraordinary perseverance of these and other black women professionals whose contributions, nonetheless, expanded fields of mathematics, computing, and engineering, and these American institutions, through their activities.

Shetterly's eloquent analogy, characterizing these women as "*individual islands*" with "*unique, rich biodiversity*" in and of themselves beautifully captures the value of their stories. She also extends this analogy a step further, insisting their micronarratives have wider "relevance for the ecosystems everywhere," if we acknowledge, as I believe most will, that "no man (or woman) is, in fact, an island" [9, p. xv]. The micronarratives Shetterly gathers here are inextricably entangled with official, engrained ideas and standardized accounts of Cold War America and the scientific efforts so central to this moment, and to our own. And together, Shetterly's interwoven micro and meta-narratives retrace the threads linking these past moments to the present. These four black women at NASA, only a few of the many who worked and contributed to this history, reshaped the larger forces and aims and practices of 1950s American science and engineering, early computing, and the gendered and racialized institutions and belief systems that remain so entrenched today.

In "*studying seemingly isolated or overlooked people and events from the past,*" Shetterly, "turn[s] up unexpected connections and insights to modern life" [9, p. xv]. In fact, she acknowledges in the prologue how "*These women's paths set the stage for mine; immersing myself in their stories helped me understand my own*" [9, p. xv]. On one level, this claim seems commonsensical, repeating the familiar notion that through the reading and sharing of stories we make connections, both with our parents (at bedtime reading during childhood), with peers and cultural figures, as we get older, and through our visceral, symbolic identifications and (dis)identifications with the narrators, protagonists, and antagonists in the fictional and non-fictional stories we read and share as a culture.

In the context of telling and sharing stories about women in STEM, it is worth underscoring, as Shetterly's statement does, the importance of narratives, not only in their capacity to document or recount events and order them in time, but also to provide highly contextualized, potential models of action and promote social collaboration in this way. In *Stories and the Brain: The Neuroscience of Narrative* philosopher Paul B. Armstrong describes how this work of "*creating [and sharing] plots that simulate structures of action can have such a profound effect on our patterns of configuring the world*" [10, p. 13]. Or, as Shetterly provides her own gloss on these empowering interrelational dimensions of narrative: "*immersing myself in their stories, helped me understand my own*" [9, p. xv].

From this vantage, it may be possible to understand why the specific, contextual dimensions of micronarratives can present real-world scenarios and potential situations in which we imagine ourselves, however vicariously, as potential agents, or reluctant heroines, as taking action, and pursuing these, or alternate trajectories. As the storytelling metaphor of weaving the threads of a narrative together reminds us, narratives thrive on and are well-poised to reproduce contextual depth and complex interrelationality. This is why they are so essential to learning, from childhood onward. Listening to a bedtime story, we find ourselves identifying with the chameleons in Leo Lionni's *A Color of His Own* [11] or with the rebuffed friends in the recent Golden Book, *Grumpy Cat* [12], and we dexterously apply and adjust this scenario to our own circumstances (in any number of ways). Shared reading experiences also connect us with our parents and peers, the narratives serving as influential, culturally and historically variable, models for interpersonal ways of relating and embedding (or contesting) cultural values. They also, more simply, can help us attempt to understand and empathize with what might be going through others' hearts and minds, or our own, for that matter.

These characteristics of narrative and interpersonal micronarratives, in particular, are now recognized as a key reason that the sharing of a micronarratives, like these, is more likely to lead to behavior changes, such as quitting smoking or changing one's lifestyle, than is the mere sharing of information in the form of a statistic or other abstract medical information (such as one's statistical chances of getting lung cancer). Interpersonal identification and highly contextualized, recognizable, real-world scenarios and embodied activities, such as the ones shared in the micronarratives in this collection, engage us in impactful ways that we are cognitively attuned to. In this way, and others, the "*exchange of stories brings different worlds into relation with one another*" [10, p. 2] in all their diversity and difference.

Although these aptitudes and affordances of narrative may have been hiding in plain sight (as have these women's stories), in the context of data-rich information cultures and digital processing these specific traits of narrative are prompting important reconsiderations of the kinds of knowledges and social and cultural orientations narratives excel in transmitting and transforming, especially now that they can be deployed alongside data-driven, computational pattern-recognition and processing. The micronarratives included here frequently employ such "mixed-methods," combining narrative explanation with good old-fashioned data-based scientific research methods and macro and micro-analysis.

With the characteristics of micronarratives and narrative modes of knowledge and community building I've described here in mind, I'll end by underscoring how the specific micronarratives gathered here might be read with a careful attention to their ability to register changes in women's experiences and contributions to mechanical engineering today, as well as to reveal shifts that may be uneven, even idiosyncratic, yet nonetheless are contributing to changing patterns in these areas of Mechanical Engineering and changing "structures of feeling" in STEM disciplines and industries, more widely. These women engineers, and the micronarratives about their educational, personal, scholarly, and professional trajectories, which they share with such humility, humor, generosity, and insight, creatively engage narrative methods as an invaluable means of building knowledge and enhancing understanding at cognitive, social, and emotional levels, at once. As current, real-time accounts of their experiences as women in STEM, their stories about living and working in Mechanical Engineering today actively assess and reevaluate the inherited, official story of Women in STEM (which I've been spelling with a capital W to underscore its grand, macro-scope and, as a result, its necessary generalization of actual women in their fantastically invaluable diversity). They turn attention to the specific, local stories or *micronarratives* of women in STEM, in the plural. They remind us that these individual and decidedly (inter)personal stories are, in fact, what gives the larger, official discourse or grand narrative of women in STEM its force and what renews or refreshes its continued purchase on reality and its power. Gathering together and sharing these micronarratives, it is hoped, will, in turn, put pressure on our own understanding of women in STEM writ-large, introducing new concerns, priorities, approaches, and realities to its future iterations as story and in research, professional, and educational practices, alike.

Their stories remind me to relish and repeat (and repeat and repeat ad infinitum) the field-changing research, teaching, and professional *contributions* of women mechanical engineers, recognizing the positive inroads and alternative paths so many women engineers have made in these pivotal areas of energy and environment, in addition to their innovative, studied approaches to methods of collaboration, education, professionalization, and industry. As the examples of Kate Gleason, Edith Clarke, and Mária Telkes in the first chapter make clear, women's contributions are frequently unacknowledged, unseen, and untold, rather than actually absent, the result of gendered value systems and an incredibly persistent, patriarchal undervaluation of women's work and accomplishments. The very real absences of women from STEM are historically compounded by this disregard, if not active erasure of their many, many contributions.

Stories or narratives, both nonfictional and fictional, are a primary way that we produce meaning, identify patterns, and consolidate more complex and varied experiences and transmit these across generations, cultures, and places – that's why they are so essential to human understanding. Yet precisely these aptitudes and inclinations to identify and consolidate recognizable, repeatable, memorable patterns in coherent, linear patterns can serve simultaneously as a limit and liability of narratives in their capacity to calcify, to oversimplify and mislead, introducing blindspots and limits that soon stand in the way of our perception of changing realities. To end, I'll suggest that these seemingly contradictory influences or affordances of narratives are, in fact, best understood not in terms of an either/or, a gain or loss, but, instead, as a both/and, or pivot point. Armstrong thoughtfully situates stories at the crux of our need to cognitively balance our ability to identify patterns, while remaining open to new information and change, describing the complex dynamics of exchanging stories as a "two-way, back-and-forth interaction" between the story and reader that can keep "our cognitive processes from congealing into rigid habitual patterns," "by holding open their capacity to be reshaped and re-formed" [10, p. 180]. They involve us in the "play of configuration and refiguration," which can "loosen the habitual, ideological hold of any particular set of narrative patterns on our individual and social minds" [10, p. 180]. With this heady, dynamic combination of pattern-making and open-ended play in mind, I think we can begin to see why stories like the ones gathered here are so crucial to navigating social, cultural, technical change of the kinds we are currently facing; they change some of our customary ways of configuring the world and encourage us to seek out new patterns and acknowledge important changes in patterns that we habitually take for granted.

The micronarratives gathered here offer multiple, thoughtful, and curious views that open onto the *both/and* dynamics of changing stories in Mechanical Engineering and STEM today. They remind us, as readers, of the unbelievable potential, growth, and tangible accomplishments women mechanical engineers are leading in areas of energy and environment, allowing us to see emergent patterns that are rarely seen as such. Equally, their stories draw our attention and energy to the continued limitations of these gendered and racialized patriarchal domains of research, institutions, and belief systems, and the cost of these reductive, binary strictures to scientific inquiry, practice, and knowledge, and to us all, as a society in great need of more complex ways of understanding the world.

References

1. J. F. Lyotard. *The Postmodern Condition: A Report on Knowledge*, 1st ed. Minneapolis, MN, USA: University of Minnesota Press, 1984.
2. E. Carriere. "Re-enchanting Research: Using Narrative in Anti-Racism Education," in *Proc. 1994 Adult Education Conference,* Knoxville, Tennessee, May 20-22, 1994.
3. D. Haraway. *Simians, Cyborgs, and Women: The Reinvention of Nature*, New York, NY, USA: Routledge Press, 1990.
4. M.C. Reaves. *Filling in the Pieces: Women Tell Their Stories of the Twentieth Century*, San Diego, CA: Cognella Press, 2022.
5. R. Williams. *Marxism and Literature*, revised edition, Oxford, UK: Oxford University Press, 1978.
6. G. Bateson. *Steps to an Ecology of Mind: Essays on Anthropology, Psychiatry, Evolution, and Epistemology*, Chicago, IL: University of Chicago Press, 2000.
7. "History of WiSE: Pioneering Contributions and Visionary Leadership and Future Achievement," UM WiSE Web site, http://websites.umich.edu/~whimse/history.html (accessed 5 Oct, 2021)
8. M. Wyer, M. Barbercheck, D. Cookmeyer, H. Öztürk, M. Wayne, eds. *Women, Science, and Technology: A Reader in Feminist Science Studies*, 3rd ed. New York: NY: Routledge University Press, 2014.
9. M. L. Shetterly. *Hidden Figures: The American Dream and the Untold Story of the Black Women Mathematicians Who Helped Win the Space Race*, New York: NY: Edmund Morrow Publishing, 2016.
10. P. R. Armstrong. *Stories and the Brain: The Neuroscience of Narrative*, Baltimore, MA: Johns Hopkins University Press, 2020.
11. L. Lionni. *A Color of His Own*, NY: Dragonfly Books, 1975.
12. *Grumpy Cat*, NY: Golden Books, 2016.

Dr. Laura Shackelford is Professor in the English Department and the founding director of the *Center for Engaged Storycraft* at the Rochester Institute of Technology. She received a B.A. in English from the University of Minnesota, Twin Cities; an M.A. and a Ph.D. in English with a specialization in Literature and Science from Indiana University, Bloomington; and taught as a postdoctoral researcher in the English Department at Penn State, State College before joining RIT's College of Liberal Arts. Her research examines contemporary fiction, narrative, and emerging digital story practices with a particular interest in how literary fiction creatively draws from, and reflects on the computational, bioinformatic, and networked knowledges post-World War II information, systems, and bioinformatic sciences introduce. Speculative fiction that experiments with emerging digital media and spatial forms like the network, she finds, reminds us to attend to these socio-cultural and technological transformations with an eye to the possibilities these changing knowledges, scientific practices, and emergent social systems might introduce to the benefit of women, other minoritized groups, as well as to the entangled, nonhuman material life forms with which we share our ecosystems. Her research and writing unfolds in conversation with women's and gender studies and, especially, feminist science studies' explorations into the material technologies, social contexts, and systems of power that shape knowledges and might re-shape knowledges – and the broader social systems they help to co-realize – in important, necessary ways. She is the author of *Tactics of the Human: Experimental Technics in American Fiction* (2004) and co-editor of *Surreal Entanglements: Essays on Jeff VanderMeer's Fiction* (2021), and the author of numerous book chapters and journal articles on digital literary and narrative practices that contribute to understanding the potential and limitations of digital cultures and narrative understanding, at present. At RIT, she teaches contemporary fiction, speculative and science fiction, narrative theory, storytelling across media (print fiction, graphic narratives, interactive narrative, digital games, animation, film, photography, digital literary arts), and women's and gender studies courses.

Recent research and activities in support of the *Center for Engaged Storycraft* have strengthened her abiding interest in mobilizing interdisciplinary encounters across the humanities and sciences using story as a common thread, whether these activities center on literary arts and culture, engineering education, narrative medicine and health communication, writing, or creative research. Recent interdisciplinary projects include co-designing and facilitating a national KEEN *Engineering Unleashed* faculty development workshop to support engineering faculty in pursuing story-driven engineering education in their classrooms. Dr. Shackelford is also the co-creator of a *Transnational Digital Creation Workshop*, which pairs RIT students in human-computer interaction, game design and interactive media, psychology, and new media technology with French graduate students and faculty in digital literature and computational arts at the University of Paris-8, Saint-Denis, guiding their collaborative work on a transnational, creative digital literary project and travelling to Paris to co-realize these projects on site.

In Pursuit of an Inclusive Learning Environment in Engineering

Margaret Bailey

In this chapter, I recommend that students and young engineers train themselves to see, or be open to potential areas of opportunity, whenever and wherever they might arise. The chapter offers what I hope is an interesting "What if?" speculation for older and younger mechanical engineers – What if we focus our energies on creating solutions more systematically, growing *diversity of thought* on our teams and paying attention to the non-technical aspects of engineering, and that we deliberately teach students and young engineers supportive habits of mind and skills from the start?

Like several of the authors in this volume, I've had a diverse and tenured background within engineering. First as an engineering student at Penn State and then a practicing professionally registered engineer, which led to becoming a professional engineer and then a graduate student, which led to a PhD from the University of Colorado at Boulder and becoming a faculty member within mechanical engineering at two very different institutions.

A great deal has changed from a technological standpoint since I started as an undergraduate engineering student at Penn State in 1983. When I began my studies, I didn't own nor use a desktop computer. When I completed my engineering undergraduate degree, I had taken one computer science course where I learned to program in FORTRAN but never used computing again in any of my other classes. I still remember how exciting it was to set aside my typewriter and use an Apple Macintosh classic cube computer at Pattee Library on the Penn State campus to create my resumé before graduation. Today, mechanical engineering (ME) students use some form of computing in nearly every class they take, and their educational benefits are significant. The changes that have occurred from a technological standpoint since I graduated college have been life-altering, and engineers have played a primary role in how these technological advances have manifested. Considering the

M. Bailey (✉)
Department of Mechanical Engineering, Rochester Institute of Technology,
Rochester, NY, USA
e-mail: mbbeme@rit.edu

M. Bailey, L. Shackelford (eds.), *Women in Mechanical Engineering*, Women in Engineering and Science, https://doi.org/10.1007/978-3-030-91546-9_3

accelerated pace and continuing nature of these changes, those of you who are just beginning your engineering coursework or profession should expect even more profound impacts of technology in the course of your career or lifetime and ideally, many of you will be a part of this development. Within engineering, technological advances present opportunity areas for engineers which are embraced and proactively taken on by those in the field, in contrast to other highly aligned areas which seem to be ignored.

Ironically, in other areas that align with, closely relate to or even support engineering work, the advancements made within engineering practices, education, and in the profession itself seem to have lagged behind. These areas are what I'll call opportunity areas for engineers, and although there are some companies, agencies, and universities where significant advances have been made in these aligned areas, in general, there is a great deal of opportunity for more widespread advances. Based on my experience, three of these opportunity areas are:

1. Creating innovative solutions, rather than problem-solving
2. Embracing *diversity of thought*
3. Seeking out new tools and methods with a curious and open mind to support non-technical opportunity areas

I imagine what it could be like if engineering students were deliberately taught these non-technical topics in engineering curriculum to begin building foundational knowledge and capacity in these areas, and where they learned in vibrant, inclusive, effective learning environments and then transitioned into professional work environments that had similar characteristics. While the idea of advancing significant culture change across the entirety of our work and learning environment may seem somewhat galactic in scale and leave you with a sense of "what can I do" about impacting such change, it is important to realize that most major change starts with you, and a few others like you. Margaret Mead, a prominent anthropologist and activist during the 1960s and 1970s, captured this best: "Never doubt that a small group of thoughtful, concerned citizens can change the world. Indeed it is the only thing that ever has." I'm hopeful that reading this chapter will invigorate and energize those of you who are already members of small groups of people working towards advancing engineering to become more creative, innovative, and inclusive. For others, perhaps reading this chapter will give you something to consider and strive towards when the opportunity presents itself.

Creating Innovative Solutions, Rather than Problem-Solving

There are important and numerous distinctions between work that focuses on solving problems and the process of creativity, or innovation, which drives significant change and advances. The distinction between problem-solving and the creative, or innovation, process is simple, yet at the same time it can be profound. We can easily recognize this distinction when made consciously aware of it, but we tend to navigate absent of being intentional of its implications: Perhaps this is due to the intense focus we give to the short-term versus the long-term, the urgent versus the important, or the simple versus the complex. Robert Fritz describes the difference between solving a problem and creating a solution in his book *The Path of Least Resistance*::

> When you are solving a problem, you are taking action to have something go away: the problem. When you are creating, you are taking action to have something come into being: the creation. [1 , p. 11]

Take, for example, the simple and common practice people have towards their weight. Many people simply would like to "lose weight." They see their overall physical nature as overweight and therefore want to lose weight – they see weight as the "problem" and as such by losing weight want to make it "go away." Most realize that after such attempts this is not an ideal or sustainable solution. Conversely, a vision of creating a healthy lifestyle comprised of eating healthy, physical exercise, personal growth, quality relationships, and building self-esteem all lead to a sustainable solution for what was originally considered a problem. Losing weight is a quick fix, short-term, simple approach aimed at solving a problem. Creating a healthy lifestyle is a "systems approach" which takes more time and effort and results in creating an improved overall environment and lifestyle which can last a lifetime. This simple example illustrates the differences between the two approaches and, more importantly, the difference in what remains at the end of the process: in one case, the problem may have been solved, albeit temporarily and certainly not at the root cause level, in the other approach, a new and sustainable lifestyle and environment has been created. Something new has come into being [1].

A Technical Example

Consider the significance of this distinction (application of the creative/innovation process) in the music industry, in just the last 20 years. In 1999, a college freshman at Northeastern University in Boston, named Shawn Fanning, was looking for a way to readily share music. Using compression algorithms designed for storage of library material, he developed code that compressed, stored, and made it easy for the sharing of music. He dropped out of college and founded Napster, a company that despite legal issues with copyright regulations can be credited with redefining the music industry. By combining the features of existing programs (Internet Relay Chat's instant-messaging system, MS Windows file-sharing functions, and the searching capabilities of various search engines) to create new software that included all of those features [2]. Until then, company giants such as Panasonic and Sony dominated the industry. The hundreds of engineers who worked for these companies focused on problem-solving issues such as miniaturization – what seemed like significant breakthroughs involving transitioning from albums to the cassette and ultimately to the CD. This was simply a form of miniaturization. While these companies generated significant revenue, as they sold each newer compact product of music, it left the "business model" of music unchanged. However, the solution developed by the founder of Napster completely disrupted the music business model. Industry response to Napster was to force it to shut down with injunctions regarding copyrights. It did not, however, stem the evolution of digital music storage, and with the resolution of copyright issues, it led to innovations like the Apple iPod. Today, it is rare to buy a CD, or even an entire album of music, rarer still to listen to it on a traditional stereophonic system. In 20 years, the entire industry was redefined by a creative solution devised through the work of a creative young person, and then a small group who focused on sharing music more readily, and this became their vision.

Unlikely Partners Experience

In the Napster example, it's interesting that Shawn Fanning was 18 years old when he began working on a vision that ultimately redefined the music industry. I am often inspired and humbled by the intelligence, focus, drive, and openness of engineering students in my classes or who work in one of my funded research projects. Over the years, I recognize increased abilities of my students to operate at the connection between two different disciplines or where "disciplines converge." An example of disciplines converging is architectural engineering, which is a discipline which exists between the intersections of architecture and engineering; my undergraduate and graduate degrees are in this area. This program engaged both the analytical left brain and creative right brain. Students' growing abilities to manage converging disciplines is exciting because this is where innovation and creativity are boundless. I hope that in the future, we will see the emergence of more non-conventional degree programs, such as engineering with psychology or computer science with fine art or biology with organizational behavior. But even in single discipline courses, elements from very different disciplines can be introduced. One of my favorite past educational experiences was when I team taught a series of classes within a modern poetry class with a faculty member from English, Professor Anne Coon. Our collaboration helped to establish the Unlikely Partners project at RIT [3]. We focused on how electrification (or the distribution of electricity), which began in the 1890s in New York City (NYC), impacted the language used by poets, especially those living in the city or in the areas which were greatly affected by the coal mining needed to keep up with electricity demand. As a thermodynamics professor, in the modern poetry course, I taught non-technical students how coal is used to create electricity and the environmental impact associated with early coal mining.

Because poetry is like a mirror into our society's soul, we explored what was happening within the lives of the people within NYC and the coal country of PA as these rapid changes unfolded. For example, as more wires were hung in NYC in order to meet demand for lighting in a market with little to no regulation or electricity codes, sunlight actually began to be blocked on some city streets, that's how many wires were present overhead. As workers sought to keep up with demand, some were electrocuted because of faulty electrical lines and an overabundance of them, and people within the city were horrified by the results. These unfortunate incidents eventually led to adoption in 1915 of the country's first electrical code, the New York City Electrical Code [4]. My colleague from English explored how these changes influenced the language used by poets, and how terms such as "incandescent," "illumination," and "electrification" began to appear in their work. Exploring this topic in such a convergent manner between thermodynamics and modern poetry was an exceptional learning experience for all of us. It changed the way I taught the concept of electrification in Thermodynamics to ME students. In addition, the experience helped to shape how I've created courses related to thermodynamics, sustainability, and public policy [5–7] as well as features within the thermodynamics textbook that I co-author [8].

An Application in Culture – West Point

Creating innovative solutions to improve your health or redefine the music industry or enhance the learning experience for college students are offered in this section to demonstrate the possibilities that become available through creating innovative solutions rather than through solving a problem. My final example focuses on creating a culture within a large, complex organization. The design of the actual environments where we learn and work is rarely ideal and often challenged by our behaviors, processes, and habits as opposed to created through intentionality of how it can be a highly effective, productive, and synergistic system or place for us to influence both what we do and how we do it – collectively. What we each bring to the environment, including our passion and intelligence, can be leveraged to affect the creativity of the group. Interestingly, this distinction, once recognized, helps not only advancements by the group in their area of focus but can provide significant benefits to all aspects of one's life. It enables the synergy of "one plus one equals three (or more)," in learning groups, work environments, technological breakthroughs, and relationships.

I first realized the significance of working within an intentionally designed environment during the years that I served on the faculty within the ME Program at the United States Military Academy (USMA) at West Point. This distinctive environment was intentionally designed to create a culture based on a set of well-established values and it has been in existence since 1802. The 5 years that I spent at West Point provided me with a unique and enlightening experience, as one of the few non-military personnel in such a role and someone who had previously been exposed only to civilian university culture. All new instructors in the ME program at USMA went through a six-week teaching "boot camp" experience over the summer before they began teaching. This was helpful to the department because of their high annual turnover rate; 60% of the faculty are active duty military who have completed an MS degree at a civilian university, and upon graduation, they teach for 3 years at

USMA before rotating back into the regular Army. Therefore, every year there is a relatively large cohort of new faculty within the ME program.

During my tenure at USMA, I learned a great deal about teaching, assessment, and working on effective teams, while I conducted undergraduate research in advanced thermodynamics. I also had the opportunity to create the first Society of Women Engineering student chapter at West Point, and I served as its first faculty advisor. Due to the unique context of West Point, I was able to deepen my technical knowledge in thermodynamics while learning from military officers who possessed deep knowledge in technical military applications. West Point provided me with a unique and excellent insight into *leadership*. I understood the intuitive importance of leadership prior to this, but as I became acclimated into the USMA environment, I began to recognize and experience leadership attributes first-hand in my colleagues, department head, and students. My West Point experience helped me to deepen my understanding of *inclusion* and how it "looks" when it works and when it does not. It is important to note that this unique environment "works" for the vast majority of people who work and study there, but it doesn't work for everyone.

Upon arrival at USMA, I was the only woman in my teaching boot camp cohort and the only civilian. After I got over the initial culture shock, I started to become aware of the strength of the values, mission, and purpose that has been created and established there. The culture is purposeful and intentional. It is the by-product of the creative process that starts with a vision and is fulfilled through alignment at all levels such that everyone there both understands and follows accordingly to make it what it is. It is an example of the power of leadership around purpose, mission, and vision. West Point is able to achieve this through its long history, focus on leadership and mission, and its military orientation, and it would be impractical and ill-advised to assume that it could be replicated at a civilian university. However, elements of the West Point model could be adopted and would sit in stark contrast to everyday typical learning and working environments, many of which have mission and visions, but where little intentionality and alignment to ensure such "ideal" environments actually exist.

I continue to serve on the USMA ME program's external advisory board and witness how the culture continues to evolve. When I worked there, I was one of ten civilian women on the faculty of about 600 – 90% men, 85% active-duty military – and I could recognize the need for improved gender inclusion within the environment. Because that organization is military, a needed culture change initiative would need to originate with military leaders. As a civilian, junior faculty member, I did not have the positional power to initiate this type of change but I did have the desire to begin taking a more active role in shaping my academic work environment and the culture if needed. My tenure at West Point and the years spent as an engineering student (BS and PhD) and practicing engineer before graduate school are the experiences that shaped and fueled my passion for organizational transformation and change for creating more effective, inclusive, and vibrant learning and working environments.

Embracing *Diversity of Thought*

When working in a group, embodying the desire to achieve *diversity of thought* among group members is strengthened by an understanding of what this means, what it looks and feels like, its advantages and how each member can work towards achieving this state. It is also key to recognize that the concept of *inclusion* is a prerequisite for *diversity of thought* and that both are prerequisites for exceptional or "vibrant" working and learning environments for the group, where the collective intelligence of the group far exceeds the sum of individual member intelligence. Ideally, the group is within an overall organization that also offers a vibrant or highly effective working and learning environment. I will share with you some examples of how I've worked towards creating and promoting vibrant working and learning environments – the work is challenging, strategic, and requires a high level of *systems thinking* in creating innovative solutions.

In 2003, I joined the faculty at the Rochester Institute of Technology (RIT) as the inaugural Kate Gleason Endowed Chair and Associate Professor within the ME Department. I was thrilled to be joining the Kate Gleason College of Engineering, the only college of engineering in the US at the time named for a woman (see the first chapter in this collection – "Energetic Trailblazers: Kate Gleason, Edith Clarke, and Mária Telkes" – for more information on Kate Gleason). In my new role, half of my time was dedicated to addressing gender diversity opportunities within my college. I began to research why engineering continued to lack gender diversity and associated implications of the underrepresentation. I'll share thoughts on this before exploring how I've worked towards creating inclusive learning environments.

In the US, women are under-represented in the field of engineering. This issue is widely known and its causes have been studied and analyzed for over three decades. Governmental agencies like the National Science Foundation (NSF) and National Institute of Health (NIH) recognize this under-representation and its implications from a self-interest perspective, as described in this quote from a 2000 report to the US Congress.

> For the US to remain competitive in a global technological society, it must take serious steps to encourage [women and minorities] to enter [science, mathematics, engineering, and technology] fields… It is time for our nation to examine and reaffirm its policies of equal opportunity and access for all. [9]

NSF, NIH, and many other government and private organizations have funded projects that aim to improve understanding of the problem and to create research-informed solutions to address the issue. However, there has been slow progress in increasing the representation numbers. As of 2018, women have accounted for more than half (56%) of domestic undergraduate enrollment at all institutions, this trend has remained steady since 2000. Looking at graduate student enrollment in 2018, compared to 77% of the graduate students in psychology (increase from 70% in 2000 and 52% in 1980), only 25% of the graduate students in engineering (increase from 19% in 2000) and 32% of the graduate students in physical sciences and computer sciences (up from 28% in 2000) were women. Representation of women within the labor force has increased more significantly. In 2019, women constituted about half of the population employed in science and engineering (up from 23% in 2000). However, in comparing unemployment rates by gender, women's rates are higher than those of men are in each major age grouping in both 2000 and 2019 [10, 11]. Because this problem is persistent and complex, researchers study various aspects such as underlying factors of the issue and the solutions that work as well as how to best measure and define success.

Forming a Vision Around Equity, Inclusion, and Diversity

Over my career, I have existed in the midst of this journey. I've experienced what it's like to be a woman undergraduate and graduate student in engineering, what it's like to be a professional engineer in the private sector, and what it's like to be in academe as an engineering professor. I have enjoyed learning how to become an engineer, how to practice engineering, and how to teach and conduct research in engineering. However, I realized very early in my undergraduate years that the reason that I was one of the only women students in my classes and the reason why I had no women STEM professors was not because of some inherent lack of intelligence or abilities in math or science but due to gender. There were environmental factors that led me to question whether women belonged.

In retrospect, while I was an engineering student and in the private sector before graduate school, I realize that I survived and thrived in engineering in part because I learned to accept elements of the culture that served me well while ignoring or downplaying the parts of the culture that didn't, when possible. I did not think that I had the ability to change the culture, but I did have the power to decide to stick it out. I recall a few times when it was impossible for me to ignore what was happening to me, and I found navigating these issues in the professional work environment much easier than as a graduate student. In my second year of graduate school, I faced a significant issue with a faculty member following the birth of my child, which led to my questioning if I really wanted to be in academe. This experience showed me many ways in which the environment needed to change in order to become inclusive for someone like me. My learning environment at that time was the opposite of a place where I could thrive. Luckily, I had received an NSF Graduate Fellowship that allowed me to transfer to another university where I could conduct innovative experimental research that was fascinating. My family and I moved halfway across the country with an infant to allow me to continue my studies. At the University of Colorado at Boulder, I found myself in an academic environment where I could thrive and where there was very little about the culture that I had to ignore. My course work in the areas of renewable energy and pattern recognition and research into artificial intelligence, building controls, and thermodynamics were amazing. It was an ideal learning environment for me and I loved living in Boulder. My decision to remain in academe after challenges encountered in grad school turned out to be a very good decision. But sometimes it was best for me to move on.

My decision to leave West Point after 5 years on their faculty and join RIT in 2003 gave me the opportunity to actively pursue and lead various change initiatives with the goal of promoting a more inclusive working and learning environment at my university. Appointments over this time have included: Founding Director of AdvanceRIT and PI of the NSF ADVANCE Institutional Transformation (IT) and IT Catalyst grants (2008–2020), the Co-Chair of the President's Commission on Women (2007–2020), the inaugural Faculty Associate and the Senior Faculty Associate to the Provost for Women Faculty (2010–2018), the Founding Director of WE@RIT (2004–2011) and as mentioned previously, the inaugural Kate Gleason Endowed Chair (2003–2009). The NSF ADVANCE program is a federally funded program created to address the under-representation of women faculty in STEM disciplines. As of fall 2018, the NSF has invested $315M in ADVANCE projects and awarded over 70 Institutional Transformation grants [12]. NSF ADVANCE IT grants are inherently large in scope with significant budgets (over $3M) and lengthy with timelines of at least 5 years. Under my leadership, RIT received an IT grant in 2012 as detailed in the following sub-section. IT grantees "develop systemic approaches to increase the participation and advancement of women in academic STEM careers" [13]. Activities of ADVANCE funded projects include mentoring programs, professional development workshops, work to promote equity in academic workplace policies and procedures, and workshops to support the creation of a more inclusive climate and culture for women faculty [12, 14].

In these roles, I have gained significant insights that shape my vision and strengthen my belief in the importance of equity, inclusion, and diversity within our working and learning environments. Each is beneficial in very practical ways to the mission of universities and research backs this up. Recent research shows that differences in how people perceive, analyze, and apply information, also known as cognitive diversity, relates directly to other types of diversity including differences in gender, social class, and race [15]. Ample research has shown that work environments that achieve high levels of perceived equity, inclusion, and diversity are more productive, innovative, and creative. Research has also shown that people working in diverse groups are better prepared to explain their perspectives as well as more open-minded, diligent, and active in their listening because they cannot assume that everyone has the same viewpoints [16]. In an inclusive environment, people feel safe to share ideas; there is a stronger sense of belonging and a sense that people can be authentic. Most importantly, the outcomes a group can achieve are likely to be better, more significant and impactful than in the case of a non-inclusive one. This type of environment is difficult to attain within a large, complex organization like a university without deliberate and consistent effort from many people, who together serve a collective role in moving the needle on equity, inclusion, and diversity. I will share with you some history and insights gained based on a series of interrelated projects that allowed me to work with and lead a group of faculty from across the university and administration to transform our university into a more inclusive environment. This dedicated group that I created and led sought to transform a university towards a more inclusive environment.

The WE@RIT, EFFORT@RIT, and AdvanceRIT Journey

The journey at RIT that I will share with you began in 2004 at the campus coffee shop when I assembled a group of women faculty and administrators, and we brainstormed the term WE@RIT [17] to represent a program that we imagined could be created within the college of engineering to address the severe under-representation of women engineering students. At that point, the college enrolled less than 10% women undergraduates, lower than the national average. We also knew that women engineering students were consistently succeeding at rates higher than their male colleagues in regards to GPA and retention. I continued to lead this motivated and energized group, and we created a strong organizational structure that is still vibrant and thriving today within the college [18–25]. WE@RIT motivated the creation or strengthening of similar organizations in other colleges within RIT, including Women in Technology [26], Women in Science [27], and Women in Computing [28]. A great achievement for our college is that the gender profile of our incoming engineering undergraduate class now typically exceeds over 25% women, surpassing national benchmarks.

Around this time, in the late 2000s the NSF ADVANCE program [29] announced a new grant called the Institutional Transformation (IT) Start grant (later renamed IT Catalyst), which would provide up to $200K to support a university in conducting a self-study to explore the status of women STEM faculty on their campus. The IT Start program was created to help universities position themselves to submit a successful NSF ADVANCE Institutional Transformation (IT) grant proposal with much larger funding available (over $3M). At RIT, a group of women faculty and administrators already engaged in addressing gender diversity among students through their involvement in creating WE@RIT, WIC, and WIT collaborated and submitted a funding proposal for an IT Start grant. I led this effort as the principal investigator (PI), and we were successful and received funding in 2008 (NSF Grant No. 0811076); we called this project, EFFORT@RIT [30]. Over the next 3 years, we worked in close collaboration with the human resources to create a database to study faculty employment patterns by gender and gender equity in salary and start-up packages, analyze faculty climate survey responses using a survey that we created and administered (first of its kind at RIT), and conduct a thorough benchmarking of other universities [31–33]. Our research revealed barriers for women faculty in regards to climate, work/life integration, and career navigation [34]. Most importantly, the study revealed opportunities for enhancing our academic organization towards a more inclusive environment.

I worked with this same group plus a few more, and we took these opportunities and shaped them into a detailed, comprehensive plan to motivate important organizational change on campus. A shared passion to expand the safe places on our campus where authentic dialogue could occur regarding gender equity and the inclusion of those who are different in some way from the majority, led us to embark on a large, IT project called AdvanceRIT [35], funded by NSF in 2012 (NSF Grant No. 1209115). Interestingly, RIT is the only university that has successfully received both the IT Start/Catalyst and the full IT grants. I was PI on this nearly $4M effort aimed at transforming campus culture to become a place where more women faculty will join and thrive throughout their careers. Reshaping institutional culture in order to bring about a more inclusive and equitable working and learning environment was a strong motivator for the group, who engaged in questioning, analyzing, and re-envisioning our environment. We imagined a workplace where we could thrive, and we set out to create it.

AdvanceRIT focused on the goal of increasing the representation, retention, and career advancement of women faculty on our campus. The grant's social science researchers examined the unique challenges experienced by women faculty of color and Deaf and Hard-of-Hearing faculty, and we refined interventions to address the needs of these key sub-populations. The program influences long-term changes that transform culture, promote inclusion, and expand the representation of women on campus and among leadership. At the end of the NSF funding, RIT established a permanent ADVANCE office within the Office of the Provost. With the support of hundreds of people on campus, AdvanceRIT is a success in many ways, and the creation of this program is a significant achievement. In leading this project, I took a *systems* approach in creating new programs and refining existing policies, procedures, and practices [36–52]. While leading this effort, I found that patience was key because changes within higher education are evolutionary at best. The project also raised the consciousness among the faculty, staff, and administration regarding campus culture and approaches that are helpful in raising levels of inclusion, including bystander awareness/action workshops. Finally, the project created change agents among our faculty, staff, and administration, which has led to a more diverse representative voice involved in decision-making and improved assessments of diversity, marginalization, and privilege to help the university continue to move towards equity.

I'm very proud of all of our team's accomplishments. Key project achievements focus on faculty advancement and research success, faculty recruitment and retention with a focus on culture change, refinement to policy, practices, structures, and strengthened institutional collaboration in support of faculty equity, inclusion, and diversity. For a detailed list of outcomes, refer to Appendix I at the end of this volume. In August 2021, RIT was one of ten universities recognized by the National Institute of Health in receiving their new Prize for Enhancing Faculty Gender Diversity in Biomedical and Behavioral Science [53], which is in large part due to the impact of the AdvanceRIT grant on the RIT campus.

From a student's perspective, there are now more women faculty in the classrooms and labs to serve as important role models and to offer unique and different perspectives along the way. Therefore, the seeds that were planted in my mind as an undergraduate engineering student, sitting in 5 years of intensive engineering classes at Penn State as I wondered "where are all of the women on the faculty?"

grew into shaping me as a person who questions things and dreams up possible solutions.

From a faculty member perspective, prior to AdvanceRIT, the capacity to discuss work environment matters, specific to inclusion, diversity, conflict resolution, behavior issues, and open dialogue, was very limited on our campus. In the AdvanceRIT effort, the research team facilitated both cultural and environmental transformation initiatives and interventions. The group collaborated effectively with the faculty and all levels of the administration, using a multi-faceted approach addressing political, structural, symbolic, and human resource aspects of the organization. We embraced the desire to achieve *diversity of thought* among the research team members and our collaborators, key partners, and program participants. We addressed this deliberately and gained proficiency over time, while serving as role models for others on campus. We demonstrated that inclusion is a pre-requisite for *diversity of thought* and that both are pre-requisites for an exceptional or "vibrant" working environment where the collective intelligence of the group far exceeded the sum of individual member intelligence.

Helpful Tips, Tools, and Techniques

In this section, I will share just a few of the many tools that I've used to support my work, starting with some engineering tools that I have used to address non-technical challenges. As a ME, I study complicated systems like chillers within refrigeration systems, air conditioning systems, and power plants using thermodynamics to find ways to improve efficiency. I often work on *solving inverse problems* – given lots of data, what can be learned about the health or status of a system and where improvements can be made. With the help of NSF ADVANCE funding, I've described how I applied this technique to another type of system, focusing on my university rather than a power plant. I studied the status of RIT in regards to the environment for women using data to inform the study, starting with a focus on women students and then expanding to include faculty. Extending this approach to a non-technical context made sense.

Similarly, I've often used and extended the *engineering design process* to serve as a framework, as I've created new programs and initiatives focused on non-technical areas like attracting, retaining, and developing college students and faculty who will work and learn within an inclusive academic environment. My activities have often included the need to change culture and many aspects of the engineering design process that have been crucial in addressing these types of efforts. These include:

- Building the design team in order to promote *diversity of thought*
- Establishing team dynamic expectations and "ground rules" for behaviors
- Articulating the need and developing a shared understanding of this need
- Conceiving possible solutions and evaluating likely outcomes

- Defining what success will look like and how it will be measured
- Delving into detailed concept design
- Studying data based on how the created solution is doing
- Refining the designed solution based on data where you repeat many of the previous steps

Each one of the bullets requires a great deal of time and effort, and I've often relied on internal and external experts with expertise outside of engineering to support these efforts.

The first bullet above regarding *building a team* that can achieve *diversity of thought* is not a simple task, and it requires deliberate effort. At the start of the AdvanceRIT project, I asked each member of the core team to complete a Myers-Briggs personality test, [54] to help identify where opportunities might exist for the team. For example, this can help to raise the mindfulness of team members to make space in conversations for the quiet introvert(s) on the team or the quiet extravert(s) who may need it, in order to hear their voices. At RIT where we have many faculty, staff, administrators, and students who are deaf and hard-of-hearing, we also deliberately create our collaborations with accessibility in mind. Luckily we had accessibility experts on our team, who helped us become mindful of taking turns while speaking (often passed a stuffed toy to the person speaking) to reduce the instances where people interrupt each other. Sign language interpreters can better keep up with conversations when there are fewer (or no) interruptions. This led to the creation of a tip sheet to help us effectively communicate with people who are deaf and hard-of-hearing that we disseminate broadly [55]. Work in this area influenced the design of workshops, modeling the importance of making space to hear all voices.

As the team became much more mindful of how to promote *diversity of thought* among our members, we decided to create and adopt a set of ground rules that we often visited and refined as needed over the years. This led to the creation of a ground rules tip sheet for faculty and student teams that we distribute widely [56]. Several workshops have also been conducted with organizations to help them begin to use and adapt this practice, to help enrich their interactions and communications.

I found that leading an effort that transforms an organization to become more inclusive and equitable is extremely challenging, sometimes even overwhelming. I needed to help create ways to enhance creativity, patience, motivation, excitement, and endurance among members of the project's team, which included nearly a hundred faculty, staff, and administrators. A shared vision and common values were essential to the group. The 20 page-funding proposal that we created and submitted to NSF took about 6 months to create and vet. We based the proposal on 3 years of past research and we had a strong foundation. However, the months that we spent creating the institutional transformation strategy for change allowed us to begin reimagining our work environment to become more equitable and inclusive. After a month of this reimagining, the evolving solution became so complex that we needed a tool to help us wrap our minds about the overall concept.

I found the ideal tool in the book *Reframing Organizations*, by Bolman and Deal, where they offer four frames or lenses through which individuals experience and view their organizations [56]. These frames – human resources, political, symbolic, and structural – can also inform the strategic approaches created in the institutional transformation process [57]. We adopted the approach and created a strategy to make progress in all four of the frames in order to influence long-term changes that will transform the culture, promote inclusion, and expand the representation of women. Use of this multi-frame approach improved the team understanding of the organization and enabled the group to create interventions within the overall strategy based on thinking which employed all four of the cognitive lenses.

Conclusion

In this chapter, I've illustrated a connection between *systems thinking* in engineering and the value of similar *systems thinking* in non-technical areas. I've also identified three supporting opportunity areas within engineering. The first two involve improving engineer's ability to recognize the distinction between the creative process and traditional problem-solving, and with that understanding extend creative solutions beyond technical areas to thoughtfully shape the level of inclusion within the overall environments in which we work and learn in order to promote *diversity of thought*. The last area supports the first two in its focus on expanding our ability to seek out new tools and methods with a curious and open mind.

There are important and numerous distinctions between efforts that focus on solving problems and the process of creativity, or innovation, which drives significant change and advances. Differences include: the process followed for each, time required, mindset and behaviors needed, amount of mental agility required, ability to share or relinquish control, and level of *systems thinking*. Ideally engineers will be able to recognize when one approach is advantageous over the other given the context and be able to move from solving problems to creating solutions when needed while becoming comfortable and proficient in applying the process of creating solutions to both technical and non-technical challenges. As an additional bonus, when mastered, it can also help shape many aspects of your personal life.

When working in a team or group, it is essential to embody the desire to achieve *diversity of thought* among group members. It is key to understand what this means, what it looks and feels like, its advantages and how each member can work towards achieving this state. Recognize that the concept of *inclusion* is a pre-requisite for *diversity of thought* and that both are pre-requisites for exceptional or "vibrant" working and learning environments for the group where the collective intelligence of the group far exceeds the sum of individual member intelligence. Ideally, the group is within an overall organization that also offers a vibrant or highly effective working and learning environment.

Engineers will benefit by being open to and curious about new tools needed – knowledge, habits, behaviors, attitudes, and skills – to support your engineering work, like those required to support the two opportunity areas listed. It will be helpful to recognize that acquiring this new non-technical knowledge is not simple and it will take time and can be challenging to learn and master especially if you are in an environment where leaders and those in power do not emphasize the importance of these topics or serve as effective role models. Some of these non-technical topics are complex but engineers are used to dealing with complexity in technical areas and therefore should be able to translate this ability to other areas. Perhaps some engineers do not know what they know about embracing and understanding complexity. Instead, within many engineering environments, the treatment of these topics is superficial or even absent as if these things will organically occur and are simple to master. This brings Einstein's quote to my mind, "Everything should be as simple as it can be, but not simpler."

Over my career I gained momentum when I began to see how the value of *systems thinking* that I learned in engineering began to inform my later thinking about systems change in organizations like universities. These two areas began to synergize and revealed new opportunity areas for my own research and administrational efforts. I gained more motivation through my passion for creating new approaches to improving the inclusivity of learning environment for students and working environment for faculty and staff. My efforts are made possible by funding programs that focus on increasing the representation of women in the STEM workforce. However, the intended environmental improvements that I create benefit everyone, not only women. It is crucial that universities work deliberately towards creating inclusive campus cultures. Transforming the culture within a university alters its fabric. It is complex work and creative engineers who believe in the importance of inclusion and *diversity of thought* are well suited for leading these efforts, especially those who lead and create highly functioning teams and who are open to learning about new approaches and tools while relying on key experts to fill in knowledge gaps. I have been able to use my passion, strengths plus some luck, and a great deal of privilege to do this type of work. I see similar assets in other engineers and hope that more will become engaged in these types of efforts in the future.

For young engineers and engineering students, you should know that these changes are underway and you should keep an eye out for these areas of opportunity and other areas of growth and innovation that are sure to come in the next 10 years. In preparation for these changes and based on my past experiences, I recommend that you develop a mindset that allows you to stay curious, identify and pursue areas of opportunity, wherever you may find them, and to look beyond problems to create expansive solutions.

References

1. R. Fritz, *The Path of Least Resistance*. New York, NY, USA: Fawcett Books, 1989.
2. K. T. Greenfeld. "Meet the Napster - Shawn Fanning was 18 when he wrote the code that changed the world. His fate, and ours, is now in the court's hands." Time Magazine. 02 Oct.

2000, http://content.time.com/time/subscriber/article/0,33009,998068,00.html (accessed Jul. 29, 2021).
3. "Program helps satisfy need for creative outlet." RIT News. Rochester Institute of Technology. https://www.rit.edu/news/program-helps-satisfy-need-creative-outlet (accessed Aug. 1, 2021).
4. *Electrical Code Revision Cycle Handbook*, New York, NY, 2018. [Online]. Available: https://www1.nyc.gov/assets/buildings/pdf/electrical_code_handbook_2018.pdf#:~:text=In%201889%2C%20regulations%20governing%20outdoor%20wiring%20were%20issued.,the%20United%20States.%20This%20modern%20code%20provided%20a
5. M. B. Bailey, "Enhancing life-long learning and communication abilities through a unique series of projects in Thermodynamics course sequence," in *Proc. 2007 ASEE Annu. Conf. and Expo.,* Honolulu, HI, Jun. 2007.
6. M. B. Bailey, "Studying the impact on mechanical engineering students who participate in a unique series of projects in thermodynamics," in *Proc. 2011 ASEE Annu. Conf. and Expo.,* Vancouver, B.C., Canada, Jun. 26–29, 2011.
7. R. M. Muckel and M. B. Bailey, "Combining thermodynamics and public policy: Exploring the benefits of this alternative instructional method," in *Proc. 2014 IEEE Frontiers in Education Conf.*, Madrid, Spain, Oct. 22–25, 2014.
8. M. Moran, H. Shapiro, D. Boettner, and M. Bailey, *Fundamentals of Engineering Thermodynamics*, 9th ed. Hoboken, NJ, USA: John Wiley & Sons, Inc., 2018.
9. "Report of the Committee on Equal Opportunities in Science and Engineering to the United States Congress," CEOSE 2000, p. 41. [Online] Available: https://www.nsf.gov/pubs/2002/nsf02107/nsf02107.pdf 3 May 2021
10. National Science Foundation, "Women, Minorities, and Persons with Disabilities in Science and Engineering," Arlington, VA: National Center for Science and Engineering Statistics (NCSES), Directorate for Social, Behavioral and Economic Sciences. National Science Foundation. NSF 21-321. Tables 2-1 & 3-1, Apr. 29, 2021. [Online]. Available: https://ncses.nsf.gov/pubs/nsf21321
11. National Science Foundation, "Women, Minorities, and Persons with Disabilities in Science and Engineering," Arlington, VA: National Center for Science and Engineering Statistics (NCSES), Directorate for Social, Behavioral and Economic Sciences. National Science Foundation. NSF 21-321. Table 9-9, Apr. 29, 2021. [Online]. Available: https://ncses.nsf.gov/pubs/nsf21321
12. J. DeAro, S. Bird, and M. R. Shermaine, "NSF ADVANCE and gender equity," *Equality, Diversity and Inclusion: An International Journal*, vol. 38, no. 2, pp. 131–139, 2019, doi: https://doi.org/10.1108/EDI-09-2017-0188.
13. National Science Foundation, "Women, minorities, and persons with disabilities in science and engineering," 2019. [Online]. Available: https://ncses.nsf.gov/pubs/nsf19304/
14. J. Yen, E. A. Riskin, C. Margherio, J. H. Spyridakis, C. M. Carrigan, and C. Ana Mari, "Promoting gender diversity in STEM faculty through leadership development," *Equality, Diversity and Inclusion: An International Journal*, vol. 38, no. 3, pp. 382–398, 2019, doi: http://dx.doi.org/https://doi.org/10.1108/EDI-09-2017-0181.
15. S. E. Page, *The Diversity Bonus: How Great Teams Pay Off in the Knowledge Economy*, vol. 2, Princeton, NJ, USA: Princeton University Press, 2017.
16. K. W. Phillips, "How diversity works," *Scientific American*, vol. 311, no. 4, pp. 42 – 47, Oct. 2014.
17. "WE@RIT." Women in Engineering at RIT. Kate Gleason College of Engineering. Rochester Institute of Technology. https://www.rit.edu/engineering/women-engineering (accessed Aug. 1, 2021).
18. E. Dell, E., M. B. Bailey, S. O'Hurley, and E. Kohl. "WE-IMPACT- Women in engineering - Improving program assessment tools for outreach and retention programs," in *Proc. 2011 ASEE Annu. Conf. and Expo.,* Vancouver, B.C., Canada, Jun. 26–29, 2011.

19. E. DeBartolo, M. B. Bailey, W. Scorse, S. Gillow, and R. Liccion, "An evaluation of the effectiveness of traveling engineering activity kits (TEAK) in pre-college classrooms," in *Proc. from Int. Mechanical Engineering Congr. and Expo. 2009,* Lake Buena Vista, FL, Nov. 2009.
20. M. B. Bailey, E. DeBartolo, J. Mozrall, and J. Olney, "Creating a comprehensive women in engineering organization using a managed resource strategy," in *Proc. 2008 ASEE Annu. Conf. and Expo.,* Pittsburg, PA, Jun. 2008.
21. E. DeBartolo and M. B. Bailey, "Making engineering appealing: Programs for grades 6-12 girls," *J. of Engineering Education*, vol. 23, no. 5, pp. 853–860, 2007.
22. M. B. Bailey and E. DeBartolo, ""Heat transfer" Traveling engineering activity kit: Designed by engineering students for middle school students," i in *Proc. from Int. Mechanical Engineering Congr. and Expo. 2007,* Seattle, WA, Nov. 2007.
23. E. DeBartolo and M. B. Bailey, "Traveling engineering activity kits (TEAK) – Energy and the environment: Designed by college students for middle school students," in *Proc. 2007 ASEE Annu. Conf. and Expo.,* Honolulu, HI, Jun. 2007.
24. M. B. Bailey and E. DeBartolo, "Creating a community for women engineers at RIT," in *Proc. 2005 ASEE Annu. Conf. and Expo.,* Portland, OR, Jun. 2005.
25. E. DeBartolo and M. B. Bailey, "A continuous series of outreach programs to recruit young women to engineering," in *Proc. 2005 ASEE Annu. Conf. and Expo.,* Portland, OR, Jun. 2005.
26. "Women in technology." College of Engineering Technology. Rochester Institute of Technology. https://www.rit.edu/engineeringtechnology/wit/ (accessed Aug. 1, 2021).
27. "Women in science." College of Science. Rochester Institute of Technology. https://www.rit.edu/science/immerse-yourself/women-science (accessed Aug. 1, 2021).
28. "Women in computing." Golisano College of Computing and Information Sciences. Rochester Institute of Technology. http://women.rit.edu/ (accessed Aug. 1, 2021).
29. National Science Foundation, "ADVANCE: Increasing the participation and advancement of women in academic science and engineering careers," NSF Wide Programs ADVANCE Program Information 2018. [Online]. Available: https://www.nsf.gov/funding/pgm_summ.jsp?pims_id=5383
30. "EFFORT@RIT website." Grant Research History. Advance RIT Program. Rochester Institute of Technology. https://www.rit.edu/advance/our-archives (accessed Aug. 1, 2021).
31. C. M. Marchetti, M. B. Bailey, S. Mason, S. Baum, and M. Valentine, "Perceived levels of faculty value, influence, and satisfaction by gender, rank, ethnicity, college, and department at a large private university," in *Proc. 2012 ASEE Annu. Conf. and Expo.,* San Antonio, TX, 10–13 Jun. 2012.
32. M. B. Bailey, C. M. Marchetti, E. DeBartolo, S. Mason, S. Baum, J. Mozrall, M. Valentine, and G. Williams, "Establishing the foundation for future organizational reform and transformation at a large private university to expand the representation of women faculty," in *Proc. 2011 ASEE Annu. Conf. and Expo.,* Vancouver, B.C., Jun. 26–29, 2011.
33. M. Bailey, S. Baum, S. Mason, J. Mozrall and M. Valentine, "*RIT EFFORT Career Life Survey," EFFORT@RIT*. Rochester Institute of Technology. Oct. 2009. [Online]. Available: https://www.rit.edu/nsfadvance/sites/rit.edu.nsfadvance/files/docs/faculty%20career%20life%20survey.pdf
34. M. B. Bailey, S. Baum, S. Mason, J. Mozrall, M. Valentine, C. Marchetti, G. Williams, E. DeBartolo and S. LaLonde, "*Establishing the foundation for future organizational reform and transformation at a large private university to expand the representation of women faculty:* Report to the Rochester Institute of Technology community," [Online]. Available: https://www.rit.edu/nsfadvance/effort-it-catalyst-study May 2012
35. "AdvanceRIT program website." Advance RIT Program. Rochester Institute of Technology. https://www.rit.edu/advance (accessed Aug. 1, 2021).
36. M. B. Bailey, N. Sharma, L Verschage, and T. Conlon, "Creating an inclusive, vibrant learning environment within a large, software engineering program experiential learning experiences created for students, faculty, and senior design coaches & sponsors," presented at the *2021 Collaborative Network for Engineering and Computing Diversity*, Crystal City, VA, Jan. 2021.

37. E. Dell, M. B. Bailey, E. Litzler, M. James and E. Affolter, "The development and evaluation of an ADVANCE professional development series to promote institutional transformation," *ADVANCE Journal*, vol. 1, no. 2, 2019.
38. C. Marchetti and M. B. Bailey, "The importance of "how" in faculty salary equity studies," presented at the *2019 Leadership in Higher Education Conf.*, St. Louis, MO, Oct. 2019.
39. M. Bailey, M. Valentine, E. Dell and C. Marchetti, "Final outcomes report: Rochester Institute of Technology: Advance RIT," NSF ADVANCE Program, Feb. 28, 2019. [Online]. Available: https://ssrn.com/abstract=3344659
40. M. B. Bailey, L. Manning, E. Dell, C. Marchetti and M. Valentine, "Reimagining and refining campus cultures: Promoting an inclusive learning environment at RIT," presented at the *SIGITE 2017: The 18th Annu. Conf. on Information Technology Education/Research Proc.*, ACM, Rochester, NY, Oct. 6, 2017.
41. C. Marchetti, M. Valentine, S. Mason, M. B. Bailey and E. Dell, "No more duct tape! Institutionalization of ADVANCE initiatives," in *Proc. 2017 ASEE Annu. Conf. and Expo.*, Columbus, OH, June 25–28, 2017.
42. E. Dell, C. Marchetti, K. Crawford, M. Bailey, M. Valentine and S. Mason, "Career navigation initiatives for women stem faculty in support of institutional transformation," in *Proc. 2017 ASEE Annu. Conf. and Expo.*, Columbus, OH, June 25–28, 2017.
43. S. Mason, E. Dell, R. Garrick, M. Bailey, C. Marchetti and M. Valentine, "Advocates & allies: Engaging more men in institutional transformation," in *Proc. of 2016 ACM & Anita Borg Grace Hopper Celebration of Women in Computing*, Houston, TX, Oct. 19–21, 2016.
44. S. Mason, M. Bailey, M.D. Sorcinelli and S. Wadia-Fascetti, "Advancing diversity and inclusivity in STEM education," presented at the *SIGITE 2016: The 17th Annu. Conf. on Information Technology Education Proc.*, Boston, MA, Sep. 30, 2016.
45. E. Dell, S. Mason, M. Bailey, C. Marchetti, M. Valentine and K. Crawford, "*Connectivity* at RIT-Developing and delivering an effective professional development workshop series for women faculty in STEM," in *Proc. 2016 ASEE Annu. Conf. and Expo.*, New Orleans, LA, Jun. 26–29, 2016.
46. C. Marchetti and M. Bailey, "Into the light: Diffusing controversy and increasing transparency in the AdvanceRIT salary equity study process," in *Proc. 2016 ASEE Annu. Conf. and Expo.*, New Orleans, Jun. 26–29, 2016.
47. C. Marchetti, S. Mason, M. Valentine, E. Dell and M. Bailey, M., "Collaboration: The key to leading institutional transformation," in *Proc. of 2015 ACM & Anita Borg Grace Hopper Celebration of Women in Computing*, Houston, TX, Oct. 14–16, 2015.
48. S. Mason, M. Bailey, C. Marchetti, E. Dell, A. Rommel, M. Valentine and L. Clayton, "AdvanceRIT *Connect* grants: Driving momentum for disruptive change for women STEM Faculty," in *Proc. 2015 ASEE Annu. Conf. and Expo.*, Seattle, WA, Jun. 14–17, 2015.
49. M. Bailey, C. Marchetti, S. Mason, E. Dell and M. Valentine, "Institutional transformation guided by a multi-frame organizational analysis approach," in *Proc. 2015 ASEE Annu. Conf. and Expo.*, Seattle, WA, Jun. 14–17, 2015.
50. S. Mason, C. Marchetti, M. Bailey, D. K. Crawford, S. Foster, E. Dell, L. Clayton, S. Baum and M. Valentine, "Institutional transformation at a large private technical university: Reporting on the accomplishments of a combined data driven top down and bottom up approach," in *2014 IEEE Frontiers in Education Conference Proc.*, Madrid, Spain, Oct. 22–25, 2014.
51. S. Mason, C. Marchetti, M. Bailey, S. Baum and M. Valentine, "Faculty awards at a large private institution: An indicator of evolving university values?" in *Proc. 2014 ASEE Annu. Conf. and Expo.*, Indianapolis, IN, Jun. 15–18, 2014.
52. S. Mason, M. Bailey, l. Clayton and C. Marchetti, "AdvanceRIT: Transforming an institution using data to inform creative interventions," n *Proc. of 2014 ACM & Anita Borg Grace Hopper Celebration of Women in Computing*, Phoenix, AZ, Oct. 8–10, 2014.
53. "NIH office of research on women's health statement on announcement of winners of the NIH prize for enhancing faculty gender diversity in biomedical and behavioral

science." National Institutes of Health. https://orwh.od.nih.gov/sites/orwh/files/docs/NIHPrizeGenderDiversity508C_0.pdf (accessed Sep. 22, 2021).
54. "Myers-Briggs personality test." The Myers & Briggs Foundation. https://www.myersbriggs.org/my-mbti-personality-type/take-the-mbti-instrument/ (accessed Aug. 1, 2021).
55. "AdvanceRIT tip sheets." Advance RIT Program. Rochester Institute of Technology. https://www.rit.edu/advance/overview-of-resources#tip-sheets(accessed Aug. 1, 2021).
56. L. Bolman and T. Deal, *Reframing Organizations: Artistry, Choice, and Leadership,* 6th ed. Hoboken, NJ, USA: John Wiley & Sons, Inc., 2017.
57. A. Austin, S. Laursen, A. Hunter and M. Soto, "Organizational change strategies to support the success of women scholars in Science, Technology, Engineering, and Mathematics (STEM) fields: Categories, variations, and issues," in *Proc. Annu. Conf. of the American Educational Research Association*, New Orleans, LA., Apr. 2011.

Margaret Bailey, P.E., began her journey as an engineer upon entering the Pennsylvania State University in 1983. After graduating with a B.S. in Architectural Engineering (1988), she worked as a building HVAC design engineer before successfully passing the professional engineers exam in 1993. That same year Bailey began studies toward a Ph.D. and graduated from the University of Colorado in Boulder (1998). Her dissertation involved experimental and analytical research, which gave her the opportunity to address an important question that she had been contemplating, based on her time as a practicing engineer. Could an artificial intelligence-based fault detection system observe the behavior of building mechanical systems when no one was present and alert the building engineers when a fault was likely occurring? Bailey worked under Professor Jan Krieder's advisement while developing this artificial neural network-based system that successfully detected faults within mechanical equipment.

In 1998, she joined the faculty in the Department of Civil and Mechanical Engineering (ME) at the United States Military Academy (USMA) at West Point where she taught courses in thermodynamics and created an active research program. Due to her passion focused on increasing the representation of women within engineering, she was actively involved in the USMA Margaret Corbin Forum, which is dedicated to improving diversity-related issues at the USMA, and she ensured the creation of the SWE USMA Student Section and served as its first advisor. Although she left USMA in 2003 for an ME faculty position at Rochester Institute of Technology (RIT), she has served since 2009 as a member of the USMA Mechanical Engineering Program External Advisory Board. Bailey is now a Professor of ME within the Kate Gleason College of Engineering, RIT. Past appointments include the Founding Director of AdvanceRIT Program (2012–2020), the Co-Chair of the President's Commission on Women (2007–2020), the inaugural Sr. Faculty Associate to the Provost for Women Faculty (2010 –2018), the Founding Director of WE@RIT (2004–2011), and the inaugural Kate Gleason Endowed Chair (2003–2009). She teaches courses and conducts research related to thermodynamics, engineering and public policy, engineering education, and gender in engineering and science. Bailey is the co-author of a widely used engineering textbook, *Fundamentals of Engineering Thermodynamics*, and she has published over 90 refereed journal and conference proceedings papers.

Bailey served as the Principal Investigator (PI) or co-PI on several National Science Foundation funded efforts, including the RIT ADVANCE Institutional Transformation (IT) grant (2012–2019, PI), the ADVANCE IT-Catalyst grant (2008–2011, PI), the ADVANCE Partnership grant focused on faculty salary equity practices and study (2021–2026, co-PI), and a PFE:RIEF grant focused on "Understanding the Relationships Between Gender, Self-Efficacy, and Extracurricular Activities in the Professional Formation of Engineers" (2021–2023, co-PI). The projects that she has led have resulted in impactful new programs, practices, and policies as well as a dedicated ADVANCE unit within the Office of the Provost and the women in engineering center, WE@RIT, within the Kate Gleason College of Engineering. Both of these units have received national recognition. In August 2021, RIT was one of ten universities awarded the National Institute of Health (NIH) Prize for Enhancing Faculty Gender Diversity in Biomedical and Behavioral Science, which is in large part due to the impact of the RIT ADVANCE program. Dr. Bailey is a recipient of the national Maria Mitchell Women in Science Award, the RIT Edwina Award for Gender Diversity and Inclusiveness, the RIT Isaac L. Jordan Sr. Faculty Pluralism Award, and the WEPAN 2008 Women in Engineering Program Award in recognition of the WE@RIT program's offerings and impact for diversifying engineering.

Section II
New Perspectives

Educating the Next Generation of Mechanical Engineers in Fluid-Thermal Sciences

Francine Battaglia, Lu Chen, Mirka Deza, and Bahareh Estejab

The Nexus of Education

Our story begins with the aspirations of one person that brings together the lives of three more. Collectively, they pursue careers educating the next generation of engineers. These experiences document the evolution of mentoring and elucidate how diversity and inclusivity can be achieved. The next section presents how four women perceived the world and themselves. They experienced challenges and questioned whether their ambitions were achievable. That is nothing new and everyone faces doubts. But what binds them is mentoring and the chain reaction as it is passed from one to another. Each woman, armed with their doctorates, now educate, and do so differently but with the commonality of their experiences. The stories are assembled to provide insights on how to educate and prepare undergraduate students for engineering careers, graduate school, industry, and academia. While the techniques may not all be novel, it is the advice and revelations that are most important. The thread connecting each story is having someone believe in you.

F. Battaglia (✉)
Department of Mechanical and Aerospace Engineering, University at Buffalo, The State University of New York, Buffalo, NY, USA
e-mail: fbattagl@buffalo.edu

L. Chen
Lam Research, Fremont, CA, USA
e-mail: Lu.Chen@lamresearch.com

M. Deza
Department of Mechanical Engineering, Iowa State University, IA, USA
e-mail: mdeza@iastate.edu

B. Estejab
Department of Mechanical Engineering, Manhattan College, New York, NY, USA
e-mail: bestejab01@manhattan.edu

M. Bailey, L. Shackelford (eds.), *Women in Mechanical Engineering*, Women in Engineering and Science, https://doi.org/10.1007/978-3-030-91546-9_4

The Value of Mentoring Aspiring Engineers

Enter Francine Battaglia. In seventh grade, I announced that I wanted to be an engineer. I had the aptitude and it sounded cool. But I didn't really know what it meant to be an engineer. That is where my first role model emerged: my mother. She was determined to help me learn about engineering and found a local after-school program with the US Corps of Engineers and enrolled me. My cursory understanding of engineering through the program didn't waiver, and I applied to universities convinced I wanted to be an engineer.

I attended the University at Buffalo (UB) while living at home. Being local, I often spoke with my high school chemistry teacher (who earned his doctorate when I was a student in his class) to share my experiences. One day he asked if I ever thought of becoming a teacher, and I laughed, immediately retorting that there was no way I could stand in front of people and talk. Obviously, he had the last laugh when he learned years later that I attended graduate school and planned to become a professor. I won't discuss the trials and tribulations of being one of the few females in my mechanical engineering classes; I tried not to think about it and pushed forward, hoping to persevere. Was it difficult? Yes. But I didn't want my gender to be thought of as an excuse, and definitely not a weakness. I was fortunate to have wonderful people mentor me, and I never thought twice that they were men. Their advice and encouragement shaped the person and engineer I've become, and now I get to do the same.

As a tenure-track assistant professor at Iowa State University (ISU), I was somewhat naïve. It was not about what I needed to do in my job, but what I *could* do in my job: be a mentor and inspire the next generation of engineers, especially women. It never crossed my mind that I had another role to play. In fact, one of the first two graduate students whom I hired was a woman. It was a very easy decision as I was looking for a student with a background and interest in computational fluid dynamics (CFD). I took it for granted that I had hired a female, because in my mind, I was looking for the best candidate.

As the years progressed, I realized I could make a difference in the lives of young women, and this notion strengthened when I moved to Virginia Tech (VT) as an associate professor. Mirka Deza and two other graduate students followed me from ISU to VT. Deza would be the second female student to pursue a doctorate under my supervision. She also shared a commonality with my first female doctoral student. Both women worked in industry before pursuing doctorates and had a different appreciation for their education. Something else was happening at VT that I hadn't seen before, the number of females in my undergraduate courses was uncommonly large. I am not sure how long it would have taken me to notice that more women were in my classes, but it became apparent when they had to get my permission to force-enroll because my section was at full capacity. My new role started to emerge: encourage, support, council, and inspire these young women who needed to see someone like themselves break the glass ceiling.

Enter Mirka Deza. I was born and raised in Lima, Peru, where from an early age my parents taught me the value of education. They made huge sacrifices to give my siblings and me opportunities they did not have to succeed. As a result, we became engineers. My older sister is a civil engineer with a PhD, designing and analyzing bridges in Minnesota. My brother is a mechanical engineer and leads new technologies in a copper/zinc mine in Peru. My younger sister is also a mechanical engineer and the owner of an engineering project company in Peru specializing in process automation.

As a woman in mechanical engineering, my journey has not been a stranger to challenges. I was one of the few female undergraduate students in my major at the Pontificia Universidad Catolica del Peru. I felt that I had to work harder than the rest to be considered "one of the guys." My first teaching experience was working as an undergraduate teaching assistant for a thermodynamics course. I graduated with a strong academic background from a university highly ranked internationally. But I longed for a female engineer role model; sadly, even when I worked in industry, I was not able to find a mentor.

In 2001, my husband found a consultant job in the USA, so when we moved, I decided to get a master's degree with the intent to return to industry. As a graduate student at ISU, I experienced being a minority again, this time as an international student and as the only female pursuing a graduate degree while having her first child. I once again put in extra work to succeed academically while taking care of my baby daughter; thankfully, I had the support of my husband and our parents. During this time, I met wonderful professors (some of them are my colleagues now) and finally found a mentor in my heat transfer professor, Dr. Francine Battaglia.

Seeing the impact that her dedication had on her students made me realize the critical role of a mentor. Battaglia was such a great influence in my professional career that my family and I moved to Virginia so I could continue my doctoral degree at VT under her guidance. Besides teaching me the fundamentals of CFD and numerical modeling, she passed onto me a passion for the fluid-thermal sciences, and taught me how to write papers, present at conferences, and teach. She provided me with an opportunity to teach fluid mechanics with her, observed my teaching, and provided feedback and invaluable advice. My skills improved and made me an effective teaching professor. Battaglia was very supportive of my family responsibilities, especially when my daughter was in primary school and halfway into my PhD studies, my son was born. She helped me plan and prepare for my preliminary exam and worked with my schedule. Years later, I wanted to be not just a teaching professor, but a mentor to all my students, and especially a role model to minority students in engineering. Living and working in this wonderful country helped me recognize the intrinsic value that diversity of experiences and ideas brings to the learning and working environment.

Enter Bahareh Estejab. I vividly remember the day I received my acceptance letter to one of the most prestigious universities in my birth country of Iran. I had worked toward this goal for years and was overjoyed with a great sense of accomplishment. I owed it all to my family's support and encouragement, especially my first inspirational role model, my mom, who believed in me and taught me that I can do anything. During orientation, as I walked through the hallways, classrooms, and libraries, I looked around eagerly absorbing my surroundings, thinking it was just the beginning and I would be an engineer very soon.

Unlike segregated girl-boy schools before college in Iran, all students sat together in college classes. During my first freshman semester, less than 5% were female students and the faculty were male. Excited to begin the new chapter of my life, none of this bothered me, nor even surprised me! My second semester math class was taught by a very knowledgeable and helpful professor, who was also a female. She talked about her research passionately, and my dream of working in industry transformed. I wanted to be a professor like her, pursuing research and training young engineers. During my sophomore year, I was determined to find engineering classes taught by female faculty. Shockingly, there were no female faculty in the Mechanical Engineering Department, despite its large size. The lack of female role models was discouraging, and I could not see myself in academia anymore.

After graduating with my bachelor's degree, I worked in industry as an engineer and gave up my dream to become a professor, until a few years later the thirst for learning pushed me toward graduate school. Since I wanted to explore the world and experience other cultures, I decided to attend the University of Kentucky (UKY). Immigration was indescribable. It was as if you were born again as an adult, where everything was new and different. Everyday tasks were an obstacle for an immigrant: grocery shopping, driving, visiting the doctor were all challenging. (Grocery shopping would take hours until I learned "new food words" and recognized labels.) Despite years of learning English before coming to the USA, it was hard to understand native English speakers at first. There were times when I felt helpless and lost until I overcame these obstacles and became comfortable in my new country.

Lexington, KY, was a nice, quiet city and UKY was a great place to begin my graduate studies. My advisor, Dr. Sean Bailey, was a knowledgeable and kind professor who helped me grow as a researcher. However, I was still missing a female professor role model that I could look up to professionally. After graduating with my master's degree, I was torn between finding a job and pursuing a PhD. I felt one step closer to my ultimate career goal of being a professor, but the vision had not fully returned. Upon further contemplation, I applied to PhD programs just to have the option.

I was accepted in two of my top choices that were perfect for a graduate student at the beginning of her journey. Although both universities were appealing, VT offered something I needed desperately, the vision of my future career and the presence of female faculty. This need helped me overcome the daunting decision to move 500 miles and endure a long-distance relationship with my husband. My PhD advisor, Dr. Battaglia, was a role model and mentor. I learned how to manage classrooms, conduct research and in general be an academic without sacrificing who I was as a female in a male-dominated major. I continued at VT as a postdoctoral researcher for 1 year while raising my daughter who was born a few months after I defended my dissertation. My dream of being a faculty member was restored, and I was ready to start my career in academia.

Enter Lu Chen. My path towards engineering began when I realized I enjoyed math and science. Growing up in China, I was captivated by how engineering contributed to society. Engineering changes our lives in every aspect, from modern homes, bridges, space travel, vehicles, and the latest cellular technology. Realizing that engineering can be a driving force to change the world, I was convinced that it would be very rewarding to be a part of contributions that advance society. With my growing curiosity and passion, I went to Southeast University, one of the top engineering colleges in China.

My first role model that broke the clichés of women was a female physicist, Dr. Chien-Shiung Wu, who graduated from my undergraduate college. She had made significant contributions in nuclear physics by developing the process for separating uranium into uranium-235 and uranium-238 isotopes. To me, she was the "Chinese Madame Curie" and inspired me to break the invisible barriers and pursue my interest in engineering and research.

However, the journey was not very easy for me. I found it difficult to speak up and lead projects in all-male groups. When the project teams met in the male dormitory because it was easier for the other team members, I felt unimportant and dejected, and my confidence waned. To make matters worse, there were concerns female students would not be able to handle heavy-labor engineering lab tests. The negative experiences during my undergraduate and first-year master programs were not as rewarding as I expected. As a female seeking to find my place in engineering, I desperately needed mentoring and guidance. After sharing my struggles with one of the professors in the undergraduate college, he suggested that I seek opportunities abroad and change my study focus to computational analyses. Finding that overseas universities had more female professors, I contacted Dr. Battaglia about graduate school opportunities and she encouraged me to apply.

I officially started my journey abroad to pursue a doctorate at Virginia Tech in 2013. During the next 4 years of graduate school and 1-year postdoctoral experience, Battaglia's everyday demeanor answered my fundamental question: women *can* lead in engineering. I was very impressed by her leadership in the profession and realized that it was possible for women engineers to lead in scientific research. At that moment, the answer I sought for years was vividly standing in front of me and I felt that I could become a leader, too.

Battaglia was not only a role model but a mentor and taught by example. During my studies, I learned how to write concisely and deliver clear messages to an audience. Her work ethic was a constant motivator to work harder and prepare myself to be a successful researcher. In addition to learning from Battaglia's expertise in CFD, the 5-year experience taught me to hone my strengths as a woman. For example, women are very diligent and attentive to details, helping us create work plans and priorities, multitasking, and exploring small nuances that could turn into valuable findings. I developed many good habits from her lessons, which continue to benefit my career and personal growth. I was fortunate to travel abroad and find a great role model, which made me realize how important it is to have successful women engineers around you for inspiration and motivation. Now I am part of a "special" group, and I can carry forward what I learned to become a leading expert in my field and inspire future women engineers.

Exit doctoral recipients. After graduating from Virginia Tech and parting ways from their advisor, Deza, Estejab and Chen took different paths, each one with the conviction that they could inspire and prepare new generations of mechanical engineers. Deza started a postdoctoral position and then became an assistant teaching professor focusing on the education of undergraduate students in one of the largest mechanical engineering programs in the USA at ISU. Estejab also held a postdoctoral position and is now a tenure-track assistant professor at Manhattan College pursuing her research interests and guiding undergraduate and graduate students. Chen followed Battaglia to UB to work as a postdoctoral researcher before finding a modeling engineer position at Lam Research, where she balances the different demands of working in industry and coaching engineers and graduate students, and most recently, raising her newborn daughter. Battaglia, having returned to her alma mater of UB, was given opportunities to lead and mentor her colleagues. In 2019, she was appointed acting associate dean for faculty affairs in the School of Engineering and Applied Sciences, and in 2020, assumed the role as department chair in Mechanical and Aerospace Engineering.

These women now share how they continue the tradition of mentoring and how they have adapted their skills to educate future engineers.

Preparing Undergraduates for a Career in Mechanical Engineering

After graduation, Dr. Deza worked as a postdoctoral research associate at ISU investigating numerical models for natural ventilation and taught a freshman class. Being an instructor was a rich learning experience, but she had to adapt to new ways by which undergraduate students expected delivery of the class. Now as an assistant teaching professor at ISU, Deza teaches core fluid-thermal classes from freshman to senior levels and continues to evolve her teaching methods.

One effective classroom tool is active learning, where students solve problems and receive direct supervision and feedback, reinforcing problem-solving methodologies. Other classroom techniques such as individual and small group brainstorming help further the learning process. Providing active learning opportunities in the classroom can increase student confidence. Using active techniques underscores how students learn differently and how expectations have changed. Filling a blackboard with equations and words is truly becoming a thing of the past.

The popularity of the flipped classroom (lecture notes delivered as a short video and exercises solved in the classroom) inspired Deza to use a combination of effective teaching methods and engaging engineering content. For the pedagogical aspect, implementing effective methods needs a flexible approach for instructing and supporting students with different learning styles, majors, and academic levels

[1, 2]. For the academic aspect of teaching, it requires integrating research applications and current technology associated with the course content. Providing students with various learning methods, studying habits, and test-taking strategies [3, 4] are ways in which Deza helps improve their educational experience. An example Deza uses to expand beyond the classroom is traveling with the students to visit a local power plant.

In spring 2020, when the classroom moved from in-person to online teaching, the endurance and creativity of college professors were put to the test. For many female professors, caring for their families added an extra load to an already challenging situation. Faculty adapted quickly to learn basics of online teaching while helping students adapt and navigate widely different situations and resources [5–7]. Deza made the transition as smoothly as possible by offering synchronous classes and the opportunity to interact with one another while living in isolation. Student evaluations revealed how much they valued being able to see their professor and peers, ask questions and work in groups. To make online delivery successful, Deza exploited technology by using one device to monitor and answer questions immediately and another device to write class content and problems, while recording the class.

Deza strives to teach her students more than technical content by instilling personal and professional values such as integrity, hard work, appreciation for an inclusive learning environment, and the love for learning. For Deza, the most rewarding experience is hearing from students that found the same passion for the energy fields that she found years earlier. To know she inspires struggling students that were planning to leave the engineering program but remained fills her with joy. Deza finds that encouragement is especially important for minority students to stay motivated. She imparts her personal experiences to encourage them to be the best in their fields so they can grow professionally, earn promotions or recognitions they deserve, and become role models to new generations.

Educating and Training Graduate Students

After working as a postdoctoral researcher, Dr. Bahareh Estejab began her career in education and joined Manhattan College as an assistant professor teaching undergraduate and graduate classes, while conducting and supervising research in the fluid-thermal sciences. Teaching core courses in fluid mechanics, thermodynamics, and special topics such as compressible flow, she endeavors to integrate new technologies to make classes more enjoyable and appealing to students. Estejab's acquired knowledge in the discipline of pedagogy is from VT courses she completed to earn the "Future Professoriate Graduate" certificate, seminars, and workshops. Coupled with her teaching experiences, she knew that the classroom should be consonant with the students' generation to encourage learning. Knowing that students today are fascinated by social media, Estejab joined the trend, creating an Instagram account for each of her classes. One activity is that students upload a picture to the class Instagram page of a real-life event and write a paragraph explaining how the entry relates to the classroom material. Doing so helps students conceptualize the practical and everyday use of theories taught in class. She emphasizes real-life and industrial examples in her lectures and provides students with a list of beneficial and related educational videos to watch. Estejab has found that building a link between theory and its potential application in industry, while creating an enjoyable classroom environment, motivates students.

Estejab brings to the classroom her expertise and contemporary research in renewable energies with a focus on wall-bounded two-phase flows encountered in many industrial applications. Despite their widespread applications, there are still many unknown key issues about the flow behavior such as the flow regime, instabilities, and pressure changes. Estejab has the rewarding experience of training students in computational methods to accurately predict the behavior of two-phase flows. While supervising graduate and undergraduate research, one of the more subtle goals is teaching them to conduct research correctly and professionally. A technique Estejab finds effective is providing students enough freedom to find their path. She offers students a vision of the research and its ultimate goal upfront rather than giving them step-by-step instructions. Estejab cultivates students' potential and capabilities by getting to know them, talking to them on a regular basis, and empowering them to find their true abilities. From her experiences, Estejab appreciates the value of having moral support during difficulties in research and thus supports her students professionally and personally.

Beyond maintaining responsibilities as a faculty member, being a female professor in a male-dominated major carries other unofficial responsibility. Estejab has become a role model for female students and guides them on their journey to become successful and confident engineers. The key is helping them realize their abilities and capabilities. It is very easy for female students to feel unappreciated, and encouragement goes a long way. Estejab's role as a mentor is gender inclusive, and she has also found that she can help male students understand unintentional biases and appreciate the equality between genders in STEM majors.

Preparing for a Career in Industry

After holding an appointment as a postdoctoral researcher, Dr. Lu Chen embarked on the next leg of her professional journey by joining a semiconductor company, which develops next-generation equipment for chipmakers. The question they seek to answer is how to engineer powerful memory chips that are incredibly small at the nanoscale? Chen's work focuses on integrating fluid-thermal modeling into multidisciplinary engineering to provide hardware design guidance for precise process control so that billions of memory cells can be built atom by atom. Different from focusing on one research topic when Chen was a graduate student, industrial work has short timelines to solve problems. The shift from theoretical study to application modeling also brought new requirements for using computer-aided design (CAD) software to defeature/build real-world engineering drawings, and familiarity with advanced meshing tools. The industrial modeling work involves a variety of topics with a limited time window to find a solution. Chen expanded her expertise from syngas combustion studied during her doctorate to conjugate heat transfer, compressible fluid flow, and gas-surface reactions. It is a testament that education is a degree in life-long learning.

At first, Chen's new role in industry was overwhelming, as it required new skills. However, her education equipped her with capabilities and training to deal with challenges and difficulties. Quickly developing and executing work plans, Chen was able to complete tasks while learning required skills and adapting to the industrial work-pace. In addition, the work required multidisciplinary knowledge and interfacing with cross-functional teams. Collaborating and interacting with engineers from chemical, electrical, and material science fields, as well as experimentalists, designers, and process engineers, has been an invaluable resource for broadening her knowledge in other engineering fields.

After 3 years, Chen's productivity earned her a promotion to senior engineer and the added responsibility to coach new employees. In her secondary role as a mentor, she is in a position to strengthen the confidence of other young men and women who want to succeed in industry. She currently coaches two engineers but from different backgrounds: one employee has 5 years of industrial experience as a modeler and the other is a recent graduate student with a solid theoretical background but almost no experience with modeling software packages. Engineers with different backgrounds provide a more diverse work environment where they can apply their talents. However, they still require on-the-job training to understand the company culture and product development, and to revisit fundamentals to deliver quick answers and solutions within a limited timeframe. This requires the engineer to be a self-motivated learner and prepared to solve the next industrial problem. Students in engineering should consider learning different modeling techniques and become familiar with a variety of modeling approaches for fluid-thermal related problems. For example, familiarity with CAD packages (e.g., NX, Solidworks, etc.), modeling packages, and programming using Matlab or Python will help them quickly adapt, complete tasks, and become valuable members of a team.

A Profession in Academia

Two decades after beginning a career in academia, Battaglia continues to find mentoring opportunities that support the STEM fields. She has taught thousands of undergraduate and graduate students, and supervised dozens of students who earned MS and PhD degrees. Not only did she educate students but grew professionally as a leader, and now offers advice based on her experiences.

In Battaglia's recent career move, she returned to her hometown of Buffalo for an opportunity to be a faculty member in the department that provided her engineering foundation. It was serendipitous bringing her career path full-circle and giving back to the community that gave her so much. Three years later, Battaglia was appointed department chair in May 2020 at the height of a global pandemic. Her immediate concerns were helping faculty and students resume research activities safely with the start of summer. However, an adjacent issue emerged: the pandemic was adversely affecting productivity of faculty, especially women, in the remote environment thrust upon society. Reports indicated disparities between men and women scholars, whereby productivity of women was declining. With their families at home, women were burdened with more domestic responsibilities [8, 9].

In some engineering settings, where gender and racial inequalities are still prevalent, the pandemic brought greater challenges [10, 11]. For untenured faculty, stress mounted as they tried to sustain their research momentum. Universities responded by providing their untenured faculty the option for a "COVID-19 stop-the-clock" [11, 12]. Not so long ago there was a stigma when untenured faculty requested a "stop-the-clock" to extend their probationary period. If it was a woman, it was assumed she had given birth. Although less common, men requested a clock stop typically associated with a research set-back. No matter the reason, tenure committee discussions would focus on the extra time and speculate how this provided an advantage to the tenure candidate. The notion that a clock stop is an advantage must now change. It remains to be seen if a clock stop will really help, but don't assume it will. Take action and talk to your junior faculty to gauge how they are doing. Offer assistance and craft a plan to help them be successful.

Another way to support faculty is by showing them that they are valued. Merit raises reward faculty for their productivity or outstanding contributions if money is available. Alternatively, nominating faculty for local and national awards can provide them additional motivation, while enhancing their career profile. The key is taking time to learn about your faculty and their professional interests and remember that women and minorities might be overlooked otherwise. As a starting point, nominating a colleague for professional awards or election to fellow membership provides them with international recognition, and can be used to leverage other award nominations.

There are two final recommendations for sustaining a career in academia. The first is to schedule time every week where you can focus on *your* work. Covet those hours and do not fill them with meetings; time for yourself during normal business hours becomes a pleasure and can increase productivity. This connects to occasionally saying *no*, that is, no to too many service requests. The recommendation to "just say no" is especially important for underrepresented minorities whose presence may have a secondary purpose to ensure a diverse committee. While selection of such individuals brings a richness of knowledge and experience, it can unfairly overburden an individual who fears that saying no might hurt their career. Of course, say *yes* when you find passion with the work or enjoy the benefits of being part of a team that can make an important impact. After all, we choose careers in academia because we want to make a difference.

Concluding Remarks

A career in mechanical engineering is rewarding. Find what motivates you, and you can accomplish anything. Four women found their career paths through strong mentorship, family and professional support, and a yearning to have a positive impact, whether it be in research, education, service, or all three. The exercise of drafting our stories provided a revelation that may not have been recognized otherwise. The collection of experiences unveiled how important it is to have a nurturing work environment. Although Battaglia did not have a female academic role model, she became one. At first, it was not obvious or conscious, but eventually she realized females were seeking her advice. Through her compassion and understanding, she helped these women see themselves in a "mirror of success."

For Deza, she found the support she needed. Not only is she a role model for engineering students but is an example that it is possible to pursue your professional goals while raising a family. Both of her children love engineering, and her daughter is pursuing a mechanical engineering degree. Estejab saw that possibility, too, when considering graduate school, and knew that she could thrive although separated by distance from her husband. She fulfilled her desire to become a professor and returns that passion and enthusiasm to her students every semester as she helps guide their professional life. For Chen, she desperately needed a strong female role model, and noting the diversity of Battaglia's research team, also found an extended family. As an engineer in industry, she applies her knowledge and innovation to change future technologies and empower team members to develop their skills.

Together, these women will forever be connected personally and professionally, and are poised to lead future generations of women and men in engineering. The intangibles are who you are and how you project yourself. Hold up a mirror and study the reflection.

References

1. R.M. Felder and R. Brent, *Teaching and Learning STEM: A Practical Guide*. San Francisco, CA, USA: Jossey-Bass, John Wiley and Sons, 2016.
2. S. Lord, M. Ohland, and M. Prince, *35th National Effective Teaching Institute Workshop*, American Society for Engineering Education, Philadelphia, PA, August 1-3, 2018.
3. R.M. Felder, "Memo to Students who are disappointed with their last test grade," *Chem. Eng. Edu.*, vol. 33, no. 2, pp. 136–137, 1999.
4. R.M. Felder and J.E. Stice, "Tips on Test-taking," *Chem. Eng. Edu.*, vol. 48, no. 1, pp. 57–58, 2014.
5. P. Semingson, P. Kavathas, and S. Silverman, "Basics of Teaching and Learning," *BOLT 2020*, Center for the Integration of Research, Teaching, and Learning, July-August 2020.
6. L. Hassal, "Applying the Quality Matters Rubric (APPQMR)," On-line Workshop, June 2020.
7. M. Lamm, "CELT Teaching and Learning Academy," Center for Excellence in Learning and Teaching, Iowa State University, Ames, IA, USA, Aug. 2020–May 2021.

8. C. Flaherty, "No Room of One's Own," *Inside Higher Ed*, April 21, 2020. [Online]. Available: https://www.insidehighered.com/news/2020/04/21/early-journal-submission-data-suggest-covid-19-tanking-womens-research-productivity
9. P. Vincent-Lamarre, C.R. Sugimoto, and V. Larivière, The decline of women's research production during the coronavirus pandemic, *nature index*, May 19, 2020. [Online]. Available: https://www.natureindex.com/news-blog/decline-women-scientist-research-publishing-production-coronavirus-pandemic
10. J.L. Malisch, B.N. Harris, S.M. Sherrer, K.A. Lewis, S.L. Shepherd, P.C. McCarthy, J.L. Spott, E.P. Karam, N. Moustaid-Moussa, J. McCrory Calarco, L. Ramalingam, A.E. Talley, J.E. Cañas-Carrell, K. Ardon-Dryer, D.A. Weiser, X.E. Bernal, and J. Deitloff, "Opinion: In the wake of COVID-19, academia needs new solutions to ensure gender equity," in *Proc. of the National Academy of Sciences*, vol. 117, no. 27, pp. 15378–15381, July 7, 2020, doi:https://doi.org/10.1073/pnas.2010636117.
11. S. Weissman, Universities Are Freezing Tenure Clocks. What Will That Mean for Junior Faculty of Color? *Diverse: Issues in Higher Education*, May 9, 2020. [Online]. Available: https://diverseeducation.com/article/175846/.
12. R. Khamis-Dakwar and J. Hiller, The Problems with Pausing the Tenure Clock, *Inside Higher Ed*, July 7, 2020. [Online]. Available: https://www.insidehighered.com/advice/2020/07/07/response-pandemic-better-alternatives-pausing-tenure-clock-should-be-considered.

Francine Battaglia is a Professor and Chair in the Department of Mechanical and Aerospace Engineering at the University at Buffalo (UB). She joined UB in 2017 after working 10 years at Virginia Tech and 8 years at Iowa State University. Dr. Battaglia is Director of the Advanced Simulations for Computing ENergy Transport (ASCENT) laboratory (formerly CREST), which solves problems in the fluid-thermal sciences using computational fluid dynamics and developing computational models. Current research interests include single- and multi-phase turbulent reacting flows with applications in gasification processes, alternative energy production, building energy utilization and particulate transport related to pollution and viruses. Dr. Battaglia has published over 135 refereed journal and conference proceedings papers and co-authored a textbook on natural ventilation. To date she has supervised a very diverse research team (with underrepresented minorities comprising 25–30% of the team) of 25 M.S. and 13 Ph.D. students, eight postdoctoral research associates, and currently supervises two M.S. and five Ph.D. students. She is a member of the American Institute of Aeronautics and Astronautics, elected Fellow of the American Society of Mechanical Engineers (ASME) in 2009, and Fellow of the American Society of Thermal and Fluids Engineers in 2019. Dr. Battaglia is serving as the Editor-in-Chief of the ASME *Journal of Fluids Engineering*. She earned a Ph.D. from the Pennsylvania State University (1997) and B.S. (1991) and M.S. (1992) degrees from UB.

Lu Chen is a Senior Modeling Engineer in the semiconductor company, Lam Research Corporation. Dr. Chen joined Lam Research in 2018 after working as a Postdoctoral Research Associate at the University at Buffalo (UB). She graduated with a Ph.D. in Mechanical Engineering from Virginia Tech (2017) and received a B.S. from Southeast University (2012). During Dr. Chen's Ph.D. and postdoctoral studies, she worked in the Computational Research for Energy Systems and Transport (CREST) lab at Virginia Tech and (renamed) ASCENT at UB, focusing on numerical modeling of turbulent combustion flow. During her Ph.D. at Virginia Tech, she also worked at the Virginia Tech Transportation Institute (VTTI), focusing on creating a fluid-structure interactive model to predict vehicle hydroplaning on roadways and develop mitigation strategies. Dr. Chen was Virginia Tech Student Campus Champion in the National Science Foundation Extreme Science and Engineering Discovery Environment (XSEDE) Program from 2015 to 2017, serving as a university representative using high performance computing for computational fluid dynamics modeling. Dr. Chen published multiple refereed journal and conference proceedings papers before joining Lam Research. Currently, Dr. Chen focuses on integrating fluid and thermal modeling to provide guidance for semiconductor hardware design and process control. Dr. Chen expands her expertise to conjugate heat transfer, compressible fluid flow under vacuum conditions, and particle trajectory analyses.

Mirka Deza is an Assistant Teaching Professor in the Mechanical Engineering Department at Iowa State University (ISU) and previously a Postdoctoral Research Assistant at the ComPM Lab working on numerical simulations of natural ventilation. She obtained her Ph.D. (2012) from Virginia Tech where she worked in the CREST Lab researching the hydrodynamics of biomass fluidized beds. She received an M.S. from ISU with a co-major in Biorenewable Resources and Technology and worked for the ISU Industrial Assessment Center providing energy, productivity, and waste management audits to small- and medium-sized manufacturing facilities. She earned a B.S. in Mechanical Engineering from Pontificia Universidad Catolica del Peru. Dr. Deza teaches core courses including Thermodynamics I and II, Fluid Flow, and Heat Transfer, Problem Solving with Programming, and Introduction to Mechanical Design. She co-advises the ISU Society of Hispanic Professional Engineers (SHPE).

Bahareh Estejab is an Assistant Professor in the Mechanical Engineering Department at Manhattan College (MC), NY. She received a Ph.D. in Mechanical Engineering from Virginia Tech (VT), an M.S. from the University of Kentucky (UKY), and a B.S. from Shiraz University (Iran). She holds a graduate certificate in computational fluid mechanics from UKY and a graduate certificate in university teaching from VT. She joined MC after working as a postdoctoral researcher at VT where she focused on numerical modeling of multiple phase flows with and without chemical reactions. Her current research interests are computational fluid dynamics, renewable/alternative energy, multiphase and reacting flows, turbulent flows, wall-bounded flows, and unsteady vortex flows. Dr. Estejab has very robust experience in interdisciplinary research, and she has been an active member of Iota Delta Rho Interdisciplinary Honor Society. She has published multiple refereed journal and conference papers in different areas of fluid mechanics. Dr. Estejab teaches core courses in mechanical engineering including Thermodynamics I and Fluid Mechanics I/II, while conducting and supervising research in fluid mechanics and heat transfer. Dr. Estejab has a variety of experiences in mentoring graduate students and is a member of multiple professional and honor societies.

Circular Systems and the Culture of Collaboration

Heather Dillon, Rachel Dzombak, and Chrissi Antonopoulos

Introduction

> As happens so often in trying to understand and discuss other cultures, wording trips up specialists. When they claim to have never found "a true matriarchy," these anthropologists are envisioning a mirror image of patriarchy, a vision that ignores the differing ways in which females conceptualize and wield power. (Ryan and Jetha, p. 133)

If a friend describes a hiking trip, you might have a vision of reaching beautiful waterfalls and mountains with little effort. Or perhaps you see only the beautifully crafted photo on Instagram, where the sweat has magically disappeared, and the hiker is in a triumphant pose with a stunning view behind her. Often the career of women in science, technology, engineering, and math (STEM) fields is narrated in the same way. You hear a famous name followed by a highlight reel of her amazing accomplishments in science and engineering. Often the rocky parts of the journey are not explored. All the times she lost her way or faced a setback have been omitted from the record. However, it is exactly the challenging times, the falls, bumps, lost paths, that make her successful, foster her resilience, and ultimately lead to success.

The number of female graduates in STEM fields has continued to increase, though the number of faculty members has remained stagnant [1]. While women account for nearly half of the US workforce, they only hold 25% of jobs in STEM occupations [2]. Many factors contribute to gender disparities in STEM fields.

H. Dillon (✉)
Department of Mechanical Engineering, University of Washington Tacoma, Tacoma, WA, USA
e-mail: hedillon@uw.edu

R. Dzombak
Haas School of Business, UC Berkeley, Berkeley, CA, USA

C. Antonopoulos
Pacific Northwest National Laboratory, Richland, WA, USA
e-mail: chrissi.antonopoulos@pnnl.gov

M. Bailey, L. Shackelford (eds.), *Women in Mechanical Engineering*, Women in Engineering and Science, https://doi.org/10.1007/978-3-030-91546-9_5

Researchers have found that the lack of supportive networks and gender identity in STEM fields plays a strong role in women's decision to leave the field [1]. In addition to striving for gender equity within organizations, there is a clear need to foster supportive environments for women in STEM that focus on enhancing women's roles in technical positions.

We hope to share with the reader some of our own experiences navigating the career path as a woman in engineering alone, and also the power of tackling a career journey with great collaborators. Building trust in other collaborators has been a key mechanism for how we found new paths and peaks together. Coming together has both fostered our individual abilities and confidence as well as supported us to pursue new paths that were only accessible as a group (Fig. 1).

This chapter leverages the hiking metaphor to explore the ways that women might build supportive networks and environments. We have used the techniques we outline here to build a trusting and productive research team that operates fully remotely. Together we have navigated complex trails on our own professional journeys. We have come to believe that who you hike with on your research journey may be more important than the destination. Along our journeys, our research team has also demonstrated resilience to challenges and the capacity to navigate ambiguity, with many changed directions and research questions over the years, supported by the confidence we have in each other.

The authors include three avid hikers and researchers. Dr. Heather Dillon is Professor and Chair of Mechanical Engineering at the University of Washington Tacoma, where her team is researching thermodynamics and energy systems. Dr. Rachel Dzombak is a Lecturer at UC Berkeley's Haas School of Business and runs an independent consultancy. In both roles, she supports individuals and

Fig. 1 Hiking is a good metaphor for the complex journey that many women in STEM fields experience. (Sketch by Heather Dillon)

organizations to foster innovation and solve complex problems, including how to enable a circular economy. Chrissi Antonopoulos is Senior Analyst at the Pacific Northwest National Laboratory where she leads a team focused on innovative energy research in building science. Together, our research has focused on the circular economy, life cycle assessment, equity, and economic drivers in the field of energy and environment.

Background

Why is it worth unpacking what positive collaboration looks like? At least for the women within this collaboration, finding academic partnerships where our voices are heard and our skills appreciated, and where mutual support is freely given, had been a rare occurrence. At the same time, due to the increasing complexity of today's challenges and problem spaces, working in interdisciplinary and diverse teams is increasingly critical in science and engineering [3]. Teaming is "a way of working that brings people together to generate new ideas, find answers and solve problems. But people have to learn to team; it doesn't come naturally" [4]. The latter point is critical. We often are placed or work in teams, though few individuals have ever learned the principles of good teaming.

Teaming matters because finding solutions to the complex, wicked problems that we are faced with today necessitates bringing together perspectives from multiple disciplines and teaming across disciplinary boundaries [5, 6]. When good teaming is practiced, teams are able to leverage the available diversity of team members. Leveraging diversity entails creating space for all of the ways that individuals see and solve problems [7]. How we see problems is informed by our cultural experiences, race, gender, socioeconomic background, and much more. For example, someone who grew up in Berkeley, CA, is going to see and understand the problem of climate change differently than someone who grew up in New Orleans, LA. Furthermore, diversity also encompasses the ways in which we solve problems and ensures that solutions disseminate to diverse groups. If your team members all have a background in engineering from the same school, then you are going to be limited in the approaches you bring to the problem at hand.

Research has shown that diverse teams either significantly under- or out-perform more homogeneous teams [8]. When teams ignore the diversity that is present or treat it with stereotypes (e.g., you are the only woman on the team, you must have the best handwriting so you take all of the notes!), they tend to underperform. On the other hand, teams outperform when they collectively adopt a learning mindset [9], both about each other and the challenge at hand, which enables them to both learn and experiment with the different perspectives and heuristics that are brought by the team members.

A critical enabler of leveraging diversity on teams can be achieved by establishing psychological safety. A term coined by Prof. Amy Edmondson, psychological safety is, "a shared belief held by members of a team that the team is safe for interpersonal risk taking." Psychological safety can manifest as the willingness of a team member to throw out a wild idea or engaging in a tough conversation about a project aspect that didn't go as planned. In a study of teams at Google, psychological safety was established to be the most critical factor for team success [10]. Within organizations, the values held and the norms practiced can reduce teams' anxiety and support the tendency for risk taking.

In the subsequent sections, we detail the journey of our partnership and how we intentionally created a high degree of psychological safety through our collective values held and norms practiced. We also include a discussion of how leveraging available diversity on our team leads to a productive and supportive research program.

Our Journey Begins

In the beginning, we were all hiking alone. Heather had worked at the Pacific Northwest National Laboratory for many years and already understood the power of collaboration for scientific discovery when she met Chrissi, who at the time was a new hire to PNNL. Heather and Chrissi worked in the same office but with different project teams, and gradually became friends. As a senior staff, Heather became a mentor to Chrissi, and over time they began developing projects they could work on together. Heather knew that it is easier to motivate yourself to hike up a steep hill when your friends are counting on you. Together, Chrissi and Heather worked on renewable energy and energy efficiency projects for building science [11, 12]. When Heather took a job in academia, the two had developed a working relationship with high levels of trust that allowed them to creatively look for ways to continue collaboration.

A short time later, Heather met Rachel, a third-year graduate student at the time, at a lighting conference when they spoke on a panel together. At the time, Rachel was navigating to find a project to focus her dissertation around and struggling to find a project fit. She recalls, "at that meeting (where I had felt like an imposter), Heather acknowledged the value of my work and told me she was interested in working together. At that moment I felt very seen." A few months later, Rachel faced an intense crossroads when her graduate advisor passed away unexpectedly. She sent a panicked email to Heather asking to talk, because she didn't know anyone within her institution that would be able to step into an advising role. Heather thankfully responded quickly and invited her to join the team. Starting the hike was as easy as setting up a 30-min weekly video call.

Our paths first crossed in different ways, at a conference, in a shared office space, and we each identified that a new trail had appeared. Why did we choose to walk together? Our collaboration was sparked not because our research efforts or focal were perfectly aligned, rather because our *values* were aligned. As individuals we implicitly realized this alignment existed early on, though our values surfaced initially through conversations about our individual goals and how we might develop mutually beneficial shared goals together. We further recognized the respect and support from the start was different than other partnerships. Because of the trust these early conversations engendered, vocalizing the values and norms we individually and collectively hold has become an important practice that we have each adopted as we work with new teams.

Team Motivation and Values

In research, there are always interesting problems to study and solve, and often the most interesting engineering problems come with human obstacles. Sometimes the obstacles are senior colleagues that insist on being first author on every publication. Sometimes the obstacles are a lack of funding. We have confirmed that if you find a group of friends that enjoy working together, you can accomplish great things and tackle these obstacles together. In order to develop this chapter, we engaged in a process of reflection on our work together. We each considered the values that our work showcases and then we participated in a series of sessions to converge on the list of team values below. In sharing, we help to showcase a set of values and norms that can help guide future collaborations (Fig. 2).

Fig. 2 Finding a way past a tricky career challenge is often difficult in research. Team discussions can be a method for finding a path forward

Our team is composed of three female scientists. During our time together we have all advanced in our academic and research/industry career. At this time our work identities include a tenured professor and two research-focused industry roles. Our motivation to support one another is one of the most important facets of our partnership. Many of the core values we share were important to each of us before we worked together, but we quickly realized that reinforcing these values in the team increased our productivity.

The research obstacles we had all encountered helped us develop a shared set of values. Each of these values contributed to team resilience and trust. Below we outline the values that we hold in our research partnership, including compassion, authenticity, humility, and listening. At the core, the collaboration centers on prioritizing the humans involved over any research or academic accolade, using our commitment to each other to drive accountability, and upholding a culture of trust. In this section, we additionally outline the norms that we have in our collaboration, which we see as our values in action.

Values: a person's principles or standards of behavior; one's judgment of what is important in life [13].

Compassion and Kindness

Our team shares deep compassion for one another and the environment. We are strongly committed to using our scientific and engineering backgrounds to promote technologies, processes and frameworks to address and reverse climate change. This manifests in our dedication to each other as well as the ways in which we treat each other. Compassion also emerges from seeing what each team member cares about and then considering how we can integrate that topic into our work. Overall, we developed a shared value of compassion, based on the importance of our own beliefs and the desire to support the team.

For example, when Heather encountered bias and leadership integrity violations in her job, the team helped her courageously explore the best path forward. The team helped remind Heather she should stick to her own values, that changing oneself for an institution is not a good long-term career choice. Rachel put it succinctly, "I think at times we are more compassionate for each other than we are even for ourselves."

We also make time for kindness. Kindness might show up in small ways, like a text message before a presentation, or a gift for an engagement. And kindness is manifested every week when we make time and space for one another in our meetings. Because each member of the team can list several examples of how this type of compassion and kindness impacted their work, we included this as a shared value.

Authenticity and Respect

Our research agenda is rooted in respect for the planet and one another. We are focused on helping individuals and organizations to more effectively steward natural resources and move away from extractive practices. The same authenticity and respect naturally extended to our work with one another.

Authenticity allows us to explore the truest version of ourselves as we each change and grow over time. Freedom to cry when the big emotional blows occur is one example.

> Rachel's example: There's bringing your whole self to work and then there is bringing your whole messy self. After going through a painful breakup, I remember logging onto our video meeting and just weeping. I couldn't hold it together and was truly depressed. Heather stopped what she was doing and focused in. Really held me in that moment and normalized how I was feeling. There was time for pep talks and strategies later. At the time there was a need to just stop and let it be. That has remained consistent – though crying on the phone was a rough moment of authenticity – the rest of the spectrum of emotions is also allowed. Being really excited about a research idea, frustrated with a colleague, happy because it's warm outside – whatever it is I feel like we give ourselves the time and space to be.

This example demonstrates the trust the team has built around the space for personal sharing. We always take time to treat one another with respect, every aspect of our shared humanity deserves time and respect. We recognize that people have things going on in their lives that may affect their ability to show up on a week-to-week basis, and that is valid. After all, we are more than the sum of our dynamic parts. Moving, childcare, illness, are all examples of things that demand our attention and we respect the individual who may need to step away and prioritize another facet of life.

Humility and Equity

Another important value for the team is humility. While each member of the team is confident, there is no sense of privilege, even when the professional titles might imply it. We each work consistently to support each other to contribute in whatever way we can, even if that week contributing just looks like showing up to the meeting and providing words of encouragement to others. While humility and equity may have been important to us before we joined this team, the weekly practice of these skills has been a powerful reinforcer for each of us.

We leverage each other's strengths, ask for input, and use collective ideas to form research agendas. At times it feels like none of us care what we are working on, as long as we are working together. The good of the team is always the priority. There are no back channels supporting gossip – we each make decisions that are best for the group in that moment. There are no fights over authorship order, a unique experience when most of us have had very negative authorship issues in the past. Instead we take turns as first author, just as we take turns playing the lead on different projects. Often the author order is based on who might be leading a specific type of inquiry, or the career timing that might benefit most from a first author publication. When we reflected on our shared values, this type of equity in authorship was one of the important values of the team that was emphasized by each person.

Listening and Happiness

One of our most important shared practices is listening to one another. We don't have preconceived notions of what each other should be doing to make ourselves happy. We listen actively and trace each other's energy and see what makes that person light up – what makes them come alive. Ultimately the behavior of paying attention to each other's happiness, we think, unlocks better research because we can see what drives and motivates that person. This shared value emerged organically in our team, but it quickly became an important weekly habit.

We listen carefully to how each person is feeling and track how that is changing over time. It's not just noticing a feeling but also remembering to speak up if we're seeing a behavior pattern over a longer period of time. We push each other to grow and leverage this empathy into ways we can each take action.

> Heather's Example: I believe Rachel and Chrissi realized I needed to adjust my career long before I did. Over a year of conversations, they patiently listened as I described troubling patterns of micro-aggressions from colleagues. While they never told me what I ought to do, they did support me in determining possible paths I could take. In the end I think they understood that I needed to adjust my career in a way that would allow me to be my authentic self, and I am grateful to have a trusted team to process that with.

Trust

Each of our shared values help us to create a culture of trust to do excellent research in our team. Trust means that we can throw each other the ball and know that the other person will catch it. When someone says they're going to do something they do it, and moreover, they do it in a way that doesn't require a cross-check. Fundamentally, we work together, with an inherent understanding that our contributions will result in a strong work product. We give each other space and time; if we run into issues or conflicts, we work together to solve problem. This value emerged over time as we grew in our team.

Hiking together requires a great deal of trust in your team, particularly for long journeys. You need to trust that your friend packed the fire starter, but it is also helpful if you don't each need to carry every tool because it gets heavy. Our values help us determine the types of tools and journeys we wish to travel on together. We consider what roads we would like to explore together, and then think carefully about the skills that each of us might be able to pack for that type of research project.

Setting Norms: How We Manifest Our Values

> A norm exists in a given social setting to the extent that individuals usually act in a certain way and are often [held accountable] when seen not to be acting in this way. (Adapted from [14])

Norms describe human behavior, and the willingness to adapt, adopt and make decisions. While values lay the foundations of who we are as people, many times it's societal norms that help inform our behaviors and guide our actions in specific settings. Norms help to ensure that values are practiced and don't just live in our minds. In our collaboration, we connected on shared values and over time, worked to develop norms in our collaboration to manifest those values. Part of norms includes holding people accountable to the values that are set. By finding ways to constructively and compassionately give and receive feedback, we also encouraged each other to grow and uphold who we wanted to be within our research partnership. Our norms further connected to a broader agenda. Our behavior as a team is rooted in our strong environmental ethic and understanding of the need to solve some of the complex climate issues that face our society. As our society reaches a critical junction related to climate change, and our ability to effectively combat it, norms play a crucial role. From an environmental perspective, social norms are necessary for society to focus on the common good and find resolution to the world's environmental problems [15].

Checking In

We start every meeting by checking in with one another and seeing what is going on in each of our lives. The unique part about being at three different institutions is that we don't see each other on a day-to-day basis. We are not in each other's work environments or social circles in a daily way. In fact, we've all only been together in person once. That doesn't stop us from recognizing that there is a human being on the other side of the computer screen that has a life behind them, and we make sure to give space to be vulnerable and share the truth of what is happening on any given day of any given week. This norm is deliberate and connected to our value of listening and compassion for one another.

Stretching

Our group overall is flexible and one of the most important norms that we have is the ability to stretch and shapeshift into what the group needs at a given moment – whether that looks like developing new expertise to fit in with an emergent research need to support one person's project, or it looks like adopting new personalities when someone needs cheering section behind them as they embark on a new adventure. Sometimes the stretch only lasts one 30-min meeting or sometimes it lasts years as we are stepping into a space to allow someone to reach a professional goal or build-out work in a new domain. We made the decision a long time ago to show up as ourselves and as what the others in the group need. This norm is closely tied to our value of authenticity and respect.

Advocating

Our group is a strong sounding board for each other. Each of us has faced a professional dilemma and over the years there have been many. We each take the approach of first helping each other unpack the situation and spend time to develop the team's perspective on it. For example, when one of us stepped into a new role and was marginalized by their boss, we brought the situation to the group to discuss. At first, the predominant narrative in the situation was about being overwhelmed and frustrated; action took a backseat to emotion. Once we gave space to the big emotions, the conversation transitioned to a focus on what needed to be done. We helped her prepare to advocate for herself (and of course she already had everything she needed within herself) and to craft her point of view on her arguments and her approach to the conversation. This norm is connected to our value of listening and happiness.

Positioning

One of our norms we exhibit is quite subtle. We are good at positioning each other's work because, despite a high level of general confidence in ourselves, we can each see each other as bigger and brighter than we see ourselves. Rachel is excellent at providing big-picture connections to the most complex environmental system. Chrissi is great at positioning our work as interesting and fundable to people who want to see things happening in the world. She is also amazing at building partnerships. Heather is incredibly impactful about packaging the contributions we have made or work we have collaborated on, and she enjoys art so we often have whimsical figures to include in our publications. This norm is closely connected to our value of humility, each of us believes in the greatness of the others.

Work Norms

A simple but impactful norm is that we do not have one-hour meetings. Sometimes they only take 15 min. Sometimes they take 2 h. The format and the structure of the meetings doesn't matter. Getting to the work matters and celebrating each other as individuals matters. And one feeds the others. Leaving space for the sparks and then trusting that the work products will emerge has proven to be an excellent and productive strategy. This norm is closely tied to our value of trust, we each trust that the work will get done even if we don't meet for a specific time period.

Our norms have provided a concrete way for us to practice the values our team has developed. Whenever we do broaden our collaborations to a bigger group of people, the strength of the norms protects the integrity of the team. Several of us have also used the norms from this group in other teams and found them to be widely beneficial to any gathering.

How the Journey Has Changed Us

The journey of our collaboration started in 2012. As a research team that valued one another above academic outcomes, we have exceeded our own expectations for productivity. In the beginning, Chrissi was on a trail focused on energy-efficient buildings, Rachel was studying the circular economy, and Heather was looking at energy-efficient lighting. Over time, we were intentional about sharing different aspects of our background and skillsets, and through focused conversation and active listening, synergies emerged that led us to collaboratively create new pathways in our research.

Our team was initially focused on lighting with ties to the circular economy, in part due to questions that emerged from Rachel's work as a graduate student [16]. Exploring the circular economy in the context of our collaborative partnership created new focus for Rachel's dissertation and a collection of new publications on the end of life implications of lighting products [17–19]. As we continued our journey together, Chrissi contributed her perspective of the circular economy from an economics lens and Rachel integrated her work in design studies leading to a publication on assessing the waste burden implications of technological change [20]. Drawing on Heather and Chrissi's previous work to view the circular economy from the broader context of building science also enabled us to integrate perspectives on exergy [21]. Throughout all of this, we had conversations about what new paths we could explore together.

This team has changed how we individually think about research careers. When most people finish a PhD program, they believe they will specialize in one topic forever. Instead, in this working group, we have found that the topic of research is significantly less important to sustaining interest in research. What we do care about is supporting the career aspirations of friends and colleagues. When we tackle a new topic as a group that might be a bit outside one person's comfort zone technically, we embrace it! We do so with eagerness, knowing that friends will help guide our growth in a new topic and do so with generosity (Fig. 3).

> Heather's example: I learned geographic information system (GIS) mapping last summer. This is a new skill set for me that I now feel confident and excited to expand. Chrissi's mentorship in this area helped me explore this in the context of other technical work I was

Fig. 3 Taking time to reflect on your own journey is a helpful way to stay motivated

doing for one of our projects. I am growing because we are doing things I would never have tackled alone. It is exciting!

Chrissi's example: Working in collaboration with Heather and Rachel has helped me realize my own personal and professional goals. Their support and encouragement helped me take the big leap to pursue a PhD and they have been there for me throughout the process. Having women in my life with similar goals and journeys has helped me gain confidence as a researcher and adds perspectives to my ideas that can only come from a trusting environment.

Conclusion: View from the Mountain

We have discovered that building a research culture that is based on shared values and norms contributes to a highly trusting and productive team. Working with other confident and kind women has made each of us more successful. Our team has published nine papers in 8 years, and we are still going strong with several new works pending. We have climbed far more mountains together than we ever could have climbed alone. We encourage women who are launching research careers in STEM, particularly focused on the environment, to consider ways to formalize norms and values for teams using the example we have outlined. We have found the trust that comes from this type of collaboration leads to high productivity. We have also found that cross-discipline collaborations that stretch us are powerful for tackling complex energy and environment challenges. Our model of collaborative research may be easiest to replicate in environment fields where principles of shared responsibility are often valued.

Creating a positive culture of collaboration can be transformative for every person in the team and highly productive scientifically. We hope that whatever research team you become a part of might be strengthened by embracing a new type of scientific culture, one that values each researcher's journey more than the scientific method. Focusing on supporting another person is significantly more motivational than a stated goal that you might wish to write a specific number of papers this year. Rather, the focus of a career could become, how many people might we help succeed in new things this year? Your research matters, but your colleagues matter more!

References

1. B. J. Casad *et al.*, "Gender inequality in academia: Problems and solutions for women faculty in STEM," *J. Neurosci. Res.*, vol. 99, no. 1, pp. 13–23, 2021, https://doi.org/10.1002/jnr.24631.
2. D. N. Beede, T. A. Julian, D. Langdon, G. McKittrick, B. Khan, and M. E. Doms, "Women in stem: A gender gap to innovation," 2011. https://doi.org/10.2139/ssrn.1964782.
3. A. C. Edmondson and T. Zuzul, "Teaming Routines in Complex Innovation Projects," in *Organizational Routines: How They Are Created, Maintained, and Changed*, Oxford: Oxford University Press, 2016, p. Part 1, Section B:8.
4. A. C. Edmondson, *Teaming: How Organizations Learn, Innovate, and Compete in the Knowledge Economy*. John Wiley & Sons, 2012.
5. A. C. Edmondson and J. F. Harvey, "Cross-boundary teaming for innovation: Integrating research on teams and knowledge in organizations," *Human Resource Management Review*, 2016.
6. H. Rittel and M. Webber, "Dilemmas in a general theory of planning," *Policy Sci.*, vol. 4, no. 2, pp. 155–169, 1973, https://doi.org/10.1007/BF01405730.
7. S. E. Page, "Making the Difference: Applying a Logic of Diversity.," *Acad. Manag. Perspect.*, vol. 21, no. 4, pp. 6–20, 2007.
8. K. Srikanth, S. Harvey, and R. Peterson, "A Dynamic Perspective on Diverse Teams: Moving from the Dual-Process Model to a Dynamic Coordination-based Model of Diverse Team Performance," *Acad. Manag. Ann.*, vol. 10, no. 1, pp. 453–493, 2016, https://doi.org/10.1080/19416520.2016.1120973.
9. M. F. Ercan and R. Khan, "Teamwork as a fundamental skill for engineering graduates," in *Proceedings of 2017 IEEE International Conference on Teaching, Assessment and Learning for Engineering, TALE 2017*, Jul. 2017, vol. 2018-January, pp. 24–28, https://doi.org/10.1109/TALE.2017.8252298.
10. C. Duhigg, "What Google learned from its quest to build the perfect team," *The New York Times Magazine*, 2016.
11. H. E. Dillon, C. A. Antonopoulos, A. E. Solana, and B. J. Russ, "Renewable Energy Requirements for Future Building Codes: Options for Compliance," Richland, WA, 2011.
12. C. A. Antonopoulos, M. C. Baechler, and H. E. Dillon, "A Drop in the Bucket or a Pebble in a Pond: Commercial Building Partners' Replication of EEMs Across Their Portfolios," 2014, [Online]. Available: https://www.aceee.org/files/proceedings/2014/data/papers/3-292.pdf.
13. "VALUE | definition in the Cambridge English Dictionary," *Cambridge English Dictionary*. https://dictionary.cambridge.org/us/dictionary/english/value (accessed May 31, 2021).
14. R. Axelrod, "An evolutionary approach to norms," *Am. Polit. Sci. Rev.*, pp. 1095–1111, 1986, https://doi.org/10.1017/S0003055400185016.

15. A. P. Kinzig *et al.*, "Social norms and global environmental challenges: The complex interaction of behaviors, values, and policy," *Bioscience*, vol. 63, no. 3, pp. 164–175, 2013, https://doi.org/10.1525/bio.2013.63.3.5.
16. The Ellen MacArthur Foundation, "Towards the Circular Economy," 2013.
17. H. E. Dillon, C. Ross, and R. Dzombak, "Environmental and Energy Improvements of LED Lamps over Time: A Comparative Life Cycle Assessment," *LEUKOS J. Illum. Eng. Soc.*, pp. 1–9, Mar. 2019, https://doi.org/10.1080/15502724.2018.1541748.
18. R. Dzombak, J. Padon, J. Salsbury, and H. Dillon, "Assessment of end-of-life design in solid-state lighting," *Environ. Res. Lett.*, vol. 12, no. 8, 2017, https://doi.org/10.1088/1748-9326/aa7ab1.
19. R. Dzombak, E. Kasikaralar, and H. E. Dillon, "Empirical Characterization of Technology Management within the Street Lighting Industry," 2019.
20. R. Dzombak, C. Antonopoulos, and H. E. Dillon, "Balancing technological innovation with waste burden minimization: An examination of the global lighting industry," *Waste Manag.*, vol. 92, pp. 68–74, 2019, https://doi.org/10.1016/j.wasman.2019.04.037.
21. H. E. Dillon, R. Dzombak, and C. Antonopoulos, "Exergy analysis of heating options for buildings in the Pacific Northwest," *Int. J. Exergy*, vol. 30, no. 3, pp. 275–293, 2019.

Dr. Heather Dillon is Professor and Chair of Mechanical Engineering at the University of Washington Tacoma, where her research team is working on renewable energy systems, solid-state lighting, energy efficiency in buildings, fundamental heat transfer studies, and engineering education.

She is the Chair of the Council on Undergraduate Research Engineering Division and recently served as the Fulbright Canada Research Chair in STEM Education at the University of Calgary, Alberta. Before joining academia, Heather Dillon worked for the Pacific Northwest National Laboratory (PNNL) as a senior research engineer working on both energy efficiency and renewable energy systems, where she received the US Department of Energy Office of Science Outstanding Mentor Award. During her time at the University of Portland, she received the Provost's Award for Outstanding Undergraduate Research Faculty Member (2017) and the Outstanding Scholarship Award (2020) for research excellence.

Dr. Rachel Dzombak is a strategist and educator working to frame and solve complex problems and train others to do the same. She teaches at the University of California (UC) Berkeley's Haas School of Business, helping students to use both design and systems thinking to solve problems that matter. Rachel additionally works as Digital Transformation Lead at Carnegie Mellon University (CMU) Software Engineering Institute. Through her work at CMU and in her independent consultancy, she supports a range of organizations working to shift business models and grow innovation ecosystems. Rachel holds a Ph.D. in Civil and Environmental Engineering from the UC Berkeley.

Dr. Chrissi Antonopoulos is a Senior Analyst in the Energy and Environment Directorate at the Pacific Northwest National Laboratory (PNNL). Her research is focused almost exclusively on residential buildings, where she leads PNNL's Research Dissemination portfolio, which includes development and maintenance of the popular residential construction website, the Building America Solution Center. Chrissi also supports research projects focused on the building science of thermal enclosures, indoor air quality, and assessment of socio-demographics and equity. Chrissi has published studies outlining opportunities for embodied carbon and circular economy perspectives related to building technologies, and socio-demographic analysis of energy use in residential buildings. Chrissi has MS and Ph.D. degrees from the Toulan School of Urban Studies and Planning at Portland State University.

How to Stop Imposter Syndrome from Sabotaging Your Career

Toni Crowe and Stephanie Slocum

Imposter syndrome is a feeling that you don't belong and don't deserve to be here. Many women of notable achievement also have high levels of self-doubt, even when there is substantial evidence to the contrary ... What we call imposter syndrome often reflects the reality of an environment that tells marginalized groups that we shouldn't be confident, that our skills aren't enough, that we won't succeed—and when we do, our accomplishments won't even be attributed to us. Yet imposter syndrome is treated as a personal problem to be overcome, a distortion in processing rather than a realistic reflection of [work environments] [1].

Overworked, struggling with self-doubt, feeling like you don't belong at work, and don't deserve to be here, we've been there. A 2020 Women in Technology Report notes that 79% of technical women experience imposter syndrome [2]. Experienced executive-level women aren't exempt, with a reported 75% struggling with imposter syndrome during their careers [3].

The resulting lack of confidence and feelings of fraud can derail even the most technically gifted women, stalling not just *your* progress but any progress made on normalizing women leaders in technical fields.

Toni Crowe and Stephanie Slocum connected based on the shared goal of normalizing professional and technical women as leaders at work. These women came from hugely different backgrounds. Stephanie is White, Toni is Black. Toni started in the Chicago projects, while Stephanie experienced a middle-class childhood. They are different generations, with disparate work histories and ethnic backgrounds, yet both have struggled with imposter syndrome. Both are proof that the sooner imposter syndrome is addressed, the sooner engineers lean into their unique professional and personal skills.

T. Crowe (✉)
Just One, New York, NY, USA
e-mail: tonicrowe@tonicrowewriter.com

S. Slocum
Engineers Rising, State College, PA, USA
e-mail: stephanie@engineersrising.com

M. Bailey, L. Shackelford (eds.), *Women in Mechanical Engineering*, Women in Engineering and Science, https://doi.org/10.1007/978-3-030-91546-9_6

This chapter will uncover why and how women in engineering experience imposter syndrome and explore what they can do. You'll learn through a duo of engineers that statistically should have never met: Two successful engineers turned leaders then authors, who met when writing their books.

What Is Imposter Syndrome, and How Do I Know I Have It?

"Imposter Syndrome" was introduced in 1978 by psychotherapists Pauline Clance and Suzanne Imes [4]. They found women with notable achievements struggled with self-doubt:

> *Despite outstanding academic and professional accomplishments, women who experience the imposter phenomenon persists in believing that they are really not bright and have fooled anyone who thinks otherwise.*

In her book, *The Secret Thoughts of Successful Women: Why Capable People Suffer from the Impostor Syndrome and How to Thrive in Spite of It* [5], Dr. Valerie Young shares six self-assessment questions to determine if you have imposter syndrome:

Do you chalk your success up to luck, timing, or computer error?
Do you believe "If I can do it, anybody can?"
Do you agonize over minor flaws in your work?
Are you crushed by constructive criticism taking it as evidence of your ineptness?
When you succeed, do you secretly know you fooled them again?
Do you worry that it's only a matter of time before you are found out?

Clance and Imes noted that environment, combined with a strong motivation to achieve, contributed to an individual's experience of imposter syndrome.

Imposter syndrome is incorrectly viewed as a "fix the woman" problem instead of a "fix the culture" problem. Individuals view high levels of self-doubt as a self-esteem issue. Managers see a lack of confidence as an inherent trait instead of a trait the work environment is causing. Each woman must figure out how to help herself.

This chapter uses a question-and-answer format with practical advice and exercises you can apply immediately. The questions speak to women engineers at all career levels. We offer a roadmap for anyone who wants to support others in their career success journey, framed in the authors' first-hand experiences.

Question 1: At What Times in Your Career Have You Experienced Imposter Syndrome and How Did You Overcome It?

Stephanie

I felt imposter syndrome for the first time during an internship interview in college. I sat in the lobby, waiting to be called into the conference room, fidgeting with the hem of my new tan suit I had bought for the occasion. When I heard my name, as I stood my heel caught on the chair leg. I nearly did a faceplant on the floor. Taking a deep breath, I walked into the conference room, attempting to regain my composure. I sat, silently berating myself for my clumsiness, until the interviewer interrupted my self-flagellation with his first question.

Since then, imposter syndrome consistently flares when I'm in new situations or stretching beyond my comfort zone. That is especially true when I am the only woman in the room or feel I have to put on a "tough woman" facade to be respected.

Examples of ways this has shown up for me:

- Looking at my resume as if it were someone else's → "I can't believe I've done all these things. I'm not this accomplished; it must have been luck."
- Walking out on the site/plant floor → "I don't have enough expertise to belong here."
- Before giving a public presentation → "Who am I to be speaking? There are people out there who have more expertise."
- Before a networking event → "I don't belong here, I'm an introvert."
- Before an interview → "Am I qualified for this position?"
- Before a meeting → "I'll only speak up if directly asked. No one cares what I have to say."

Toni

"This is a tough place for a woman. I've been put down, pushed aside, knocked out. The truth is I have had to fight my whole life because of who I am, who I love, and where I started. But I didn't let anything get in my way." Sharice David spoke these words during her 2018 campaign for Congress. The words resonated with me immediately. She told of her time in the political world, but she might also have been speaking of my time in corporate America [6]. Professional and technical women, particularly in men-dominated fields, don't have a tranquil relationship with success. My time working in corporate America exacerbated my imposter syndrome.

As an ambitious Black woman, I was triggered quite a bit. It did not matter how difficult the mission, nor how successful the completion—it came back to me. Anyone questioning the success of my work triggered anxiety. One time was when I was the engineering team leader, we were working on a nuclear energy update project. The project involved castings, which are complex metal pouring. From start to finish, a casting takes 30 weeks. My team reduced the time from 30 to 25 weeks.

The day the parts were due (but did not arrive), my boss stormed into my office. The pieces were delayed by 1 day. His face was red; he was breathing hard with his hands clenched into tight fists. I thought he might punch me. The man climbed on my desk. He kicked the phone off the desk, then my paper clips, then my calendar. While he was standing on my desk, I wondered if he knew I was a fraud.

I didn't allow anyone to see my genuine emotions when experiencing imposter syndrome. What my boss saw was that I appeared calm; I had on my game face. He told me to get my "stuff" together and left my office. Even though my team saved the project with innovative techniques, I spent the next three hours agonizing over my team's performance. My boss had behaved brutally, but all that I could think of was that I was a fake.

It took a great deal of self-examination to realize nothing was wrong with me. Imposter syndrome had me on the ropes until I could gain control of my emotions.

Exercise 1. The Game Face

Every woman needs a game face. Never let your peers or boss know you are experiencing imposter syndrome. This is the face that I show when I am retreating behind a façade.

I went for "the interested and intelligent look." Practice every morning. Your goal is to hide chaos or distress as they trigger you. *It is not required to be interested and intelligent; you only need to look interested and competent.*

Upon your imposter syndrome triggering,

1. Immediately put your face into your game face.
2. Look intelligent and interested.
3. Take one deep breath, breathing from the abdomen.
4. Invoke a sentence that captures your desired state. Do not speak it. Mine is four words. "I handle my business."
5. Control runaway emotions by grabbing them in your mind and putting them in a box.
6. Casually look the provoker in the eye. Be cool.
7. Handle your business.

Steps 1 through 7 should take 10 seconds. The Game Face is a game-changer.

Question 2: What Role Does Confidence and Experience Play in Imposter Syndrome?

Stephanie

Until women in technical roles are normalized, confidence and experience help but do not eliminate imposter syndrome, I experienced imposter syndrome at all levels of my engineering career.

Imposter syndrome has lessened with more experience. The strongest flare-ups occur in the following circumstances: (1) new situations, (2) situations where I am stretching myself/taking a career risk, (3) situations when I am afraid of being judged harshly, and (4) a defensive reaction has been triggered.

I was elected to a board-level position, and was one of only three women. At my first meeting, a new committee proposal focused on justice, equity, diversity, and inclusion (JEDI). I wrote several questions about the proposal as the presentation proceeded. The presentation finished. The Chair of the Board asked if there were questions.

No one spoke. In those 10 seconds of silence (which felt like 10 min to me), I felt tightness begin in my chest and feelings of shame creep into my body. My inner dialogue filled with doubt: "*No one else is speaking up. Who are you to ask questions, given it's your first meeting?*"

Recognizing the signs of an imposter syndrome trigger, I took a deep breath, asked myself, "How best can I serve this group at this moment?" and asked my questions. The result? A 10 min discussion about both the committee's proposal and about advancing JEDI initiatives across the organization. It resulted in an email from the Board Chair afterward saying he appreciated my thoughtful questions.

In my experience and in coaching hundreds of women in engineering, I find new situations like the board trigger imposter syndrome. You get stuck in your head, ruminating over all the things you don't know. You feel like everyone else knows more. Those doubts and ruminations translate into outward behaviors that make it appear that you aren't leadership material.

This shows up in so many ways: Projects just don't go as well for you and you don't know why. You hesitate to volunteer for an exciting new project, so someone else gets it. You avoid active networking and challenging conversations. You say "yes" to everything instead of the things that matter.

Combined, this creates a self-perpetuating loop of more doubts, more rumination, and more imposter syndrome. It makes you feel terrible, and who wants to stay in a career that makes you doubt everything about your capabilities? Based on the women I have talked to who have exited engineering entirely, low confidence and high imposter syndrome levels are leading indicators she will leave her profession. My anecdotal discussions are not isolated. According to an NSF report, "Stemming the Tide: Why Women Leave Engineering," "feeling a lack of confidence in their ability" is one of three primary predictors that a woman will leave engineering (the other two: excessive work responsibilities and lack of role clarity) [7].

Confidence is a muscle we can build. Psychologists have studied what it takes to build confidence and have boiled it down to three things: competence, compassion (with self-compassion being primary), and courage (Fig. 1).

Experience is only one part of "competence." We understand what success is when you tie your shoes. (Your shoes are tied.) Do you similarly know at work when you are successful on a project, on a weekly and daily basis? You need feedback to tell you how you are doing, often the missing link between getting experience and feeling confident about that experience.

When expectations are unclear, or there is no (or unfavorable) feedback, you are left to "guess" how you are doing. This leads to doubts, rumination, getting stuck in your head, and imposter syndrome. The solution: Proactively get more feedback. Ask your manager what project success looks like and how it will be measured. Drill those measurements down into your day-to-day work so you know EXACTLY what metrics define your success.

Here's an example of what one woman I mentor said after having that conversation with her manager:

	Competence	Compassion	Courage
DEFINITION	Able to effectively complete work tasks. Encompasses both a) Informational competence: having the technical skills needed; and b) Relational competence: having sufficient team and people skills needed to complete work	Ability to understand the emotional state of another person or oneself (1) Compassion requires 3 components and applies to both self and others: a) Serious troubles b) Troubles are the result of unjust fate c) We are able to picture ourselves in the same predicament (2)	Mental or moral strength to venture, persevere, and withstand danger, fear, or difficulty (3) In the context of work, it is calculated risk-taking aligned with a sense of purpose (4)
SIGNS YOU'VE GOT IT	• Stretching yourself to learn • Viewing feedback on your work as a gift and regularly asking for it • Proactive in figuring things out for yourself • Ask for network, peer, mentor, and sponsor support at regular intervals • You believe you can find the answers even if you don't know them in the moment • You have sufficient resources, training, and support to complete your work • Awareness of your own strengths and weaknesses	• Do not feel pressure to be "perfect" at work • Have cordial work relationships, even with those different from you • Set clear boundaries and say no to projects or activities that drain you • See failure as a learning experience • Rebound quickly from setbacks • In work conflict situations, you focus on solving the problem, not "fixing" or blaming the person • Prioritize responsibilities so your most important work gets done, and avoid ruminating (much) over the rest	• Connect with your work to a sense of internal purpose (intrinsically motivated) • Speak up for yourself and others • Face work conflict head-on and resolve it immediately • Take calculated career risks • Apply for stretch roles • Find a way to align your work with your personal values • Are outward-focused and connected to why your work matters in the world • You are afraid sometimes, but still take action
SIGNS OF STRUGGLE	• Worry about looking bad or making a mistake • Hide mistakes or blame others for them • Not asking for and/or receiving feedback, so you're not sure how your work is perceived • View "working harder" as the solution for everything • Ask for help only as a last resort • Constantly asking for validation from your manager • You think you should know more and/or everyone knows more than you • You have insufficient resources, training, and support to complete your work	• Afraid of making even a small mistake • Struggle with stress, overwork, burnout, and/or "busy-ness" • View perfectionism as a good thing • Defensive, point fingers, or jump to conclusions when things don't go your way • Have difficulty staying calm in work conflict situations • Say yes to everything • View failure as a permanent setback • Holding yourself and others to impossibly high standards and judge harshly when mistakes are made • Struggle with self-care	• Stay silent when you want to speak up • Avoid work conflict at all costs • Stay stuck in roles you don't love because they feel safe • See your work as something dictated to you by your manager • Feel your work doesn't matter • Afraid of asking for what you need to thrive at work • Often rationalize and/or complain about why you aren't doing the work you really want to be doing • View your career path as "options available" instead of imagining your own future

1 American Psychological Association. APA Dictionary of Psychology.
2 Lonczak, Heather, PhD. 20 Reasons Why Compassion is So Important in Psychology. 01.09.20. https://bit.ly/3kZgbN4
3 Mirriam-Webster Dictionary
4 Reardon, Kathleen. Courage as a Skill. Harvard Business Review. January 2007. https://hbr.org/2007/01/courage-as-a-skill

Fig. 1 Components of professional role confidence

> *Having this conversation illuminated so many things and made me more confident. My manager and I are now on the same page, I know what's expected, and I know where I'm headed.*

The second component of confidence is compassion, specifically self-compassion. Imposter syndrome and perfectionist tendencies go hand in hand. When you beat yourself up about mistakes (or you are working with someone who is hypercritical and you internalize that criticism), lack of self-compassion causes low confidence.

Self-compassion is a struggle for myself and every high achiever I know. Short bursts of mindfulness practices built into your day are the best tool to address it. Specifically, 10 min per day of meditation, 10 min of journaling, and a 10 min gratitude practice. I do this right before starting my work, and I "beat myself up" much less since I started this practice.

The last component of confidence is courage. Practically, courage means you default to action. Stop thinking (or planning or researching, as I tend to do!) and MOVE your project forward. Two questions I use that immediately spur me into action when confronted with the feelings associated with imposter syndrome:

- "How do I want to show up in this situation?"
- "How can I best serve the group/project/greater good at this moment?"

Understanding the "why" which triggers your action is essential. Many women—as I do—have more courage when supporting and advocating for others than ourselves. That's why the second question above works so well in tapping my inner courage.

Toni

Daniel Goldman, an internationally known psychologist and author of *New York Times* bestseller *Emotional Intelligence* [8], has identified a group of competencies predicting when a leader will be an outstanding performer. The five competencies are self-awareness, self-regulation, motivation, empathy, and social skill [9]. When these distinguishing traits of leadership are combined, they are called Emotional Quotient (EQ) or Emotional Intelligence (Fig. 2).

Women engineers in technical situations often have not developed the level of EQ (Emotional Intelligence) to protect themselves from the imposter syndrome effects.

	DEFINITION	HALLMARKS
Self-awareness	The ability to recognize and understand your moods, emotions, and drives, as well as their effect on others	• Self-confidence • Realistic self-assessment • Self-depreciating sense of humor
Self-regulation	The ability to control or redirect disruptive impulses and moods. The propensity to suspend judgement, to think before acting.	• Trustworthiness and integrity • Comfort with ambiguity • Openness to change
Motivation	A passion to work for reasons that go beyond money or status. A propensity to pursue goals with energy and persistence.	• Strong drive to achieve • Optimism, even in the face of failure • Organizational commitment
Empathy	The ability to understand the emotional makeup of other people. Skill in treating people according to their emotional reactions.	• Expertise in building and retaining talent • Cross-cultural sensitivity • Service to clients and customers
Social Skill	Proficiency in managing relationships and building networks An ability to find common ground and build rapport	• Effectiveness in leading change • Persuasiveness • Expertise in building and leading teams

Fig. 2 The five components of emotional intelligence at work

A study by the Goleman Company reviewed Star Performers in Fortune 500 companies. They examined the competencies of Emotional Intelligence.

The study found that a star performer's IQ was not as crucial as they promoted a person through the Leadership Chain. EQ was the leading predictor of their success in becoming a superior performer in meeting objectives.

Unlike IQ, we can teach EQ. There are several training programs where a participant can gain the required knowledge. While increasing your emotional intelligence (EQ) command will not guarantee that they will promote you, the training will facilitate a deeper understanding of what it takes to succeed as a high potential leader.

Question 3: Why Do You Think Imposter Syndrome Is Particularly Acute When Transitioning from Technical to Management Roles? What Can You Do to Combat This?

Toni

The reward for great technical work is promotion into a position where you understand nothing. As a Design Engineer, I did an outstanding job. Because I was excellent at designing, they gave me a small team to manage. Since I had not managed before, I was lost. Sink or swim was not unusual treatment, yet men did not seem to have the same doubts about being a new leader.

I headed to my boss's office to tell him I no longer wanted the promotion. He had me wait a moment while another supervisor stopped by and said, "I'll be waiting in the lab with my group for you to help me with the overtime announcement." The boss said, "No, I'm not coming. Just do it as I showed you."

It had not occurred to me to ask my manager for help. I asked the boss if he could show me how to manage timecards. My imposter syndrome calmed down. Asking for help was not a weakness, but a normal part of learning (Fig. 3).

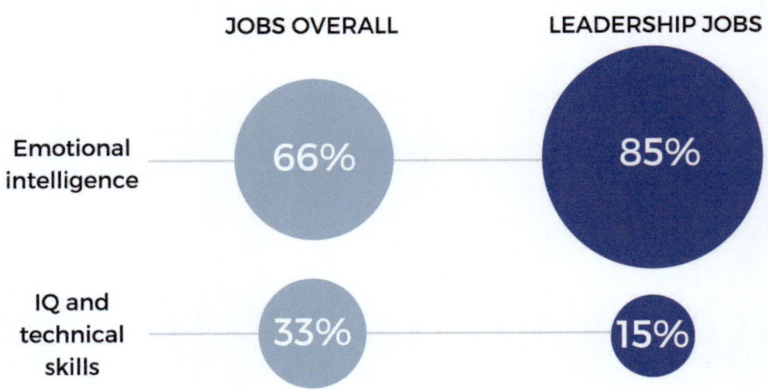

Fig. 3 The ingredients of star performances: distinguishing competencies

Star players use their IQ 33% and their EQ 66%. However, once those players took leadership positions, the balance changed. IQ was no longer a significant factor in their success; it was needed only 15% of the time. Instead, EQ was an essential factor 85% [10].

Self-regulation is the ability to control, manage, or disrupt impulses or moods. It encompasses the power to suspend judgment—to think before acting. Self-questioning causes hesitation and procrastination. The knowledge that imposter syndrome is at work will drive behavior change.

An exercise that identifies self-awareness issues is writing the thoughts in your head around an action down. Pay attention to the voices in your head: Capture them on paper. Are they mostly negative?

Take action with these three steps:

1. Forgive yourself: Everyone experiences indecision.
2. Reexamine goals: Have you set up unattainable actions?
3. Find someone to share your fears and hopes.

Stephanie

Toni's story of the transition from technical to management is very similar to my own. I was promoted to a position with a management title, but no defined change in responsibilities or additional training. This was the norm at the firm where I was working, and it took several years for me to become comfortable in that role.

Looking back, I wish I had purposefully sought mentors, as well as proactively asked for training. For me, this transition created a perfect storm for imposter syndrome. I didn't have the self-awareness or EQ at the time to recognize the problem, and I considered leaving the field entirely.

It turns out more than half of women who are promoted or transitioning to new roles experience imposter syndrome during that transition [11]. Fear of failure, commonly triggered during those transitions, is correlated with increased incidences of imposter syndrome [12]. To this day, new situations and situations where I am the only woman consistently trigger imposter syndrome, irrelevant to my level of experience on paper.

In addition to practicing self-awareness, you can reduce incidences of imposter syndrome during role transitions by:

- Seeking mentors inside and outside of your company and get extra support at home.
- Asking for clearly defined expectations and success metrics.
- Collecting (measurable) feedback on successes and areas for growth.

In future role transitions, I learned to purposely seek people who had been in those roles before and ask their advice about common mistakes new people made in that role and what the best people in that role did. I also learned to proactively seek mentors and a better support network outside my company to understand my own experiences better.

Question 4: What Are Signs That Imposter Syndrome Is Sabotaging You at Work?

Toni

There are four warning signs that imposter syndrome is sabotaging your career:

1. Worry about meeting the expectations of others.
2. Apprehension that people will find out you are not all that they make you out to be. You fear that "everyone" will find out you are not as good as they thought.
3. Attributing your success to "luck." The careful preparation for meetings and research, the diligent checking and rechecking of your work, and your team's work are dismissed. The meticulous preparation means nothing: it is merely "good fortune."
4. Others take credit for your work, including the people that work for you. They don't worry that you will ever claim the credit for what you have accomplished. They offer your ideas as theirs.

If even one of these situations is part of your work experience, then imposter syndrome is holding you back.

I was the Director of an Aerospace company. Five manufacturing plants were under my team. The CEO invited me to a corporate meeting to discuss closing one of my plants. The month before, I presented a plan to my boss to merge the five plants into four. Employees would transfer, but I would sell the building and land, providing a bump to our profit.

Being a paranoid engineer, I went to the meeting room the day before to check the setup. To my horror, I found another member of the team was presenting my charts. I sat there for quite a while, thinking about how unworthy I must be if my boss had given my presentation to someone else. Every reason that I could think of why I didn't belong overcame me. Then, I took a few deep breaths and went home early.

That night, I prepared a set of custom emotional responses for myself. I worked hard to ensure that I was ready to deal with whatever happened at the meeting. I would be tough to ignore as the only woman *and* the only minority in the room.

When it was time for my information to be presented, I acted. After my peer stopped talking, I added additional facts and figures regarding the data not on the prepared charts. After all, I did the research, so I knew the backstory, equipment, and products.

When my peer challenged me, no amount of poking or prodding would get me to blow my stack. Eventually, it was apparent that it was my presentation. If I had allowed my imposter syndrome to remain in charge, that meeting would not have led to my next promotion.

Stephanie

Signs that imposter syndrome is sabotaging your career include a general lack of confidence. This can manifest at work as follows:

- Challenges in speaking up.
- Not going for a promotion or stretch project because you don't feel 100% qualified.
- Walking into a meeting and immediately looking for a seat far away from the "important people."
- Brushing off complements.
- Perfectionism: Nothing is ever "good enough."
- You must have all the answers to be competent in your role.
- You hate asking for help and instead spend hours researching (also known as wasting time when asking someone was much faster!).
- Trouble delegating or micromanaging.
- Overwork and exhaustion.
- Saying yes to everything/not setting boundaries.

The most common of these are the last three. I see overwork and struggling to set boundaries at all experience levels, which results in trouble delegating and significantly contributes to keeping women stalled in middle management instead of advancing to executive levels.

Signs that imposter syndrome is sabotaging you show up daily when simple tasks that aren't important in the big picture—like writing an email—take much longer than they should. For example, I wrote an email to my boss for help prioritizing my heavy workload. That email should have taken 15 minutes, or better yet a 5-minute conversation. Instead, it took over 2 hours to write. In retrospect, the *real* reason it took so long was that I was afraid of looking inept. It was a classic case of imposter syndrome.

Question 5: What Steps Can You Take to Prevent Imposter Syndrome or Stop It When It Occurs?

Stephanie

Long-term fixes and in-the-moment fixes can minimize imposter syndrome—in the long term, understanding your purpose at work (i.e., why does your work matter to YOU). Getting comfortable routinely being out of your comfort zone will reduce your incidences of imposter syndrome.

We must work on each of the three components of confidence: competence, compassion, and courage, and understand that this will not happen "naturally"—at least until women in leadership in technical roles are normalized—without practiced attention to it.

In the practical sense, starting can be as simple as asking yourself daily:

- "What did I fail at today?"
- "What did I do well today?"

I ask my three daughters these questions after school and I journal on them daily. Struggling to answer them is a sure sign we're stuck in our comfort zones AND fueling the self-doubts inherent in our society around what it means to be a girl or woman.

Consistent practice getting out of your comfort zone means that when you do stretch yourself, imposter syndrome is less likely to strike and hurt your career opportunities.

Exercise 2. Take Immediate Steps to Overcome Imposter Syndrome

Steps to overcome imposter syndrome in the moment:

1. Make a list of your imposter syndrome triggers.
2. Prepare (a reasonable amount) for new situations. Training and role-playing help substantially. Example: Practice asking for feedback.
3. Increase your energy level immediately before a potentially imposter-triggering situation:
   - Take a 5-minute break to listen to upbeat songs, take a quick walk, or meditate.
   - Write your intention before going into one of these situations: How do you want to show up? What outcome do you want from this interaction?
   - Try power poses (Google them).
   - Start a brag file of "attagirl"-type messages. Look at that file right before a potentially triggering event.
4. Plan what you will do when imposter syndrome is triggered. Practice taking a deep breath, telling yourself, "That's my imposter talking," and then asking: "How can I best serve or show up in this situation? What is the outcome I want?" That kicks your brain out of rumination mode (where imposter syndrome thrives) and into a problem-solving mode (where your inner boss lives).

Toni

Take three measures to prevent or stop imposter syndrome when it occurs. I've listed them in order of relative importance.

1. Stop it. Instead, replace negativity with positive, self-affirming talk. Each time a negative thought seeps into your mind, replace it with an affirmation. The affirmations must be something that means much to you. You must find the affirmations that lift your spirits and place you in a better frame of mine.
2. Control your words. "Words are powerful—they bring up images of success or failure and significantly affect how we approach tasks and overcome obstacles and challenges [13]." I can tell you precisely what happened at every misstep I made in my career, detail by painful detail. I cannot describe the many more successes that I had during my career. The hits' give me a fading sense of pride, but the misses still cause me pain. Analyze your successes and failures. When I started tracking my accomplishments, I found out something I had not recognized: My success was breeding success in others. I also found that, after 10 years, other than money, I had not fully defined what success meant to me.
3. Allowing others to trigger your emotions at work is a losing proposition. I invested too much in conflict management with my peers or bosses while I was in the throes of imposter syndrome. There is no success for anyone under those circumstances. When triggered, imposter syndrome is a terrible teammate.

Imposter syndrome works when one person is questioning themselves. To limit that questioning, individuals must know what success looks like for them. Once they define their success, they must visualize and hold success in their minds and hearts.

One of the consistent thoughts that imposter syndrome sufferers have is that everyone cannot be successful. No one needs to fail. Success is not a pie. There is no limit to the amount of success existing in this world. My success does not diminish your accomplishments. Your success does not diminish mine.

Exercise 3. Define Your Definition of Success

(A) Write what you believe your success looks like for your mind, body, and spirit.
(B) Examine these musings for perfectionist tendencies, then stop.
(C) Eliminate the unreasonableness and rewrite with reality in mind.

Now, close your eyes and visualize yourself as your definition of success.

When I performed this technique on myself, I acknowledged that something was missing. I was an up-and-coming executive, making six figures, traveling the world. But when I visualized my life, I saw myself with more personal time. There were interactions in my vision that I was not having with people I loved.

I lived an outdated version of success I had set years ago. This realization caused a profound change in my direction as a high-performing executive. I got out of the fast-track travel position and settled down in a less exciting homebound office position. I became more successful, though it looked like I had stepped off the fast track to outsiders.

I decided what success meant to me and stopped chasing other's versions of achievement. I encourage technical women, in particular, to make a decision early in their careers of what success means to them. Knowing your definition of success is a major quieting factor to the imposter syndrome voice in your head.

Creating a custom emotional guideline is an excellent way to deal with the changing values of success.

Exercise 4. Create Custom Emotional Guidelines

When you expect an interaction with a person, process, or team that triggers your imposter syndrome, perform the following exercise as close to the exchange as possible. This exercise takes 90 minutes—your definition of success matters in this exercise.

You'll need a timer, a pencil and paper (or a computer), and a mirror.

1. Set the timer for 5 minutes.
2. Write every interaction, verbal or nonverbal, that could trigger your emotions.
3. Allow your negative thoughts to run wild during the exercise. Write/type as fast as you can. Do not correct errors.
4. Stop when the timer rings.
5. Reset the timer for 10 min.
6. Define a reaction to every interaction you identified. Apply no censure.
7. Stop when the timer rings.
8. Take a 15-minute break. Do not skip the break!
9. Set the timer to 10 min.
10. Control your emotions.
11. Examine and prioritize the emotional triggers identified based on their overall effect on your work. Apply your definition of success. Repeat steps 11–14 until the actions match your desires.
12. Edit your answers into professional interactions.
13. Stop when the timer rings.
14. Set the timer for 30 minutes.
15. Stand in the mirror. Practice your interactions until you are confident you will regulate your responses.

Conclusion

The environmental factors that trigger imposter syndrome are not your fault. However, there are real consequences to your ability to succeed if you cannot overcome imposter syndrome. Missed growth opportunities, lower pay, lower career satisfaction, and a lower likelihood of staying in your industry are just a few.

Imposter syndrome does not need to restrict your success if you have the self-awareness to see it for what it is: A by-product of a work culture that tells you in subtle and not-so-subtle ways that you don't quite fit. Toni and Stephanie have shared their stories and a set of guidelines for identifying and addressing imposter syndrome. Repeat these exercises as needed when you call into question your competence.

Successful women still have doubts, but they must deal with imposter syndrome. The thing that sets them apart: They take action, anyway. You can choose to make healthy and empowering decisions to move forward, knowing that you are paving the way not just for yourself but for future generations of technical women.

We've given you the tools to be your ninja, ready to address imposter syndrome no matter the provocation—the next step: to be recognized as a leader at the top of your field. Now, get it done.

References

1. S. Slocum, Stephanie. *She Engineers: She Engineers: Outsmart Bias, Unlock Your Potential, and Live the Engineering Career of your Dreams*. Middletown, DE: Stephanie Slocum, 2018.
2. "TrustRadius. 2020 Women in Tech Report," TrustRadius, Austin, TX, 2020. Accessed: Feb. 2.10, 20.21. [Online].Available: at https://www.trustradius.com/buyer-blog/women-in-tech-report
3. K. Caprino, "Imposter syndrome prevalence in professional women and how to overcome it.", *Forbes*. October 22, 2020. [Online]. Available: https://www.forbes.com/sites/kathycaprino/2020/10/22/impostor-syndrome-prevalence-in-professional-women-face-and-how-to-overcome-it/?sh=2724e58373cb.
4. Clance, P. and Imes, S,. "The Imposter Phenomenon in High Achieving Women: Dynamics and Therapeutic Intervention." *Psychotherapy: Theory, Research and Practice*, vol. 15, no. 3, Fall 1978. Available: https://www.paulineroseclance.com/pdf/ip_high_achieving_women.pdf
5. V. Young, Valerie. *The Secret Thoughts of Successful Women*. New York: Crown Publishing Group, a division of Random House, Inc., 2011.
6. T. Crowe, Toni. *Bullets and Bosses Don't Have Friends. How to Manage Difficult Situations at Work*. Just One, Round Rock, TX. Toni Crowe, 2018.
7. N. Fouad and R. Singh, "Stemming the tide: Why women leave engineering," University of Wisconsin Milwaukee, WI.. 2011. [Online]. Available: https://uwm.edu/business/wp-content/uploads/sites/34/2014/10/Stemming-the-tide_NSF_Report_2012.pdf
8. D. Goldman, *Emotional Intelligence*. 10th ed. New York, NY: Bantam Books, 2007.
9. D. Goldman. (Accessed February 1, 2021). *Building Emotional Intelligence*. Goldman EI Online [Online] Available: https://golemanei.com/#online
10. D. Goldman. "What Makes a Leader?" 1998. *Harvard Business Review*, Best of HRB, 1998 , What Makes a Leader? Cambridge, MA; Daniel Goldman, 1998 (Reprint January 2004). [Online]. Available: http://media.withtank.com/f7420daead/040507_what_makes_a_leader.pdf
11. KPMG International. "Advancing the future of women in business: the 2020 KPMG women's leadership summit report," KPMG International, Delaware. KPMG International, 2020. [Online]. Available: 2.10.21 https://womensleadership.kpmg.us/content/dam/womensleadership/pdf/2020/2020wlsstudy.pdf

12. S. Fried-Buchalter, "Fear of success, fear of failure, and the imposter phenomenon: A factor analytic approach to convergent and discriminant validity." *Journal of Personality Assessment*, vol. 58, no. 2, pp. 368–379, 1992.
13. P. Clogh and D. Strycharczyk, 2012. *Developing Mental Toughness: Improving Performance, Well-Being and Positive Behaviour in Others*, London, England: Clogh, Peter and Strycharczyk, Doug, 2012.

Toni Crowe , CLA, BSEE, MSM, CM, P.E., founded Just One after a stellar 35-year career as an Operations Vice President. The business serves as both a publishing company and a consulting firm.

Toni holds an Electrical Engineering degree from the University of Illinois, Chicago, and a master's degree in Organizational Management from Maryville University. She is certified as a Professional Manger by the Institute of Certified Professional Managers at James Madison University. She trained in Lean techniques in Japan under Sensei Yuzuru Ito of the United Technologies Corporation ACE program. She is a Design for Six Sigma Green Belt and a Six Sigma Leadership Black Belt. Crowe worked with the National Management Association to train and certify engineers in management skills.

Toni worked her way up the chain in the Fortune 500 companies of Boeing, United Technologies, and Honeywell, ending her career as the VP of Operations of a division of Ultra Electronics, a multibillion-dollar UK aerospace company.

Toni has written six books, two of which have won Reader's Choice Gold Awards. The winners were (1) her business book, *Bullets and Bosses Don't Have Friends* and (2) *Never a $7 Whore,* the first book in her memoir. Toni is a prolific writer on the Medium.com platform with articles from management advice to cat humor.

Toni served on the Board of Directors of Matarah Corporation, the Hartford Connecticut Big Brothers/Big Sisters, and the Dunedin Fine Arts Center. She served as a member of the Round Rock, Texas Chamber of Conference.

Stephanie Slocum , BSAE, MSAE, LEED, is the founder of Engineers Rising L.L.C. and author of *She Engineers: Outsmart Bias, Unlock Your Potential, and Create the Engineering Career of Your Dreams*. She is the Fearless program's creator, the first-of-its-kind virtual mentoring and leadership development program designed exclusively for individual women in engineering who are struggling with career paths as one of the very few women on their teams or in their firms.

Stephanie shines a light on barriers to the retention of women in engineering and provides practical training, inspiration, and mentorship through her online platform and programs. She is a recipient of the 2020 Women of Technology Award for empowering and supporting women in their engineering careers.

Stephanie volunteers extensively with the American Society of Civil Engineers (ASCE). She currently serves on the Board of Governors of their Structural Engineering Institute and is the

Business Practices Committee's acting Chair. She served on the Task Committee on the Code of Ethics, is a member of the NCSEA SE3 committee on engineering retention and engagement, and spoke at the 2019 National SWE conference and the Women in Transportation (WTS) Summit on the topic of building professional confidence.

Before founding Engineers Rising, she worked in engineering building consulting for 15 years. She holds Bachelor's and Master's degrees in Architectural Engineering from the Pennsylvania State University and has held a Leadership in Energy and the Environmental Design (LEED) accreditation since 2004.

From Brazil to the World: The Journey of a Fluid Dynamics Experimentalist

Tamara Guimarães

When planning and designing a fluid dynamics experiment, we usually try to find the most efficient way to reproduce a process or to operate a machine. The less energy you use to achieve an end goal, the more efficient you are. This concept can also be applied to our everyday lives. But a lot of times in life, unlike in fluid dynamics experiments, the most straightforward and efficient path may not be the way to go.

Looking back at my journey and where life has taken me—see Fig. 1—since I first heard about this profession called "Engineering" in high school, it may seem that I did not take the most efficient way to get to my career goals. And that is absolutely correct. Just like in an experimental investigation in fluid dynamics, there were several moments in my life in which I faced questions, frustrations, failed hypotheses, and misleading conclusions, and I had to take a step back and reassess my strategies. In addition to that, I also faced some of the many challenges of being a woman in a male-dominated field such as engineering: sexism, imposter syndrome, and self-doubt.

In the next pages, I will talk about my journey as a woman and an immigrant in engineering all the way until preparing to start working as an Assistant Professor at Penn State University, and how each step of the way was influenced by the very few women I could look up to and the few men who were willing to offer me support. With this, I hope to be able to help future generations of female engineers and scientists to not feel so alone, and understand that they can and should go after the career they want, because we should all be pursuing our dreams, and the path can be much easier and more enjoyable when we feel seen, heard, and represented.

T. Guimarães (✉)
Department of Mechanical Engineering, Pennsylvania State University,
State College, PA, USA
e-mail: guimaraes@psu.edu

M. Bailey, L. Shackelford (eds.), *Women in Mechanical Engineering*, Women in Engineering and Science, https://doi.org/10.1007/978-3-030-91546-9_7

Fig. 1 Places that have been part of my (very much not straightforward) journey

From Journalism to Engineering

I was born in São Bernardo do Campo, a city in the metropolitan area of São Paulo known for being the heart of the automotive industry in Brazil. My grandfather worked for General Motors and Volkswagen before I was born. Both my parents worked long hours, so I was raised mostly by my grandmother, who would help me and my brother with our homework, even though she herself had only finished elementary school. She was also the cook, the manager of the household, my right-hand (wo)man, and definitely the first example of a strong female leader that I had in my life.

When I was around 4 years old, she gave me a cheap little toy car that she bought at the farmers market. The 50-cent toy soon became my favorite, even though I had many dolls with which I did not even once play. My parents and teachers could have then noticed that I was more interested in "boy" toys, such as cars, puzzles, and Lego blocks, and that I had the potential to be an engineer, and then could have helped me find the easiest way to get there. But traditional gender roles were (and still are) really hard to overcome in Brazil. So for the longest time, I did not even think engineering was an option for a woman. I thought I wanted to be a journalist.

I have always been very communicative. I was that kid at restaurants who would run around striking up conversations with complete strangers. So, in elementary school whenever my teachers asked what I wanted to be when I grew up, "a journalist" was my go-to answer. To that, my math teacher would always reply: "What a waste!" That reaction outraged me. I didn't understand what that meant. It always sounded very offensive to me. What did it mean for it to be a waste? What was it that she saw in me that made it such a waste for me to choose a career in journalism?

In Brazil, everyone who wants to pursue a higher education degree must take an entrance exam which is offered only once a year. I picked a high school that would prepare me for that exam. In my first year, I had a Physics teacher who was phenomenal. His classes were fun, engaging, exciting, but right off the bat he made a clear statement that girls did not do well in his class, because "*girls are not cut out for sciences*"[1]. That statement bothered me so much that one day I raised my hand in class and told him I would ace his exam. He told me that if I did that, he would dress up as a woman for the next class. Well, I did ace his exam, and while wearing a white top and a pink skirt, he asked me what I wanted to be when I grew up. I defaulted to my usual answer: journalist. And he finally finished that statement for me: "What a waste, you should be an engineer!" That was the first time someone said that to me. But that didn't go much further. What does an engineer do? Why should I be an engineer? What kind of engineer? How do I prepare for that?

I asked my mom for advice. She has a degree in electronics technology and was leading her own company as a newly divorced mother of two. She has always been the one woman among all the men, having dealt with sexism, being shut down and diminished by her coworkers and mentors her whole life, and knows how hard it can be to succeed as a woman in a technical field. She was not really supportive at first and asked me what happened to my passion for journalism – she even suggested a Law degree because I was so good at arguing with her all the time. We ended up finding a good compromise: I enrolled in a Technical Drawing class to see if I would be interested in (and if I could handle) something very technical that would be a necessary skill if I really wanted to be an engineer.

So there I was, at 16 years old, for the first time being one of the few women among all the men. The class was aimed at machine operators who worked for the automotive industry, so there were only three women – myself included – and over fifty guys in the class. The environment was so sexist and so uncomfortable that the (obviously male) teacher would only call the guys to the board to solve problems so that the women wouldn't be on the spot and subject to harassment. I was learning so much and was so excited to show that I could do everything that they could do that I found it really upsetting to never be called to the board. I wondered if that would be my whole life as an engineer. Would I never be invited to be in the spotlight? Would there never be a chance for me to prove that I could be as good as them? However, regardless of the uninviting environment, that experience was the push I needed: I was going to be an engineer.

A Woman Among the Men

The decision to study engineering turned out to be the easiest part of the process. With a lot of study hours —and the privilege to be able to focus on only studying for a whole year—I got into a public university that had just been founded near my hometown: *Universidade Federal do ABC* (UFABC). They offered an innovative

[1] This type of statement is now widely understood to be a stereotype threat that discourages female students from choosing STEM fields.

approach for higher education in STEM in Brazil: Students would first get a Bachelor's degree in Science and Technology, and then choose a specific STEM-related major and get a second Bachelor's degree.

UFABC was also the first university in Brazil in which all professors had a PhD [1], the main goal being to promote academic research. Since I was in the top 3% of the incoming students, I was offered a scholarship named *Researching Since the First Day*, which offered financial support for research activities. As Physics had been my favorite subject in high school, I started doing research with Professor Eduardo Gregores, who is an international reference in Particle Physics, having been one of the authors of the paper that describes the Higgs Boson at CERN in Switzerland [2]. Particle Physics isn't exactly an easy subject, so I struggled for a year, and ended up giving up on that research once I failed a class directly related to my research topic for the first time in my life.

I had never gotten a bad grade before, let alone failed a class. On top of that, I was one of the few women in my classes and had to deal with professors who would make jokes during the lectures, or constantly ask us the tricky questions, putting us on the spot and hoping for us to answer incorrectly. A Heat Transfer professor once added a picture of a kitchen oven to a slide "*so that the girls would see something they were familiar with*." I had a supervisor at my first internship in my junior year who would not invite me to any of the meetings with our international customers, even though I had taught English at a language school for 5 years, but would invite the male interns who could barely speak a full sentence in English. Those were some of the moments when I really considered giving up. What if my high school teacher and these professors were right and women were not cut out for sciences? What if my mom was right and I should have studied journalism? I decided to push harder and focus on my engineering classes, which I figured would give me more insight on whether I could be an engineer or not.

Energy Engineering was my chosen major after taking some introductory classes during my third year. The course offered an interesting mix of classes in the thermal and electrical sciences, so I built a solid foundational knowledge of Fluid Mechanics, Thermodynamics, and Heat Transfer, while also learning in depth about Electrical Circuits, Power Systems, and Sensors and Instrumentation. The most fascinating part of it for me was bringing all that knowledge together into the more interdisciplinary classes, which focused on Energy Efficiency, Management, and Planning. That was also when I got closer to the other women in my class, who are still today some of my best friends. At this point, there were three to five women in each class, and about twenty-five men. But it was only then that I realized something: The women were usually in the top 5% of the class.

That was mind blowing! Even though the sample size was pretty small, that was a pattern I had not noticed before. It showed me that women are not only cut out for sciences, but we can THRIVE. So I should stop questioning myself every time I struggled, after all, I would rarely see men questioning themselves. Struggling is normal and it is part of being an undergraduate student, and it was ok for me to not do so well in an exam or in a project, but yes, I could still be an engineer. My best friends from that time at UFABC now all have remarkable careers: Maria Brasil works at an international bank in São Paulo; Juliana Pin is getting her PhD in Operations Research at North Carolina State University; and Dr. Laura Novoa got her PhD in Mechanical Engineering at University of California Irvine in 2019 and has since been working as an engineer at Tesla.

Near the end of my undergraduate studies, the Brazilian government – led at the time by the country's first female President, Dilma Rousseff – launched a study abroad program called *Science Without Borders*. They had the bold goal to send thousands of Brazilian students abroad to improve their language skills, experience different cultures, and promote international academic exchange with other countries. I had always wanted to study abroad, so I applied for that scholarship and ended up as an exchange student in Engineering at University of California Davis, which completely changed my life.

What Is an Experimental Research Lab?

Studying abroad is an experience like no other: Being away from friends and family and taking classes in a language that is not your own is not easy, no matter how comfortable you are with the language. With many other Brazilian students in the same program, a support community was quickly built, which helped us navigate the challenges. The percentage of women in the engineering classes was not much higher than in Brazil, but it felt to me that being a woman in engineering in the US was *normal* and not a joke-worthy matter, which fostered a more inclusive and less hostile environment than what I had experienced before.

The *Science without Borders* program offered a scholarship for students to take classes during the academic year and also encouraged them to find a summer internship. Since I was close to finishing my degree in Brazil, I was able to take graduate-level classes, which I picked because they were taught by professors who ran energy-focused centers on campus. I wanted to have the chance to learn from professors who were references in their research fields, but also to get to know them during office hours and potentially find an internship. The Director of the Western Cooling Efficiency Center at UC Davis (WCEC), Professor Mark Modera, taught a class called Building Energy Performance offered by the Department of Civil Engineering. His lab had just moved to a new building and they were looking for students to help them with setting up and building new experimental rigs, so I reached out to him and got my internship for the summer.

My time at the WCEC completely redefined the meaning of academic research to me, especially experimental research. Since the campus of my home university was not completely built (which is way too common for government construction in Brazil), I had very few opportunities for hands-on engineering experience. Being at UC Davis and working at an on-campus research lab was the complete opposite. I was assigned to build an experimental rig that would simulate a hot room being cooled by an HVAC unit equipped with sensors that controlled and monitored the temperature inside the chamber and in the air conditioning lines. This experiment was aimed at assessing the quality of measurements taken by HVAC technicians in the field by having them come to the lab and use their equipment to measure the parameters in our rig, which then would be compared to the built-in sensors. With this, we could quantify how the efficiency in those home systems would be affected by inaccurate or biased measurements. I worked with three female engineers at the WCEC: Theresa Pistochini and Dr. Kristin Heinemeier, who were my direct supervisors; and Mayra Garcia, a Master's student whose family had immigrated to the US from Mexico, as have many others in California.

Mayra had a similar background to mine. She also came from a different culture and was used to speaking a different language back home. Seeing her working on the next steps after becoming an engineer meant I could do that as well. I could get a graduate degree after finishing my Bachelor's, and I could do experimental research, which could be relevant to the world. Unlike in Brazil, where most research funding comes from a few governmental agencies, the WCEC had funding both from the US government and from industry partners, who had more application-oriented goals. So there was an obvious next step for me after my year abroad at UC Davis: Getting a PhD in the US.

I started looking into graduate programs all over the US when I returned to Brazil. My boyfriend at the time (now my husband) also wanted to get a PhD, so we narrowed it down to a couple of universities that offered programs in both our fields. We received offers of admission to Virginia Tech (VT), Mechanical Engineering for me and Biological Systems Engineering for him. I also received a Teaching Assistantship (TA) for the first year, so I did not have to worry about finding a research group and an advisor right away. Once again, I had to leave the comfort of my family, my friends, and my country to pursue my dream. In August 2014, I packed one suitcase and moved to Blacksburg, Virginia, to start my PhD studies. I was one of the five women in a group of twenty-six TAs for a senior lab class in Mechanical Engineering. I was the only first-year international graduate student. Once again, I felt alone and like I did not belong. So questions started popping back in my mind: How could I balance classes with being a TA and doing research? Did they really offer admission to the right Tamara Guimarães? I really felt like I was going to fail and be kicked out of the program. And I might have, if it hadn't been for the support of the two female professors who led that class and the professor who was teaching my Fluid Dynamics class, Dr. Francine Battaglia. (See "Educating the Next Generation of Mechanical Engineers in Fluid-Thermal Sciences" in this volume, by Francine Battaglia, Lu Chen, Mirka Deza, & Bahareh Estejab, to learn more about Dr. Battaglia and her mentees' approaches to engineering education and mentorship.) They assured me that what I was feeling was normal and that I should not be so hard on myself, something I still struggle with constantly, which is an example of imposter syndrome, widely recognized to be a common issue for underrepresented students during graduate studies.[2] An hour of conversation and just showing that they were there whenever I needed made all the difference to me.

---

[2] If you are struggling with imposter syndrome during a graduate program, considering seeking campus resources for help.

A month or so after starting at Virginia Tech, I had my first meeting with Professor Todd Lowe, who was looking for a PhD student for his group to work with experimental fluid dynamics. Todd was, and still is, one of my greatest cheerleaders. Both he and Dr. Walter O'Brien, my unofficial co-advisor, have led established research groups and centers at VT with funding from government agencies and from the industry. My PhD research was funded by NASA Langley and NASA Glenn and focused on experimental fluid dynamics using laser-based diagnostics, mainly particle image velocimetry (PIV).

My research was part of the NASA Environmentally Responsible Aviation (ERA) Project, a 6-year program created by NASA in 2009 to explore ways in which aviation could reduce its impact on the environment through the development of more efficient aircraft designs, reducing fuel consumption, noise levels, and nitric oxide (NOX) emissions [3]. There is extensive research in the aviation and turbomachinery community on how to improve the efficiency of gas turbines (both for flight and for ground applications, such as power generation), and even very small percentages of efficiency improvement mean millions of gallons of fuel and millions of dollars saved every year [4]. To meet the very bold goals of the project, NASA developed a novel aircraft concept named Hybrid Wing Body, which, unlike traditional aircraft configurations that consist of a tube-shaped body and wings, has a different design (see Fig. 2). There are numerous steps in the research and development of a new aircraft, involving hundreds or even thousands of people in different research groups around the world. The Turbomachinery and Propulsion Laboratory at Virginia Tech (TurboLab) was tasked with developing a 3D-printed device called StreamVane™ that would assist in the ground testing phase of assessing the impacts that such a novel aircraft concept would have in engine performance. The device can be seen in Fig. 3 installed in a research engine inlet.

My work focused on analyzing in depth how distortions to the airflow caused by that aircraft design would behave in the inlet of the engines. I performed experiments in different wind tunnels and at the full-scale engine test rig at the TurboLab using a laser-based diagnostics technique to measure the flow called particle image velocimetry (PIV), in which micron-sized particles are added to the airflow and a

Fig. 2 Hybrid Wing Body aircraft concept. Credit: NASA

Fig. 3 Laser sheet during PIV experiment in the inlet of a full-scale research engine at the Virginia Tech Turbomachinery and Propulsion Laboratory. Credit: personal archives

laser sheet is used to illuminate those particles (see Fig. 3). Using two cameras placed at an angle from the laser sheet, two consecutive pictures are taken with a predetermined time delay. By knowing that time delay and using a post-processing software to track the displacement of the particles between the frames, it is possible to calculate the three-component velocity vector field in the flow, describing the flow behavior. The main impacts of my research were applying this measurement technique to a full-scale engine experiment, which had not been done before, and describing the similarity in the flow distortion across the different scales of the experiments, showing that the StreamVane is a powerful tool that can be used for reducing costs during the development of novel aircraft concepts [5, 6].

The PhD in Engineering Experience Beyond the Lab

Getting a PhD in engineering is as much about technical developments and advancements in science as it is about the personal aspects of the academic and research world. Being in the turbomachinery field as a woman and a non-US citizen can be as frustrating as it is exciting. A lot of the funding for this line of research comes from federal agencies, so international students usually cannot work on these projects. As mentioned before, I was extremely lucky to have PhD mentors who kept that in mind and went above and beyond to ensure they were fair to all their students. We had weekly meetings to discuss my progress in the program, and they always asked me how I was feeling and made it very clear that there should be nothing making me uncomfortable. Both my advisors have daughters, which I know has helped them tackle diversity issues with different eyes. Still, I was one of the two women in my lab at the time and the only international student, and I did feel like some of the other students questioned my background and whether I was qualified to do the research. The support from my advisors and the way they believed in me—even when I didn't—helped me build confidence in my work and in the quality of my research, so I was able to stand up for myself or just ignore some of the comments that could bother me.

Attending conferences is also a great chunk of PhD work. I was lucky to present my PhD work in many, both in the US and internationally, and to network and meet people who I could potentially work with in the future. Regardless of the size of the conference, women usually accounted for about 10% of the participants. Once during a coffee break, an older man walked up to me and thanked me for bringing some color to the conference because I was wearing a pink skirt while everyone around me was wearing dark suits.[3] I *think* he meant that as a compliment, though I can't help but wonder whether he would have offered a similar compliment to a guy wearing a pink suit…

The larger engineering conferences usually promote networking events for women, which are great for meeting people, learning about their stories and struggles, or even just for feeling welcome in a room in which you do not stand out for being a woman. Getting to meet women who have established careers and entered STEM fields in a time in which it was even less common for women to pursue these careers is inspiring but can also get you really (really) angry about the struggles they had to go through. It is not uncommon to hear stories of women who dedicated an entire career to science and had their discoveries and results attributed to men in their groups or departments. Or, even in 2020, hearing from women working in the industry that their male colleagues with exactly the same job earn significantly more than them is not uncommon, even if the men do not have a PhD. One friend saw herself in that exact situation and decided to confront her bosses about it and stand up to what she believes is right, given she worked at a company that had been awarded for having strong female leadership. She was told that her male colleague had been a more aggressive negotiator and that she should not be so emotional when discussing topics such as her salary. Salary negotiations are one of the known causes of the wage gap between men and women, but there is research showing that even if that behavior is changing and women are getting better at asking for raises, men are more likely to actually get them [7].

The Next Generation of Experimentalists

I spent the summer before the last year of my PhD doing research at the *Institut für Strömungsmechanik und Aerodynamik* at the Bundeswehr University in Munich, Germany. I met the head of the institute, Professor Dr. rer. nat. habil. Christian Kähler, at a conference the year before, and contacted him as I was looking for

[3] This type of comment is sadly common for many women in STEM fields. If you experience comments like this it is okay to explain to a male colleague that comments about dress and appearance can make others uncomfortable.

opportunities for summer research. I spent 2.5 months doing experimental research in Acoustofluidics, a field that studies the effects of acoustic waves on microfluidic devices.

What I first saw as an opportunity to learn about a different field of fluid mechanics while living in Europe ended up leading me to another overseas move for a 1-year postdoctoral position once I finished my PhD. Since the field was different from my PhD topic, the research was a lot more challenging. It felt like I was learning everything from scratch, and it took me months to finally start being productive, which is one of the scariest things for women in academia. The levels of frustration I faced during this time, while also struggling with the language and cultural differences made me decide to look for a job in industry at the end of that year.

I started working as a Business Development Manager at a startup that manufactures instrumentation devices for aerodynamic measurements in May 2019. Working more closely with customers and visiting wind tunnels, experimental setups, and universities, it did not take me long to realize that I wanted to have a more direct impact on people's lives while also working on research and development. Throughout my academic life and career, I have always tried to be a role model for the younger students I interacted with, especially if they were women or foreigners. But to increase the presence of women in STEM fields, it is also extremely important that there are initiatives to recruit girls focused on elementary and middle school. I wanted to join a program in Munich that would allow me to work with girls to foster their interest in sciences. I couldn't find one that I would fit in, mainly because of the language barrier, so I started my own program, targeted at a very specific group: children of Brazilian or Portuguese parents. I teamed up with the NGO *Casa do Brasil*, and our program named *Science for Kids: Exploring a New World in Portuguese* was created. Once a month, this program would bring together groups of ten to fifteen kids, and was always organized and led by a local Portuguese-speaking scientist who worked directly with research or had research experience from a Master's or PhD. This program did not only benefit the kids, but also the scientists, who often do not have the opportunity to interact with children and present the contents of their research in a simple way, accessible for people outside the scientific community.

Getting back in touch with researchers in academia and being in this learning environment with the kids, I concluded that the next step in my career would be to go back to academia in the US. It wasn't long until I learned from someone at a conference that the Mechanical Engineering Department at Penn State was hiring new faculty members. I was going to visit them the following month for my job, and I had the chance to also meet the Department Head, Dr. Karen Thole—who describes her own journey as a woman in Mechanical Engineering in the Research/Technical section of this volume, with the chapter: "From Watching Planes in the Sky to Making Turbines More Efficient."

Until then, I hadn't been able to find a Mechanical Engineering program that I could see myself in, as most seemed like the same old traditional white male–dominated programs. It was different with Penn State. From my first meeting with Karen, I could see her commitment to diversity and how she had brought her vision to the

Fig. 4 My grandma, my mom, and 9-year-old me

department. I met several assistant professors who had recently joined the department, and about half of them were women. Just like at UC Davis when I first met a female engineer from an immigrant family working on her Master's degree, I could see myself as part of this group of young faculty members. I felt so welcome and comfortable that I even asked about the maternity leave policies of the department during my job interview, which anyone will tell you is one of THE topics to avoid during an interview!

As an Assistant Professor at Penn State, I will be able to work closely with the next generations of engineers through teaching in the classroom, in my research group, and in programs aimed at school kids. My research will focus on sensors and instrumentation for gas turbines and other applications using additive manufacturing. Collaboration between faculty members is highly encouraged at Penn State, so I will team up with professors from Mechanical Engineering as well as from other departments to develop research that will be highly focused on presenting solutions in measurement and control of aerodynamic and fluid dynamic parameters to improve the efficiency of processes and machines.

This is only the next chapter of my career at the moment, but looking back at how I got here and where I have been, there is so much I owe to the women who have been role models for me, from my grandma and my mom back when I was a kid (see us in Fig. 4) all the way to watching history being made by the first female vice president in the US, who is a black woman and daughter of immigrants. We as women still waste so much energy second-guessing ourselves and worrying we might not make the cut. Having someone to look up to and to talk to has made all the difference to me, even if it took me longer to get where I wanted. Seeing someone like me in a position of power or in a next career step made me believe that I too could be there one day. By sharing my story and showing that I, a Brazilian woman who did not know what an engineer was until entering high school, can not only be an engineer, but also teach and advise future engineers as a professor in the US. I hope to be someone that the future generations of girls and women in STEM can also look up to.

References

1. "Federal do ABC aposta em pesquisa, e 100% dos professores são doutores." Jornal da Ciência. https://web.archive.org/web/20140330220212/http://www.jornaldaciencia.org.br/Detalhe.jsp?id=79633 (accessed Jan. 15, 2021).
2. CMS Collaboration, "Observation of a new boson at a mass of 125 GeV with the CMS experiment at the LHC," Phys Lett B, vol. 716, no. 1, 2012, pp. 30-61, Sep. 2012, doi: 10.1016/j.physletb.2012.08.021.
3. J. Bonet et al., "Environmentally responsible aviation (ERA) project - n+2 advanced vehicle concepts study and conceptual design of subscale test vehicle (STV) final report," NASA, Hanover, MD, USA, CR2011-216519, 2011. [Online]. Available: https://ntrs.nasa.gov/citations/20140005559
4. T. Grose. "Reshaping flight for fuel efficiency: Five technologies on the runway." National Geographic. https://www.nationalgeographic.com/news/energy/2013/04/130423-reshaping-flight-for-fuel-efficiency/#:~:text=For%20U.S.%20airlines%2C%20domestic%20flights,(0.12%20km%2Fl.) (accessed Feb. 3, 2021).
5. T. Guimarães, K. T. Lowe, and W. F. O'Brien, "StreamVane turbofan inlet swirl distortion generator: Mean flow and turbulence structure", *J Propul Power*, vol. 34, no. 2, pp. 340-353, Oct. 2017, doi: https://doi.org/10.2514/1.B36422.
6. T. Guimarães, K. T. Lowe, and W. F. O'Brien, "Vortical flow development in round ducts across scales for engine inlet applications", Exp Fluids, vol. 60, Mar. 2019, Art. no. 52, doi: https://doi.org/10.1007/s00348-019-2702-9.
7. B. Artz, A. Goodall, and A. Oswald. "Research: Women ask for raises as often as men, but are less likely to get them." Harvard Business Review. https://hbr.org/2018/06/research-women-ask-for-raises-as-often-as-men-but-are-less-likely-to-get-them (accessed Feb. 10, 2021).

Dr. Tamara Guimarães was born in Brazil and has a Bachelor's degree in Energy Engineering from *Universidade Federal do ABC*. She received her Ph.D. in Mechanical Engineering from Virginia Tech in 2018. Her doctoral research focused on fundamental vortical flow development in the inlet of gas turbine engines. Upon graduation, she moved to Germany for a postdoctoral position at the Institute for Fluid Mechanics and Aerodynamics at the Bundeswehr University Munich, working on Microfluidics. She then worked as a Business Development Manager at a Munich-based start-up that develops flow instrumentation using additive manufacturing. She joined the Department of Mechanical Engineering at Pennsylvania State University as an Assistant Professor in 2021. Her research interests revolve around the improvement of instrumentation, sensor integration, experimental techniques, and measurements for fluid mechanics focused on gas turbine flows.

Remaining Curious: Rethinking Contributions and Opportunities as Mechanical Engineers

Kendra V. Sharp

A senior colleague recently described my career path as organic. I realized it's a surprisingly accurate description. I wouldn't say I've been tiptoeing through the daisies, but certainly the trajectory has included a sense of wandering, of non-linearity. Deviating from the standard academic track or making choices that do not appear to lead toward success by traditional metrics can feel isolating and more than a little daunting.

This chapter is aimed at those who feel like they don't fit the mold, women who are looking for nontraditional ways to contribute to advancing energy science and technology, combating climate change, or otherwise improving our environment. Indeed, many options exist for mechanical engineers to have a significant impact, whether in academia, industry, government, or other types of organizations. Through a series of lessons learned, I'll share my unique perspectives on different types of opportunities to have impact and to make a contribution, particularly for PhD degree-holders.

After completing my PhD in 2001, I was a policy fellow in the US Senate. Immediately thereafter, I launched an academic career and progressed through the ranks from Assistant Professor to Associate Professor at Penn State University, and then from Associate Professor to Professor at my current institution, Oregon State University (OSU). After my promotion to Professor, I founded and directed a humanitarian engineering program before moving into two roles in central university administration at OSU. In February 2021, I started a new position as Head of the Office of International Science and Engineering at the National Science Foundation (NSF), on loan from OSU.

The passions that have driven me throughout these various phases of my career are united by an ongoing curiosity about how to apply engineering knowledge in different work settings; a drive to be continuously challenged and always learning;

K. V. Sharp (✉)
Mechanical Engineering, Oregon State University, Corvallis, OR, USA
e-mail: Kendra.Sharp@oregonstate.edu

M. Bailey, L. Shackelford (eds.), *Women in Mechanical Engineering*, Women in Engineering and Science, https://doi.org/10.1007/978-3-030-91546-9_8

and the desire to have an impact on individuals, communities, institutions, and even national policy. I feel strongly about the importance of academic and government policymakers being well informed by science, technology, engineering, and mathematics (STEM) researchers and practitioners. A Congressional fellowship offered me the opportunity, fresh out of a PhD program, to learn about the policymaking process from a legislative viewpoint. Over time, my passions shifted as did my vision for the type of impact I could have. As a faculty member, my primary impacts were on students and my research community. As the founder of a humanitarian engineering program, my impacts expanded to include other faculty, a broader set of students, and both domestic and international partners and communities. As a university administrator with emphasis in global affairs and faculty affairs, my impacts evolved to encompass faculty across campus, and students and staff affected by our strategic priorities, policies, and processes for international and study abroad students. Now, at NSF, I've returned to a national-level position where I am engaged in developing strategies and mechanisms for international collaboration between the US and foreign funding agencies, and between US and foreign researchers.

My research contributions towards energy and the environment initially came from detailed laboratory experiments on multiphase flows in fuel cell diffusion media [1–7]. Over time, with a growing interest in engineering for global development, my research contributions evolved to estimating the resource potential in small hydropower in northern Pakistan [8–12] and studying the use of a highly efficient biomass-driven device for water pasteurization in developing world settings [13, 14]. But I have also made a number of broader, non-research contributions as a faculty member, including contributions to relevant policy dialogues, educational program design and development, and mentorship of students and faculty.

In my administrative roles, the form of my contributions continued to evolve. At OSU, instead of tracking droplet deformation, I found myself tracking risks to my institution and recommending to the University President that OSU join a global consortium of universities collaborating on providing research-based information to policymakers and the public on climate change. And at NSF, I'm tracking national-level engagements with international partners and multilateral negotiations on research and innovation policies.

In this piece, I share five lessons learned by remaining curious and seeking out opportunities that are not always visible or recognized by women mechanical engineering (ME) faculty as they move through their careers.

Contributions Come in Many Forms; Don't Discount the Broader, Non-Research Contributions

Below I describe four specific types of opportunities I've sought out to make broader, non-research contributions, ordered chronologically according to my own career progression. I've categorized these as follows: (a) opportunities in science and technology policy, (b) opportunities as an academic faculty member, (b) opportunities for international collaborations and societal impact through engineering and global development, and (d) opportunities as a university administrator.

(a) Science and technology (S&T) policy

In 2021, the need to link sound science to policy decisions has been brought into stark relief across all segments of the global population by COVID-19. But this is only one example. The need for sound science to underpin a myriad of policy decisions in the energy and environmental sector is also pressing. Serious consequences of climate change are evident globally. Eleven years ago, Dr. Bruce Alberts, former President of the US National Academy of Sciences, noted, "*The question now facing the United States is not only how to effectively reinject the facts of climate science back into the core of this particular debate, but how to ensure that good science underlies all legislative decisions* [15]." His statement rings as true now as it did then.

My own choice to pursue a policy fellowship was driven at the time simply by curiosity about the policymaking process and the factors influencing decisions that, in turn, affect everything from people's lives to the state of the planet.

For others curious about actively engaging in S&T policy, one of the best entry points for individuals with PhDs in STEM fields is the American Association for the Advancement of Science (AAAS) Science and Technology Policy Fellowships (STPF) [16]. The program is aimed at strengthening links between scientists/engineers and policymaking, by enabling scientists and engineers to work *directly* with policymakers; policymakers benefit from the fellows' relevant expertise, analysis, and science communication skills. The STPF program currently places more than 250 fellows each year across the three branches of the US government.

AAAS STPF fellowship assignments are typically for 1 year. Fellows' career stage ranges from recent PhD graduates through mid-career professionals, and even to senior professionals seeking out a unique career-capping opportunity.

While the AAAS STPF program offers an incredible opportunity for PhD-level scientists and engineers to enter the policymaking arena, it is far from the only entry point. The National Academies offer other programs for PhD students or recent graduates [17] or mid-career/senior academics [18]. The Twitter hashtag *#scipol* and the National Science Policy Network (NSPN) [19] are also good starting points for exploring options and learning about S&T policy. Table 1 highlights several S&T policy opportunities beyond the AAAS STPF.

Serving in a dedicated full-time position as a fellow is merely one way to make a S&T policy contribution. Many people contribute to S&T policy alongside their work in academia, industry, or other settings by choosing to write about policy issues or otherwise engaging with networks or publications. As mechanical engineers, we often work in the policy realm, including in energy and the environment. I highly encourage engineers at all levels to consider the societal impacts of their work, and how they might engage to ensure our federal, state, and even local policymaking processes are based on sound science.

Table 1 S&T policy opportunities beyond the AAAS STPF [17–22]

S&T policy opportunity	Additional information and key components
National Academies Fellowships [17, 18]	Christine Mirzayan Fellowship Jefferson Science Fellowship (foreign policy)
National Science Policy Network (NSPN) [19]	Annual symposium Micro-grants for student chapters Training via webinars, Science Diplomacy Exchange and Learning Program (SciDEAL), and science policy memo competition with the *Journal of Science Policy & Governance*
Journal of Science Policy & Governance [20]	Publishing platform Resource for networking and finding training and events
Government Relations Committees for professional societies such as the American Society of Mechanical Engineers [21]	Policy paper development Advocacy for priorities (currently including clean energy) Briefings or events for Congressional members and staff Expert testimony
State-level programs, such as California Council on Science and Technology [22]	State fellowship for PhD-level scientists and engineers in legislative or executive branch offices
Other opportunities	University government relations offices, think tanks, etc.

(b) Broader, non-research contributions by academic faculty

Academic faculty typically quantify their contributions through traditional metrics, such as number of journal publications, funded projects, invited talks, and graduate students advised. Hundred-page faculty dossiers contain bulleted list upon bulleted list, cataloging everything from the candidate's journal articles and grants to service on committees.

Quantitative metrics are important and, to some extent, document research productivity and scholarly impact, but they do not necessarily fully capture contributions. Many institutions allow candidates to tell their own story and highlight their key contributions, research-based and otherwise. Letters of evaluation are also requested from external experts. While not ensured to be bias-free, external letters are a useful antidote to simple quantitative metrics because writers are often able to comment on the *substance* (not just the quantity) of the candidate's contributions.

Academic faculty must meet expectations for promotion and tenure (P&T) to retain their academic positions. But quantitative output does not have to be at odds with other types of impact. Many faculty weave other types of impact into their publishable research (e.g., consideration of relevant equity issues). At some institutions, faculty may prioritize other types of impact entirely, such as commercialization and entrepreneurship, public scholarship, or curriculum innovation, if they are confident that their institution will support such choices during the P&T evaluation process.

As a faculty member, my broader, non-research contributions included organizing and offering a 3-day workshop on hydropower for Pakistani students and faculty, and enabling students to do months-long fieldwork in East Africa on the use of highly-efficient cookstoves for water pasteurization. I am encouraged that so many others are also pushing beyond research contributions. Within the energy field, alone, at my current institution, one colleague utilizes computational design tools to improve modeling and optimize wave energy converters or sustainable product design. She brings this expertise into the classroom, and when combined with her theater background, is highly engaging in large lecture settings through her style and use of humor. Another colleague focuses on the design, implementation, and assessment of household energy technologies in developing communities. Since joining OSU in 2015, she has enabled 622 student-days of research or coursework in developing countries, including through two offerings of the 10-day faculty-led field course in Guatemala that she designed.

Several of our senior mechanical engineering women faculty have made significant broader, non-research contributions too. Specific to the energy field, one spent 6 years directing the Pacific Marine Energy Center, a US Department of Energy Funded Center, to facilitate understanding and development of marine renewable energies. Under her leadership, the center was awarded nearly $50 million in federal and matching funds to launch the first US full-scale grid-connected site for wave energy.

(c) Engineering for global development: partnerships, student mentoring, and curriculum development

As my interests shifted from more fundamental to more applied research, chiefly because I wanted to see a closer link between my research and the benefit to society, I became curious about the use of small-scale hydropower for rural electrification. In high Asia, particularly in Northern Pakistan, there is an intersection of abundant natural resources for small hydropower, limited access to the national grid, and low likelihood of near-term grid expansion. This interest led to both research and broader, non-research contributions, as well as collaboration with several faculty in Pakistan. The research contribution was developing an approach for better estimation of hydropower potential for siting, investment, and improved design of micro-hydro systems in an area where access to various forms of data for sophisticated hydrological modeling is relatively limited [8–12].

Our most significant broader, non-research contributions stemmed from our impact on students and faculty. Based on our early work on the topic and partnerships we'd developed, we participated in a US Agency for International Development Partner Center (PCASE-EN) led by Arizona State University, including partnerships with the National University of Science and Technology (NUST) in Islamabad and the University of Engineering and Technology–Peshawar (UET-P). Over the course of 5 years, we hosted over 25 Pakistani graduate students as exchange visitors in our OSU mechanical engineering labs. Several have gone on to graduate programs in the US or UK as Fulbright scholars or via other means, and we've enjoyed seeing their career trajectories, whether as continuing graduate students or energy professionals in Pakistan.

Our investment of effort toward PCASE-EN led to only modest research and publication output, due to the short-term nature of the exchange and highly variable background of our exchange visitors. However, we seized many other opportunities for impact. My colleagues and I engaged in iterative evaluation of our Pakistani partners' new thermal energy engineering graduate degree program curricula. During one of my visits, PCASE-EN hosted a number of industry representatives and other Pakistani stakeholders to discuss the development of a hydropower curriculum, and organizing the 3-day hydropower workshop for students from NUST and UET-P was a highlight for me.

The draw to global development and curiosity about the use of engineering and engineering design to solve societal problems transformed my own career path, which evolved to include founding OSU's humanitarian engineering program and becoming involved with partners such as MIT's D-Lab and their broader network. Since that time, my contributions at OSU have included leading the curriculum development for our humanitarian engineering minor, sponsoring and mentoring a series of student capstone design project teams with international partners, and supporting novel interdisciplinary efforts at OSU like a dual major Master's degree in mechanical engineering and applied anthropology.

Extensive design-related collaboration with international partners has been immensely satisfying, especially in seeing the impact on students who had new opportunities to participate in international field work or collaborate with international social entrepreneurs. Importantly, as we delved further into such partnerships, it became critical to look at not only what we were doing but *how* we were doing it. How equitable were the partnerships? Were the projects focused only on student experience, or did the partners also gain tangible benefits? Venturing yet further from typical mechanical engineering research, we explored these questions [23] with partners from the Community-Based Global Learning Collaborative [24] because of their previous involvement with Fair Trade Learning (FTL) [25] and the vision that greater knowledge of FTL concepts would enhance global engineering education for sustainable development efforts writ large. Insights from our collaboration have proven very useful in thinking through *how* to approach and build partnerships with a *much* broader range of university and external partners, even now in my NSF position.

The administrative and organizational experience with the USAID PCASE-EN Partner Center, the humanitarian engineering program, and other international work served as an excellent foundation for my next transition to central university administration.

(d) Central university administration

Between being energized by international work and curious about the possibility of serving in formal administrative roles, I jumped at the chance to serve as OSU's Senior Advisor for Global Affairs. Taking on this role was distinctly rewarding, especially during the COVID-19 pandemic, a tumultuous period for international students and activities. In this role, I had several opportunities to contribute in very different ways on energy and environmental issues. Developing strategic partnerships was one element of the new university-wide internationalization and global engagement strategy that we developed. During my relatively short tenure in the position, we laid the groundwork for efforts that leverage OSU's expertise and reputation in climate science.

In one case, several well-respected and established faculty flagged an invitation for us to join the International Universities Climate Alliance [26] as founding member. This seems like a relatively easy ask, given OSU's strong reputation in the areas of climate science, climate change, and resilience. But we had a myriad of other issues to consider. For example, what if the alliance takes advocacy positions that run counter to other publicly stated positions? How much legal, reputational, and financial risk does signing-on actually entail? OSU's Office of General Counsel meticulously reviewed documents to evaluate key risks to the institution. Prior to preparing any recommendation for the Provost and/or President, we had to show clear alignment of business purpose with institutional strategy, and demonstrate sufficient support from key academic faculty and administrative leaders. As an institution, we decided that we wanted to join this alliance, and our university President signed on. The next step will be to establish how OSU faculty will be engaged going forward, and how to both leverage and support this alliance. This effort did not yield research contributions to energy and environment, but it is an excellent example of how being in an administrative position enables a dramatically different type of contribution.

For all of our international partnering efforts, one of the most interesting parts of my role was to negotiate on behalf of a large public research institution. The learning curve was very steep, especially regarding legal terminology and risks, but the staff in our Office of General Counsel have been patient and effective guides.

In 2020, I took on an additional partial administrative appointment as Associate Vice Provost for Faculty Development. I believe strongly in the importance of public scholarship, outreach, and engagement to our mission as a public land grant institution. Faculty research is all about the creation and application of new knowledge, and peer-reviewed publications establish rigor and credibility of research. However, as noted earlier in the S&T policy discussion, research expertise is invaluable when applied to policy decisions and public discourse. Curious about how to get started with public scholarship, I participated in a writing seminar offered by the OpEd Project [27], whose mission is to boost the number of underrepresented scholars emerging as thought leaders in the public sphere via authorship of op-eds and other avenues. With the emergence of COVID-19 in the US, I saw the dramatic and highly unsettling impact of rapidly escalating travel restrictions and changes in immigration policy on our international students. I subsequently worked with a mentor-editor through the OpEd Project to craft an opinion piece that was published in the *Washington Post* [28]!

I thought that other faculty would be interested in learning more about how to engage in public scholarship. The OpEd Project partners with select organizations to offer the cohort-based Public Voices Fellowship program for 6 to 12 months. I pitched the fellowship to our senior university leadership and with their support, we offered the Public Voices Fellowship in Spring of 2021. Among the 24 fellows in our initial cohort, one cluster of expertise included climate change, climate impacts, and energy.

OSU's cohort of fellows has since published pieces on a range of topics including the misrepresentation of science when scientists, journalists, and conservationists link butterflies to pollination of agricultural crops [29]; the importance of alleviating energy insecurity in the U.S. [30]; and the need for reform in faculty promotion and tenure systems to better support faculty in tackling the climate problem by encouraging rather than stymying outreach to the public and policymakers, and by encouraging high-risk research, innovation, and commercialization of climate-friendly technologies [31].

I couldn't be prouder of my colleagues' contributions to public discourse.

Above, I've described four different types of opportunities that are accessible to many mechanical engineering faculty at different career stages. But opportunities are not limited to those I have described; remaining curious and open to the full spectrum of possibilities at *every* career stage is sure to yield many ways for women mechanical engineering faculty to boost broader, non-research contributions.

Career Goals and Directions Don't Have to Be Fixed

It's all too easy to look at other women who appear to have clear and unwavering career goals and feel inadequate. They're going after their dreams! They've always wanted to be a certain type of mechanical engineer, perhaps a combustion scientist, or renewable materials expert. They're headed down a clear and seemingly preordained career path.

Even before the Instagram age, what appears to be true on the surface may not really be true. But even if it were true that these other women had clear, unwavering goals, does that imply that everyone must have clear, unwavering goals to be "successful"? Sure, eyes-on-the-prize is how athletes get to the Olympics, but allowing goals and direction to evolve rather opportunistically is an alternative approach for some people like myself.

Gilkey [32] suggests that our lives are a series of 3- to 5-year "projects" (or periods). Think about it for a minute. Can you break down your education, career, or even personal life so far into 3- to 5-year projects? Did your goals evolve over the 3 to 5 years? This is a fun and insightful exercise!

In my own case, I had some underlying and approximate long-term goals (e.g., becoming a tenured engineering professor), but upon reflection, it's easy to see how specific next steps and short-term goals very much evolved during shorter periods. In retrospect, it feels like connect-the-dots. It's now easier to see how things fit together, and how and why one experience led to the next. Even the disappointments and failures fit in, painful as it may be to revisit them. In some cases, it becomes clear that not getting a position I applied for led down a different path, one that was ultimately a better fit.

A related observation is that, during these 3- to 5-year periods, I do my best work when I focus squarely on being effective in my current position. Sure, I see other opportunities as they arise, but I'm unlikely to apply unless they strike me as simply irresistible *at that point in time*. I'm learning that, at least for me, it's perfectly acceptable to take a step without knowing what the next one will be—it's taking a step just because the step itself offers an extraordinary opportunity to be challenged and learn anew!

You Can't Say for Sure What the Future Holds, But Neither Can Anyone Else

After earning tenure at Penn State, it seemed like a change would be good for both me and my spouse. We were especially interested in moving to the western US because we wanted to be closer to the mountains, with a greater variety of outdoor activities at our doorstep. If I think of my story in Gilkey's [32] 3- to 5-year periods, the 8 Penn State years were:

(1) Pre-tenure: A time to focus on requirements to earn tenure, while also starting a family. Late nights at the office and hours of breastfeeding babies.
(2) Post-tenure: A time to start exploring new directions. Two summers in the Netherlands whet my appetite for more international collaboration.

In the second period, I started to think that I might be able to have greater impact, or at least greater near-term impact, by moving into more applied research, especially in global development. I consulted a colleague who was both well respected and a little unconventional. I knew he was very involved in faculty-led study abroad and invested in global development issues. He suggested I get more involved with engineering design, because good design is so critical to addressing global development issues. I subsequently volunteered to teach engineering design.

Around this time, I also saw an advertisement for a faculty position at OSU that seemingly offered more opportunities for collaboration and a more direct path to applied research. The timing was terrible; I was about to leave for the Netherlands to conduct research in a colleague's laboratory! I applied for the OSU position anyway, and my interview had to be sandwiched in between a quick trip to Delft with my infant to find a place to live and preparations to leave for the summer with two very young children. I was eventually offered the OSU faculty position. The negotiation process was challenged by the 8-hour time difference between the Netherlands and Oregon. One conversation stands out. On a Dutch evening, sitting in a courtyard of a lovely little hotel we were staying at for the weekend, I was trying to negotiate my new position over spotty cell service with a school head who was 6,000 miles away. My new school head was undeterred, and we agreed that I would start a year from the following January. Because we were negotiating for a move well into the future, there was plenty of time to second-guess everything.

Moving from a top-20 engineering school to much lower-ranked engineering school bucked conventional wisdom. Why on earth would one make such a move? Penn State received easily dozens of applications for each new faculty position, with candidates having increasingly dizzying records of publications and credentials. I felt confident that I could never get this job back even if I wanted it—the bar for consideration had been raised too high. Seeds of doubt were planted every time a colleague expressed bewilderment about the possible move, or if I even *perceived* a sense of bewilderment. Those seeds of doubt grew into Jack-in-the-beanstalk-sized beanstalks when a colleague I respected went so far as to say that I would never again have the same level of access to opportunities if I chose to move to OSU.

What I learned was that my colleague was (sort of) right. I wouldn't have the same opportunities. *Instead, I would have better ones (for me).* Of course, it's impossible to know what that other path (not moving) would have looked like. But what I do know is that, at the time I was making a decision, neither my colleague nor I could predict the future.

Sometimes the Move That Goes Against Conventional Wisdom Might Really Be the Right One; It's Okay to Take Risks

April 2000. I sat in a hotel conference room in a Washington, DC, ready to defend my policy paper on nuclear nonproliferation during my interview for an AAAS STPF Congressional Fellowship. I had a number of interviews for faculty positions in the final year of my PhD program. But this interview was different. There was no 45-minute academic seminar where I tried to impress people with the precision of my experiments and depth of knowledge in microfluidics. There were no lab tours where I was expected to ask clever questions or discuss my teaching interests. It was me, in a conference room facing eight representatives from federal science policy circles. They asked tough questions about current policy issues, and they expected well-articulated responses to problems that seem intractable. Various phones kept ringing during the interview, and it was a struggle to stay focused on the interview. The panel was friendly enough, and at the end they lightened the mood by explaining that while the phone ringing wasn't staged or intentional, it did provide an excellent simulation of what it's like to work in a bustling Congressional office.

I later learned that I had lucked into that interview. I was an alternate, brought in only because another candidate had withdrawn from consideration. I also later learned that I was selected! My PhD advisor is a legend in experimental fluid mechanics, and a highly trusted mentor. I was almost afraid to tell him that I was considering such a nontraditional step. Over and over, I'd rehearse the words, the justification of why this felt like a worthwhile career move. It was one of the first times I felt like I was truly stepping off the traditional track, choosing an option that might not make sense to those around me. Fortunately, he took my announcement in stride, testament to his genuine interest in seeing his mentees do what's right for them.

I wouldn't trade that year as a Congressional Science Fellow for anything. It was one of my most fun and satisfying years, and arguably one of the most formative in my career because it drove home the importance of scientists and engineers engaging with policymakers and the public. It also opened my eyes to see that academic research is not the only viable and legitimate career path for an engineering PhD-holder.

As a Congressional Science Fellow, I found myself working on big problems, with big impact. I inhabited a small cubicle with a TV, almost always on. The fellows' area in our Senate office was cozy; there were about six other fellows, all in their own small cubicles. My core focus was on S&T workforce, nanotechnology, and, after 9/11, even airport security technology.

While accepting the fellowship was one of the biggest risks I've taken, it's far from the only one. Moving from Penn State to OSU was a big risk. Having multiple children felt like a big risk. I wasn't sure my colleagues would take me seriously as an academic, and I feared the responsibility would be too much to handle as a faculty member. Transitioning from fundamental fluid mechanics to more of an applied focus, eventually moving further afield into design for global development, was a risk, as was applying for (and accepting) a central administrative position in Global Affairs. My recent move, to NSF, is a risk, both professionally and personally because we will have to navigate the impacts of being a bi-coastal family. Each time, it's been helpful to reflect on previous risks and realize that the biggest risks have offered some of the biggest rewards and opportunities to make satisfying contributions, research or otherwise.

Remain Curious and Open to Different Types of Career Opportunities, and Find the Right Mentors and Sponsors to Guide Exploration

Overall, the biggest lesson has been to remain curious and open to the many different types of opportunities around me. What it means to be an academic faculty member in mechanical engineering is changing; many more faculty are exploring opportunities to innovate and commercialize technology; deepen their commitments to equity, diversity, and inclusion; collaborate with international colleagues; or engage with policymakers and the public (including K–12). In addition to being open to the possibilities, finding good mentors and sponsors is critical, not only in terms of advancement, but also in terms of building resilience and being able to work through disappointments. But mentors and sponsors are not the same thing. In

Ibarra's words, "*While a mentor is someone who has knowledge and will share it with you, a sponsor is a person who has power and will use it for you*" [33]. Sponsorship requires a greater commitment than mentorship, and is something that evolves over time as a stronger, more trusting relationship is formed. If you're harboring an interest or curiosity toward any of these less traditional career opportunities, the right sponsor can reach out to their colleagues and potentially unlock a key invitation to participate on a committee, in a workshop, or as a collaborator in a new area.

Conclusion

I hope I've piqued your curiosity about how as engineers we can engage productively in public policy, public scholarship, international collaborations, and even academic administration. I also hope this piece has encouraged reflection on your own contributions and future goals, whether they're easy or hard to quantify. Perhaps the biggest lesson I'm learning is simply to remain curious, and to be open to the possibilities, including the unexpected. Or rather, to be open to the possibilities, *especially* the unexpected.

References

1. E. C. Kumbur, K. V. Sharp, and M.M. Mench, "A design tool for predicting the capillary transport characteristics of fuel cell diffusion media using an artificial neural network," *J. Power Sources*, vol. 176, pp. 191–199, 2008.
2. E. C. Kumbur, K. V. Sharp, and M.M. Mench, "Liquid droplet behavior and instability in a polymer electrolyte fuel cell flow channel," *J. Power Sources*, vol. 161, pp. 333–345, 2006.
3. E. C. Kumbur, K. V. Sharp, and M.M. Mench, "On the effectiveness of the leverett approach for describing the water retention behavior of fuel cell diffusion media," *J. Power Sources*, vol. 168, pp. 356–368, 2007.
4. E. C. Kumbur, K. V. Sharp, and M.M. Mench, "A validated leverett approach for multi-phase flow in polymer electrolyte fuel cell media. I. Hydrophobicity effect," *J. Electrochem. Soc.*, vol. 154, pp. B1295–B1304, 2007.
5. E. C. Kumbur, K. V. Sharp, and M.M. Mench, "A validated leverett approach for multi-phase flow in polymer electrolyte fuel cell media. II. Compression effect," *J. Electrochem. Soc.*, vol. 154, pp. B1305–B1314, 2007.
6. E. C. Kumbur, K. V. Sharp, and M.M. Mench, "A validated leverett approach for multi-phase flow in polymer electrolyte fuel cell media. III. Temperature effect and unified approach," *J. Electrochem. Soc.*, vol. 154, pp. B1315–1324, 2007.
7. S. Y. Yoon, J. W. Ross, M. M. Mench, and K.V. Sharp, "Gas-phase particle image velocimetry (PIV) for application to the design of fuel cell reactant flow channels," *J. Power Sources*, vol. 160, pp. 1017–1025, 2006.
8. T. M. Mosier, D. F. Hill, and K.V. Sharp, "30-Arcsecond monthly climate surfaces with global land coverage," *Int. J. Climatol.*, vol. 34, no. 7, pp. 2175–2188, 2014, https://doi.org/10.1002/joc.3829.

9. T. M. Mosier, D. F. Hill, and K.V. Sharp, "How much cryosphere model complexity is just right? Exploration using the conceptual cryosphere hydrology framework," *Cryosphere*, vol. 10, pp. 2147–2171, 2016, https://doi.org/10.5194/tc-10-2147-2016.
10. T. M. Mosier, D. F. Hill, and K.V. Sharp, "Update to the Global Climate Data package: Analysis of empirical bias correction methods in the context of producing very high resolution climate projections" *Int. J. Climatol.,* vol. 38, no. 2, pp. 825–840, 2017, https://doi.org/10.1002/joc.5213.
11. T. M. Mosier, K. V. Sharp, K.V., and D. F. Hill, "The Hydropower Potential Assessment Tool (HPAT): Evaluation of run-of-river resource potential for any global land area and application to Falls Creek, Oregon, USA," *Renew. Energ.,* vol. 97, pp. 492–503, 2016, https://doi.org/10.1016/j.renene.2016.06.002.
12. G. S. Warren, T. M. Mosier, K. V. Sharp, and D. F. Hill, "Small hydropower toolkit: Considerations for improving global development and an accompanying case study for Pakistan," *Univ. Pitt. L. Rev.,* vol. 80, pp. 137–174, 2018, https://doi.org/10.5195/lawreview.2018.595.
13. G. Burleson, D. Caplan, C. Mays, N. Moses, T. Navab-Daneshmand, T., K. Sharp, and N. MacCarty, "Computational modeling and empirical analysis of a biomass-powered drinking water pasteurization technology," *Energies*, vol. 13, no. 4, p. 936, 2020, https://doi.org/10.3390/en13040936.
14. G. Burleson, B. Tilt, K. Sharp, and N. MacCarty, "Reinventing boiling: A rapid ethnographic and engineering evaluation of a high efficiency thermal water treatment technology in Uganda," *Energy Res. Soc. Sci.*, vol. 52, pp. 68–77, 2019.
15. B. Alberts, "Policy-making needs science," *Science*, vol. 330, no. 6009, p. 1287, 2010. Accessed: Jun. 13, 2021. [Online] Available: https://science.sciencemag.org/content/330/6009/1287?intcmp=trendmd-sci, https://doi.org/10.1126/science.1200613.
16. American Association for the Advancement of Science, Science and Technology Policy Fellowships, https://www.aaas.org/programs/science-technology-policy-fellowships (accessed Jun. 13, 2021).
17. Christine Mirzayan Fellowship, https://www.nationalacademies.org/our-work/the-christine-mirzayan-science%2D%2Dtechnology-policy-graduate-fellowship-program (accessed Jun. 13, 2021).
18. Jefferson Science Fellowship, https://sites.nationalacademies.org/PGA/Jefferson/index.htm (accessed Jun. 13, 2021).
19. National Science Policy Network, https://scipolnetwork.org (accessed Jun. 13, 2021).
20. *Journal of Science Policy & Governance*, https://www.sciencepolicyjournal.org (accessed Jun. 13, 2021).
21. American Society of Mechanical Engineers Government Relations, https://www.asme.org/government-relations (accessed Jun. 13, 2021).
22. California Council on Science and Technology, https://ccst.us (accessed Jun. 13, 2021).
23. N. Reynolds, N. A. MacCarty, K. V. Sharp, and E. Hartman, "Using the fair trade learning framework to improve the outcomes of engagement between universities and off-campus partners: Applications and implications for program design," *Advances in Engineering Education*, accepted for publication, 2021.
24. Community-Based Global Learning Collaborative, https://compact.org/global-sl/ (accessed Jun. 13, 2021).
25. Fair Trade Learning, https://compact.org/global-sl/ftl/ (accessed Jun. 13, 2021).
26. International Universities' Climate Alliance, https://www.universitiesforclimate.org (accessed Jun. 13, 2021).
27. The OpEd Project, https://www.theopedproject.org (accessed Jun. 13, 2021).
28. V. Strauss, (introduction), K. V. Sharp, (opinion), "The Trump administration is moving to restrict international students – Why that's a bad idea," *Wash. Post*, Jun. 18, 2020. https://www.washingtonpost.com/education/2020/06/18/trump-administration-is-moving-restrict-international-students-why-thats-bad-idea/ (accessed Jun. 13, 2021).

29. E. Fleishman, “Save the butterflies – But not to save our food supply.” *Sci. Am.* (Conservation Opinion), March 31, 2021, https://www.scientificamerican.com/article/save-the-butterflies-but-not-to-save-our-food-supply/ (accessed Jun. 13, 2021).
30. E. Cotilla-Sanchez, “The value of being ordinary: Energy is not like youth sports or GRE scores,” *Visible Mag.*, Apr. 6, 2021, https://visiblemagazine.com/the-value-of-being-ordinary/ (accessed Jun. 13, 2021).
31. T. Ozkan-Haller, “To help address the climate problem, universities must rethink the tenure and promotion system,” *Yale Clim. Connect.*, May 28, 2021, https://yaleclimateconnections.org/tag/h-tuba-ozkan-haller/ (accessed Jun. 13, 2021).
32. C. Gilkey, *Start Finishing: How to Go from Idea to Done*, Boulder, CO, USA: Sounds True, Incorporated, 2019.
33. H. Ibarra, " A lack of sponsorship is keeping women from advancing into leadership,” *Harvard Business Review*. https://hbr.org/2019/08/a-lack-of-sponsorship-is-keeping-women-from-advancing-into-leadership (accessed Feb. 4, 2021).

Dr. Kendra V. Sharp is a Professor of Mechanical Engineering at Oregon State University (OSU), currently on loan to the US National Science Foundation where she is Head of the Office of International Science and Engineering. She earned her B.S. in Aerospace Engineering and Ph.D. in Theoretical and Applied Mechanics from the University of Illinois at Urbana-Champaign, and Master’s degrees in Engineering and Mechanical Engineering from the University of Cambridge and the University of California-Berkeley, respectively. Before launching her academic career, Dr. Sharp served as an American Association for the Advancement of Science/American Institute of Physics Congressional Science Fellow in the US Senate. At OSU, she previously acted as Senior Advisor for Global Affairs, where she led the development of OSU’s internationalization and global engagement strategy. She also served as Associate Vice Provost for Faculty Development. Dr. Sharp founded and directed OSU’s Humanitarian Engineering Program, and was the Richard and Gretchen Evans Professor of Humanitarian Engineering. She believes deeply in the value of interdisciplinary education and research and in using engineering and technology for positive societal impact. In collaboration with faculty, students, nongovernmental organizations, and social entrepreneurs, she works to address global issues in poverty, energy, and water through engineering design and international education. Dr. Sharp’s international research and teaching efforts include experience in New Zealand, Pakistan, the Netherlands, Ghana, India, and Thailand. Recent awards include an Erskine Fellowship at the University of Canterbury, the American Society of Mechanical Engineering’s (ASME) Edwin F. Church Medal, OSU’s College of Engineering Faculty Mentoring Award, and OSU’s International Service Award. She is an ASME Fellow.

The Changing Landscape of Mechanical Engineering: Learning to Embrace My Ecofeminist Identity Within the Elitism of Engineering

Robyn Mae Paul

Introduction

Mechanical engineering is an essential profession to our world. Mechanical engineers build infrastructure, design robots, develop energy alternatives, and create medical devices. The American Society of Mechanical Engineers (ASME) defines five core new technologies which are current areas of growth within the field: bioengineering, robotics, clean energy, manufacturing, and pressure technology [1]. Although mechanical engineering is often simply defined as the design and use of machines, or the study of objects and systems in motion, this list shows how mechanical engineering is much more than this, and is one of the most diverse and versatile fields of engineering.

So then, we must ask ourselves, what defines someone who identifies as a Mechanical Engineer? Engineering identity has been studied over the past few decades, and there is a general consensus that engineering identity is tied to one's professional identity. Beam et al. define professional identity as "how closely an individual relates to a particular field, profession, or occupation" [2]. Other scholars also emphasize the importance of professional identities considering both how people identify themselves, as well as how they are identified by others within the field [3]. This latter part of identity, where others are making judgments on who fits within the field, often comes from a comparison to the cultural norms that exist within the field [4]. Therefore, through better understanding of the professional and cultural norms of mechanical engineering, we can better understand what defines someone who identifies as a Mechanical Engineer. However, what happens when this identity and the underlying cultural norms are challenged? As we build a more inclusive culture, how do we integrate the core identity of mechanical engineering while also allowing space for a wider variety of identities?

R. M. Paul (✉)
Schulich School of Engineering, University of Calgary, Calgary, AB, Canada
e-mail: rmpaul@ucalgary.ca

M. Bailey, L. Shackelford (eds.), *Women in Mechanical Engineering*, Women in Engineering and Science, https://doi.org/10.1007/978-3-030-91546-9_9

I am a third-year PhD candidate exploring the concept of the *hidden curriculum* of engineering, which includes the invisible norms, beliefs, attitudes, and values that we all unknowingly conform to, and that are engrained within classroom and professional environments [5, 6]. Specifically, I am applying principles from ecofeminism to support the deconstruction of the hidden curriculum of engineering education. Simply put, ecofeminism recognizes the connections between the oppression of women and the oppression of nature. Further expanding, it "*is the position that there are important connections between how one treats women, people of color, and the underclass on one hand and how one treats the nonhuman natural environment on the other*" [7, p. xi].

Within our society there is higher value and power given to the male–human–culture relationship than to that of the female–biological–nature. For example, our society, particularly within technical contexts such as engineering, values problem-solving with *rational* approaches above *emotional* perspectives. We see technical problems and critical thinking as requiring rational decision-making and there is little value for using emotions and feelings to support this decision making process in engineering [8]. For example, when designing a bridge, rather than just doing a logical data analysis from stakeholder engagement surveys, it would be beneficial to put more value on factors such as previous experience (what feelings have you had on bridges in the past?) and emotional empathy (what sociotechnical considerations should be made for the local communities and ecosystems?). We also tend to associate rationality as one of the qualities that makes humans superior, whereas we see emotions as being a more feminine approach closer linked to nature. This value hierarchy operates as if it is inevitable and logical, when in reality it is an "invented superiority" [9, p. 64]. I believe that it is essential in our engineering classrooms and workplace environments that we deconstruct some of these fabricated value dualisms to promote a stronger integration of the human–nature connection, and nurture humans' ecological connectedness.

Through my work so far, however, I have learned that throwing around words like *ecofeminism* in mechanical engineering brings up a lot of feelings. Some people are excited, curious, and eager to learn more. Others are immediately defensive where they don't understand how social justice concepts such as ecofeminism have anything to do with engineering. They defend their male-dominated, and hierarchical Mechanical Engineering identity and argue that ecofeminism is not a true engineering research topic.

Through this chapter, I explore my PhD journey so far, my past journey of where I came from, my internal journey that brought me to this work, my external journal in the comments and feedback I have received, and my future journey in the importance of continuing to pursue this work. I explore the values that are core to mechanical engineering, and consider how these come into conflict with values which I believe are core to the future of engineering, particularly for addressing challenges such as climate change and achieving the sustainable development goals [10].

This chapter provides a new perspective on engineering and how I believe that through shifting some of the value hierarchies that exist within the culture of mechanical engineering, we will be better equipped to address the challenges of the future. I have had conversations on these topics with people from across career levels, including graduate students, early-career engineers, to late-career engineers. There is something different to learn from each of these conversations, and I always appreciate the wealth of knowledge and experience that people bring to understanding and uncovering engineering's value systems. Regardless of where you are in your educational journey or in your career, I hope you enjoy reading this chapter. And if you ever feel uncomfortable with my words, I hope that together we can explore where this discomfort is coming from and learn how to challenge and expand the identity that has been deeply engrained within us.

My Past Journey: Where I Came From

My engineering education and career have meandered through many different disciplines and a variety of experiences. I completed my undergraduate degree in Manufacturing Engineering with a Biomedical Specialization; I worked in industry as a Lean Project Manager for 2 years; I went back and did my MSc in Civil Engineering; I worked as a staff member in academia for 3 years as an Engineering Accreditation Specialist; and currently I am pursuing my PhD in Mechanical Engineering. This wandering journey also represents a slow increase in my self-awareness of the engineering identity I was developing.

During my undergraduate degree, I was fairly accepting of the education I was receiving, and I did a lot of "leaning in" to the culture. I knew that I was a woman in engineering and that this was an accomplishment in itself as people often reminded me of this fact, such as family members saying: "Wow, you must be a smarty pants if you're in engineering!" However, I didn't notice the microaggressions and engineering identity that was being embedded within me throughout my undergraduate degree. Thinking back to all my lab groups and design teams, I was rarely with other women. Both in my first year and my final year design team, I was with three male students, and at the time, I didn't notice the subtleties of always being the one responsible for taking meeting minutes, writing up the report documents, and organizing group meetings. I was assigned these tasks instead of the technical design tasks likely due to my gender and an assumption of my skillsets. Even when I was assigned technical tasks, this was typically on top of my more administrative tasks, often with me reassuring the others that it wasn't too much work, "Oh no worries, I'm happy to help!" This lack of awareness is common; research has shown that female students often do not "*question the profession's central narrative about itself*" [11]. I certainly did not question the gendered narrative and work distribution that was perpetuated within my design teams.

The values that I was taught during my undergraduate coursework were core to what I believed engineering to be—these being the masculine, elite, and individualistic narratives that had been promoted through my classes [12, 13]. One of the strongest values that I held core to my belief system when I graduated engineering was that the profession aspired to be *neutral* and *objective* [11]. This presumption of a neutral and objective engineering culture aligned with my apolitical approach to engineering, believing that the profession existed within a meritocracy where all groups have equal access to success and that the profession was free of politics. The emphasis on problem-solving and the application of the design process, especially, gave me the misguided impression that engineering was based on academic and technical ability in a way that was objective and value-neutral.

Immediately after graduating, I went into industry working at a small manufacturing plant as the Lean Project Manager. In this role, I was responsible for working with the employees who manufactured our products to reduce waste and increase efficiencies in their manufacturing processes. This experience was the beginning of my identity crisis and realization of my discomfort with the elitism that the culture of engineering carries. The manufacturing employees, who were all MUCH more knowledgeable than me on the manufacturing processes, assemblies, and products, often looked to me as their superior. I distinctly remember discussing ordering a new welding arm with the foreman. We were discussing different options, and then he looked at me and said, "You're the boss, whatever you want we will make it work." I felt this frustration that he didn't want to collaborate with me, and he felt that somehow 4 years of engineering education made me the better person to choose a welding arm that he would be using every day, at a company he had worked at for over 20 years. Not to any fault of his, he had also been indoctrinated to believe that I was the elite and superior person, much more knowledgeable and powerful than he would ever be, simply because I had a degree attached to my name. Although I had often doubted my long-term career as an engineer, this was the first time I can remember having a sadness that as an engineer. I would be often doing my work very isolated and it would be assumed that I would need to take an individualistic, hierarchical approach to work.

Although this example highlights my growing identity disconnect with engineering's hidden norms and values, it is not to say that much of my work as lean project manager wasn't with a collaborative approach. In this job, I often was able to truly see the value of working and designing together. For example, I ran daily stand-up meetings with the employees to gather their feedback and ideas for improvements. These meetings were modeled after the Lean Six Sigma approach, with integrated practices from Agile methodologies (specifically, the "daily scrum") [14]. I held the short 10–15-min meetings at the beginning of the day on the "gemba" (work floor), where we all remained standing, and each employee gave a quick update. Typically, the guiding questions asked employees to look back (how did I do yesterday?) and look forward (what will I do today? any issues or concerns?). During these meetings, employees were encouraged to share and discuss their ideas for improvements. Ideas ranged from very small changes such as moving the location of a supply bin so that it was in closer proximity to the work area, to very large changes such as suggesting entire changes to the manufacturing process. I felt that these collaborative discussions, although brief, were always highly valuable to the employees to learn and brainstorm together, and to feel that they were active members of the decision-making team.

At this same organization, I also coordinated a half-day project to work with a group of five employees and redesign the stockroom, where we discussed current processes, outlined challenges, and brainstormed solutions for improving the flow of the stockroom and part ordering. In planning for this brainstorming session, I worked very hard to ensure the process would be bottom-up, and the employees would be empowered to take leadership on the redesign. They made the decisions, and then they helped to implement them and iterate on them. But even in these two situations, there was always a protective bubble that seemed to follow me around, making it difficult for me to truly connect with the manufacturing employees, all of whom were men. They stepped out of my way with a gentle bow of the body, any jokes or laughter stopped when I entered the room, and they always apologized for the one wall that was decorated with magazine cutouts of swimsuit women. Even when working collaboratively with them, I never truly fit in—being a young, female, engineer made me someone different, someone they viewed as both more intellectually elite and powerful, while also being more delicate and sensitive.

After 2 years at this company, I decided that it wasn't for me. I didn't quite know why; I didn't quite have a plan on what I would be doing next, but I just knew I couldn't stay there. When I quit, I felt like a failure and I felt like I hadn't accomplished anything in the job personally or professionally. In retrospect, it was a job I really liked and excelled at, and I wonder if I had had a female mentor if I would have stayed longer. Being able to talk about my experiences with someone else might have helped me to label my feelings more easily. At the time, I didn't understand that my disconnect with the job wasn't about my own ineptitude, but rather it was that my feminine, collaborative, and empathetic identity and leadership style conflicted with the implicit value systems of engineering which emphasize a more individualistic and elitist approach to the workplace.

My Internal Journey: How I Came to This Work

Quitting my engineering job was the beginning of my transition into finding myself, my identity, and my value as an engineer. While I was unemployed, I met up with a professor who was a friend and she had an opportunity I couldn't resist. Although I never saw myself going back to school for a graduate degree, she was specifically interested in finding someone who could research our engineering faculty's leadership program to better understand the impact of the program on students who had been in their career for a few years, which eventually became the focus of my MSc thesis [15]. This field of research, known as "engineering education," was something I had never heard of, but it was a perfect blend of my two strengths: I was able to use my engineering background and the ways of thinking I had been taught through my undergraduate degree, and apply these to investigate and understand engineering education, engineering leadership, and pedagogy.

As I progressed through my master's degree, I truly felt seen and felt I was gaining a better appreciation for my engineering identity. Within engineering identity research, scholars have used the concept of *figured worlds* to conceptualize engineering identity into three areas of development:

- ***Acting as an engineer:*** This is the *making of worlds* which describes that through our actions we may shape the world we live in.
- ***Others recognizing them as being an engineer:*** This is the *positionality* of identity which acknowledges the hierarchy, social positioning, the importance of power in social interactions.
- ***Believing themselves to be an engineer:*** This is the *space of authoring* which understands how one integrates into the space/culture [4, 16].

For myself, during my MSc, I started to develop within each of these areas. By studying engineering leadership, I truly felt that I was able to *act* and use my engineering knowledge to shape and improve engineering education. As I spoke to more people across a wide variety of roles, including members from industry, professors, and students, I always received positive commentary on the importance of the work I was doing and I truly felt that I was *recognized* as an engineer. And although previously I had struggled to believe that I was a "real" engineer, I felt I had found my place in engineering and *believed* that I could do this work as an engineer. This is not to imply it was not without struggles—in fact, along with peers from across Canada, we conducted research and published a paper on the experience of engineering education graduate students. We found that students in this area often experience an identity crisis due to the epistemological differences between engineering research and educational research. Additionally, we often face critique and judgment when studying these topics from some engineering faculty members who often do not see the work and methods used as "real" engineering research (e.g., qualitative approaches are misunderstood and not valued as an accurate data analysis method by engineers who are only familiar with quantitative approaches) [17].

Fast forward a few years, and after my Master's degree I was now making the decision to continue my graduate studies and begin my PhD with a focus in engineering education (through the department of Mechanical and Manufacturing Engineering). As engineering education is not an official program at my university and there are only a few professors who study in the area, my supervisor did not have a specific focus or topic of study to give me. Rather, unlike my MSc, my area of research was left open-ended for me to choose a topic, as long as I pursued a project which investigated engineering education in some way. In my personal life, my identity was also transforming, where I had recently come out as queer and I was developing my passion for social justice work and grassroots activism. For this reason, I knew that I somehow wanted to bring a social justice or feminist perspective to engineering education in my PhD work.

Feminist thought was first brought into fields such as science and engineering by early scholars in the late 1980s including Haraway [18] and Harding [19]. These critical feminist theories have been powerful in bringing forward the invisible culture of STEM disciplines and many scholars in engineering education continue to build on this early work. Feminist pedagogies can observe the hidden ideologies in our programs and challenge the dominant assumptions of science and engineering

[20]. They make visible *"engineering's gendered boundaries,"* and consider whose identities are excluded from the universal narratives of engineering [21]. However, engineering and other technical fields are often very resistant to feminist discourses because of the presumption that engineering operates as a neutral and meritocratic field of study [11].

Meritocracies assume not to bias any group and assume that all groups have equal access to success. Engineering culture promotes the idea that a meritocracy is possible and we *believe* that engineering operates within a meritocracy. In other words, we see engineering as a value-neutral field where measurement of student success is based on purely objective assessment, and there is a positivistic approach to knowledge where there is one objective truth that we are seeking. However, although societies may aspire to be meritocracies, a purely meritocratic society is not possible, and research has shown over and over again that genetics, social class, upbringing, and just plain luck often play the most significant role in one's success in life [22]. The consequence, though, is that *this belief of meritocracy renders social justice topics irrelevant in engineering*, because any social injustices are unconnected when we believe the most talented and hard-working are rewarded [23].

As I further explored the literature on feminist thought in engineering, I was fascinated by the amount of work that had been done, with Donna Riley publishing a feminist approach to thermodynamics almost 20 years ago [24]. I was inspired and knew there was something I could do within this area of study. However, I also knew it would not be without challenges. Many of the publications talked about the difficulties of getting this kind of work recognized in engineering due to the confrontation with deeply rooted beliefs in engineering culture. For example, Beddoes [25] published an entire thesis which documented and reflected on her efforts to publish three gender-theory informed articles into engineering education journals. While I was still searching for a specific topic that I could research for my PhD, my supervisor introduced me to a professor from our Women's Studies department who teaches a course on *ecofeminism*. During our meeting it hit me like a ton of bricks—engineers have (mostly) adopted the idea of sustainability, and the "eco" lens of ecofeminism would be the perfect gateway for me to be able to talk about feminism in engineering without losing most of my audience.

Simply put, ecofeminism recognizes the connections between the oppression of women and the oppression of nature. More specifically, at its core, "*ecofeminism is the belief that coinciding ecological and feminine repressions are often more than coincidental*" [26]. Ecofeminist perspectives highlight the fact that there is higher value in our society given to masculinity, humanity, and culture as compared to femininity, biology, and nature. One of the biggest risks is that this value hierarchy operates as if it is inevitable and logical, when in reality it is an "invented superiority" [9, p. 64]. Within engineering these value hierarchies are especially prominent. For example, we emphasize the importance of innovation and technological progress. By "progress" we specifically mean that we value human and cultural progress, often with limited consideration for the impact of our progress on nature and the biological ecosystems around us. This value hierarchy is often taken even further, where we value technological solutions so strongly that we solve problems we have created with further technologyengineering education, and ensure that our future engineers understand the importance of nurturing human's ecological connectedness.

Table 1 summarizes these hierarchized dualisms from ecofeminism and which are most prevalent within engineering culture. These represent maps that have been created through a process of cultural socialization within engineering culture, where we both internalize these maps and project ourselves onto them [9]. The higher value elements on the left side of the table are seen as correlated, and we notice when something is out of alignment. For example, this is why male students are believed to be more rational, and better at technical work, whereas female students are seen within to be more emotional, and to be better at social-oriented work. Additionally, our society sees these dualisms as two extremes, rather than as a spectrum across a scale.

For decades, engineering education scholars have been aware of the technical–social dualism which is prevalent within engineering culture [28]. For example, engineering students and professors will often publicly belittle other faculties, particularly anything related to the liberal arts as they see this as "weak" and strongly in opposition to their technical engineering identity [13]. Both symbolically and in practice, this techno-social dualism is considered mutually exclusive, because "at the core of engineers' identities and engineering practice lies a sense of the technical which specifically excludes the social" [28]. This mutual exclusion not only reinforces a very masculine approach to engineering, but it also promotes a culture of power, elitism, and often arrogance [29]. The gendered techno-social dualism also reinforces the gender binary of two genders that are directly in opposition with each other and hierarchical, with man being valued higher than woman and there being no spectrum between these two discrete genders [30].

Regardless of the dualisms present, it is important to note that engineering practice is fundamentally heterogeneous and needs to integrate across these dualisms. To succeed, engineers must effectively integrate both the technical and the social to meet the needs of society. Often, this heterogeneity of engineering is used to encourage more women to come to engineering, where advertising campaigns emphasize the social elements of the technical work. However, "*the assumption that women will both be more attracted to, and have more to offer to engineering if it is defined in non-technical terms leaves intact the equation of technology and masculinity*"

Table 1 Hierarchized dualisms

Higher value	Lower value
Culture	Nature
Man	Woman
Rational	Emotional
Active	Passive
Technical	Social

Adapted from [9]

[28]. This approach of attracting more women into engineering does not address the deep-rooted educational practices and continues to promote the mutually exclusive techno-social dualism. We need to be aware of how our discourses about engineering reinforce these dualisms and more actively work to deconstruct them.

As I learned more about ecofeminism and I relished the wealth of expertise from feminist engineering scholars, my theoretical framework was coming into view for my PhD thesis. I was now feeling more confident speaking to people about the work that I would be doing. Through these conversations with others, I truly began to understand how my work would be pushing the boundaries and making people feel uncomfortable as I attempted to break down their beliefs about what is core to engineering identity.

My External Journey: What Others Think/Feel About This Work

Generally, I have been very lucky to have mostly positive and inquisitive experiences in response to the work that I am doing. People are often very curious and want to learn more, and typically they have some personal experience that they can relate to my work. Through my workshops and presentations, I have found that people appreciate being able to label their feelings. After a recent conference presentation, one of the attendees connected with me on *LinkedIn*, and talked about her experience of often feeling marginalized at STEM conferences and that my "presentation helped [her] to find the words to articulate what was happening."

Although I have not been met with direct criticism, there is certainly sometimes a feeling of discomfort which comes out in the questions being asked. After a workshop I facilitated for a national organization, one of the participants reached out to me and asked to connect so we could further discuss the topic. In our chat, he admitted to me directly that he believed engineering did exist within a meritocracy and that this was not a bad thing—that we should be rewarding effort and that it is true that not everyone can be an engineer and there are certain capabilities which are required in order to succeed in engineering. We had a productive conversation, and dove into understanding why a meritocracy is not possible, which is why believing in one is dangerous and leads to inequities in our classrooms. It was an insightful discussion for both of us. I am still learning myself how to manage these conversations, and how to articulate my arguments. Luckily this work drums up lots of curiosity in others, so I will have plenty of practice before my final thesis defense!

The biggest wins I have noticed in conversations with others is when I am able to observe small shifts in thought, where someone builds on where they are currently at and moves *slightly* toward a more inclusive lens of engineering or justice work within engineering. For example, I was once in conversation with a professor about feminism and he was arguing that feminism gives off the wrong tone because we should consider more than just the women. During this conversation, I introduced the idea of intersectionality, which can be defined as "*the interaction between gender, race, and other categories of difference in individual lives, social practices, institutional arrangements, and cultural ideologies and the outcomes of these interactions in terms of power*" [31]. He had never heard this term before, and was so excited to have the words to describe his feelings—which in essence was that much of the diversity work in engineering tends to prioritize white, western, heterosexual, cisnormative, middle-class women [25], without considering other marginalized populations.

As I continue through my PhD it will be important for me to regularly reflect and consider how I can provoke these dialogues that help nudge people to slightly expand their perspectives on what it means to be a mechanical engineer. Keeping in mind that as I continue my journey, each conversation is also a learning opportunity for me, to better understand the identity tensions that exist and to expand my own perspectives.

My Future Journey: Where Do I Go from Here

I believe that through facilitating workshops, engaging in conversations, and following-up on connections across my network, I have been able to make a difference. Although one-on-one conversations are not the most efficient way to make change, this grassroots approach provides me with the opportunity to learn and grow myself. I have been continuously learning from others, and I have been building my own knowledge to be able to better articulate myself. Slowly with time, I am gaining a clearer picture of how to make long-term, sustainable recommendations to change engineering culture. Change is often slow and difficult (especially in academia), but over time I hope I am able to continue making a larger impact, through bigger conversations, publications, and, eventually, systemic, sustainable change. This chapter is just another step in that direction. And systemically I am also making progress—our university is proposing a new program *Sustainable Systems Engineering* and they invited me to sit on the program design committee so that the program could have foundational elements of ecofeminism integrated into the program design [32].

As I continue this work, it's also important for me to consider activism burnout which comes with educational justice work [33]. Activism burnout has been shown to decline the emotional and physical health of individuals, and to cause a loss of hope in some activists. Although my supervisors have been extremely supportive of my PhD work, it is important for me to intentionally seek out mentors who can support me in my activism and feminist approach to engineering education. For every difficult conversation I have, I hope to be able to have triple the number of inspiring conversations to keep my energy and hope for a better future in this (sometimes) overwhelmingly depressing capitalist, white supremacist, and hetero-cis-normative society that we live in.

If this chapter has piqued your interest, you might also be asking yourself, what can you do? I don't expect that every engineering student, faculty member, and practicing professional become a feminist and social justice advocate in their engineering careers. However, I think we can all start somewhere. We can be more observant of the culture around us and label what we observe. We can see the dualisms that exist in our engineering culture, and break down these hierarchies to bring light to the spectrum of perspectives. We can be honest about why someone might feel a disconnect with engineering, and help them understand that it's not them—it's the system and culture that's making them feel like they are not part of the elite brand of engineering. We can resist changing ourselves to fit into this culture, and bring our own unique perspectives to the table. We can connect with and support each other on our journeys, so that we all feel a little less alone in this crazy world.

References

1. ASME. "The Future of ASME Standards." The American Society of Mechanical Engineers. https://www.asme.org/codes-standards/about-standards/the-future-of-asme-standards (accessed May 31, 2021).
2. T. K. Beam, O. Pierrakos, J. Constantz, A. Johri, and R. Anderson. "Preliminary findings on freshmen engineering students' professional identity: Implications for recruitment and retention," presented at the American Soc. For Eng. Educ. Annu. Conf. & Expo., 2009.
3. R. Stevens, K. O'Connor, L. Garrison, A. Jocuns, and D. M. Amos, "Becoming an engineer: Toward a three dimensional view of engineering learning," *J. of Eng. Educ.*, vol. 97, no. 3, pp. 355–368, 2008.
4. F. Dehing, W. Jochems, and L. Baartman, "Development of an engineering identity in the engineering curriculum in Dutch higher education: an exploratory study from the teaching staff perspective," *European J. of Eng. Educ.*, vol. 38, no. 1, pp. 1–10, 2013.
5. Villanueva, T. Carothers, M. Di Stefano, and M. T. H. Khan, "'There is never a break': The hidden curriculum of professionalization for engineering faculty," *Educ. Sci.*, vol. 8, no. 4, 2018.
6. E. Margolis, *The hidden curriculum in higher education.* New York, NY, USA: Routledge, 2001.
7. Warren, "Ecofeminism: women, culture, nature," Indiana University Press, 1997.
8. N. Kellam, K. Gerow, G. Wilson, J. Walther, and J. Cruz. "Exploring emotional trajectories of engineering students: A narrative research approach," *Int. J. of Eng. Educ.*, vol. 34, no. 6, pp. 1726–1740, 2018.
9. R. A. Martusewicz, J. Edmundson, and J. Lupinacci. *Ecojustice education: toward diverse, democratic, and sustainable communities.* New York, NY, USA: Routledge, 2014.
10. UN. "The 17 Goals," United Nations Department of Economic Social Affairs. https://sdgs.un.org/goals (accessed May 31, 2021).
11. S. Seron, E. Silbey, E. Cech, and B. Rubineau. "'I am not a feminist, but..': Hegemony of a Meritocratic Ideology and the limits of critique among women in engineering," *Work and Occupations*, vol. 45, no. 2, pp. 131–167, 2018.
12. L. Pawley, "Universalized narratives: Patterns in how faculty members define 'engineering'," *J. of Eng. Educ.*, vol. 98, no. 4, pp. 309–319, 2009.
13. J. Christman, "Where are all the women engineers ? an insider's view of socialization and power in engineering education," Ph.D. Dissertation, Dept. of Learning and Instruction, The State Univ. of New York at Buffalo, 2017. [Online]. Available: https://eric.ed.gov/?id=ED580566
14. L. Boiser, "6 Tips on Planning Effective Stand-Up Meetings," Kanban Zone. https://kanbanzone.com/2021/tips-on-planning-effective-stand-up-meetings/ (accessed May 31, 2021).
15. R. Paul, "Towards an understanding of the influence of student leadership development on early-career engineers," M.Sc. Thesis, Dept. of Civil Eng., Univ. of Calgary, Calgary, AB, 2018. [Online]. Available: https://prism.ucalgary.ca/handle/1880/106534

16. L. Tomasko, J. S. Ridgway, R. J. Waller, and S. V. Olesik, "Association of summer bridge program outcomes with STEM retention of targeted demographic groups," *J. of College Sci. Teaching*, vol. 45, no. 4, pp. 90, 2016.
17. J. S. Cicek, R. Paul, P. K. Sheridan, and L. Kuley, "Researchers explore their roles as participant-researchers in characterizing the lived experiences of graduate students in engineering education research in Canada: A collaborative autoethnography," *Canadian J. of Sci., Math. and Technol. Educ.*, vol. 20, no. 1, pp. 98–115, 2020.
18. D. Haraway, "Situated knowledges: The science question in feminism and the privilege of partial perspective," *Feminist Studies*, vol. 14, no. , pp. 575, 1998.
19. S. G. Harding, *The science question in feminism.* Ithaca, NY, USA: Cornell University Press, 1986.
20. M. Sharma, "Applying feminist theory to medical education," *The Lancet*, vol. 393, pp. 570–578, 2019.
21. D. Riley, A. L. Pawley, J. Tucker, and G. D. Catalano, "Feminisms in engineering education: Transformative possibilities," *The National Women's Studies Association Journal*, vol. 21, no. 2, pp. 21–40, 2009.
22. C. Mark. "A belief in meritocracy is not only false: it's bad for you." Princeton, NJ, USA: Princeton University Press. https://press.princeton.edu/ideas/a-belief-in-meritocracy-is-not-only-false-its-bad-for-you (accessed May 31, 2021).
23. E. A. Cech, "The (mis)framing of social justice: Why ideologies of depoliticization and meritocracy hinder engineers' ability to think about social injustices," *Philosophy of Eng. and Technol.,* pp. 67–84, 2013.
24. D. Riley, "Pedagogies of liberation in an engineering thermodynamics class," presented at the American Soc. for Eng. Educ. Annu. Conf. & Expo., 2003.
25. K. Beddoes, "Practices of brokering: Between STS and feminist engineering education research," Ph.D. Dissertation, Sci. and Technol. Studies, Virginia Polytechnic Institute and State Univ., Blacksburg, VA, 2011. [Online]. Available: https://vtechworks.lib.vt.edu/handle/10919/77992
26. J. Bile, "The rhetorics of critical ecofeminism: Conceptual connection and reasoned response," in *Ecofeminism and rhetoric: critical perspectives on sex, technology, and discourse*, D.A. Vakoch, New York, USA: Berghahn Books, 2011, ch. 1, pp. 1–37.
27. J. E. Romberger, "Ecofeminist ethics and digital technology: A case study of Microsoft Word," in *Ecofeminism and rhetoric: critical perspectives on sex, technology, and discourse*, D.A. Vakoch, New York, USA: Berghahn Books, 2011, ch. 5, pp. 117–143.
28. W. Faulkner, "Dualisms, hierarchies and gender in engineering," *Social Studies of Sci.*, vol. 30, no. 5, pp. 759–792, 2000.
29. K. L. Tonso, "Engineering identity," in *Cambridge Handbook of Engineering Education Research*, A. Johri and B. Olds, Cambridge, UK: Cambridge University Press, 2014, ch. 14, pp. 267–282.
30. Phipps, "Re-inscribing gender binaries: Deconstructing the dominant discourse around women's equality in science, engineering, and technology," *Sociological Review*, vol. 55, no. 4, pp. 768–787, 2007.
31. K. Davis. "Intersectionality as buzzword: A sociology of science perspective on what makes a feminist theory successful," *Feminist Theory*, vol. 9, no. 1, pp. 67–85, 2008.
32. R. Paul, G. Ayers, J. Bergerson, K. Black, T. Brucker, C. Dawoud, M. Eggermont, A. Knight, S. McCoy, E. Rangelova, and D. Wood, "New sustainability systems engineering program proposal – Mindset and development," presented at the Canadian Eng. Educ. Association Annu. Conf., PEI, Canada, June 2021.
33. P. C. Gorski and C. Chen, "Frayed all over: The causes and consequences of activist burnout among social justice education activists," *Educ. Studies,* vol. 51, no. 5, pp. 385–405, 2015.

Robyn Mae Paul is a Ph.D. candidate at the Schulich School of Engineering, University of Calgary in Mechanical Engineering. She is researching how principles of ecofeminism can help to deconstruct the hidden curriculum of engineering, or the unwritten and often unseen norms and values. Specifically, her work aims to model and better understand the hidden curriculum of engineering using agent-based modeling software. Robyn completed her Bachelor's in Manufacturing Engineering and her Master's in Civil Engineering, where her thesis looked at engineering leadership to explore the influence of a student leadership development program on early career engineers. Between her educational pursuits, Robyn has worked in industry as a Lean Project Manager in a manufacturing plant, and as an Accreditation Specialist for the engineering faculty at the Schulich School of Engineering. Robyn is passionate about queer advocacy and climate justice and volunteers much of her time with grassroots organizations to support causes that align with her values. During her spare time, Robyn enjoys hiking in the nearby Rocky Mountains. Most recently, at the time of this book publishing, she had a new addition to her family, and she has been spending most of her time with her daughter.

Section III
Research/Technical

From Watching Planes in the Sky to Making Turbines More Efficient

Karen A. Thole

For both propulsion and power generation, the efficiencies of gas turbine engines dictate how much fuel, predominantly fossil fuels, is used. Increasing the efficiencies of gas turbines lowers the amount of fuel which ultimately reduces our carbon footprint. How much? A 1% increase in turbine efficiencies corresponds to a reduction of carbon dioxide that is equivalent to removing 2 million cars from the road. Another illustration of the impact of increased turbine efficiencies is the recent technological breakthrough of the geared turbofan for aircraft. A plane like the Boeing 747 uses about 1 gallon of fuel every second, which corresponds to 3600 gallons per hour [1]. The highly efficient geared turbofan engines, manufactured by Pratt & Whitney, have been shown to save 100 gallons of fuel per hour, which is nearly a 3% reduction of burned fuel [2].

As illustrated in Fig. 1, the three main components of a gas turbine are the compressor, the combustor, and the turbine.

Conventional gas turbines use the thermodynamic Brayton cycle with air as the working fluid, which means that high combustor temperatures lead to high efficiencies. The heart of my work lies in the desire to keep component temperatures below their melting point as combustor exit temperatures increase. For that to occur, sophisticated strategies are needed to cool the turbine parts because the temperatures of the gases in the engine are already hundreds of degrees hotter than the melting temperatures of the turbine components. For single-aisle aircraft, gas turbines achieve approximately 50% thermal efficiencies during cruise. Two paths lead to higher thermal efficiencies:

1. Advancing materials and coatings to make sure turbine components can withstand increasingly higher temperatures.
2. Improving cooling of the turbine components.

K. A. Thole (✉)
Department of Mechanical Engineering, Penn State University, State College, PA, USA
e-mail: kat18@psu.edu

M. Bailey, L. Shackelford (eds.), *Women in Mechanical Engineering*, Women in Engineering and Science, https://doi.org/10.1007/978-3-030-91546-9_10

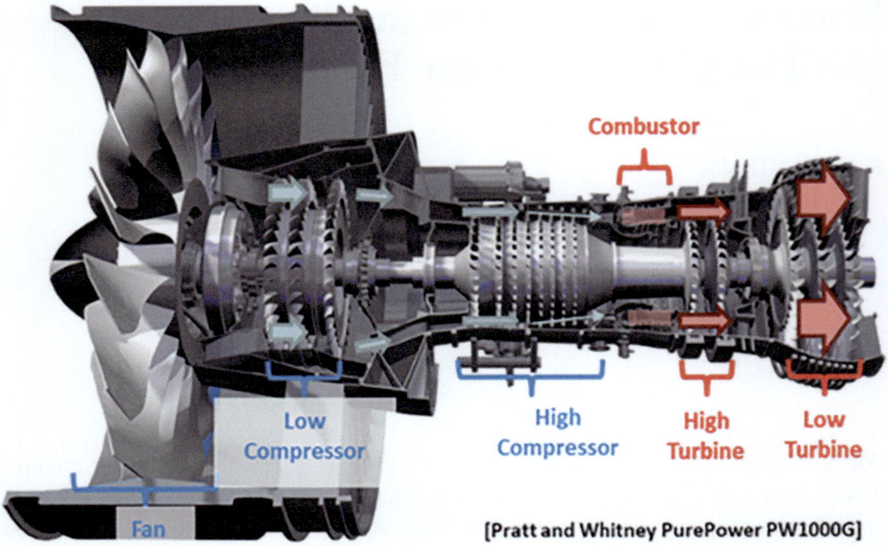

Fig. 1 Illustration of a gas turbine engine manufactured by Pratt & Whitney

It is advancing cooling technologies that I have studied since starting my doctoral work at the University of Texas at Austin.

As with so many academics, my professional path to where I am now was not straight. Seemingly small lessons that I learned along the way resulted in important turns on that path. However, my story and my first lesson began with my parents, my brother, and I watching airplanes in the backseat of our family's Ford LTD at Lambert Field in St. Louis.

My Education and my Father's Profound Influence

Lesson 1: Be Brave in Trusting your Instincts

Growing up on a farm in a southern Illinois town of fewer than a thousand people provided me with a close-knit family and strong work ethic. However, the experience did not provide me with a particularly wide vision about how I could contribute to the world. The choices seemed to be either a farmer or a teacher. Despite having only an eighth-grade education, my father was a progressive farmer. His openness to change likely resulted from his service in the army during the Korean War. Fascinated about how things worked, he drove our family on Sunday afternoons to an airport about 50 miles away to watch the planes take off and land. His fascination became my fascination.

I was the oldest child in a small family. Although farms are traditionally passed from father to son, my parents expected me to work in the fields, milk the cows, and sometimes help repair broken parts. By the time I was 10 years old, I was driving tractors and my summers were filled with hard work.

When it was time to go to college, my parents were supportive, but they had no idea how to advise me. It was my first year in college when I was walking to my apartment from chemistry that a classmate told me that I should become an engineer. After that first semester, I called McDonnell Douglas (now Boeing) in St. Louis, which was the closest large city to my farm, to ask whether they had an engineer working there. After some laughter, the operator put me through to one of their male engineers who answered my question about what an engineer's typical day was like. What I learned was that no day was "typical"—engineers were called upon to solve a host of different problems. It was then that I decided to pursue engineering. However, when I told my parents, they stated matter-of-factly that I was not smart enough to be an engineer. Despite their response, or perhaps because of it, I trusted my instincts and made my first brave move.

Engineering was challenging, but I earned my bachelor's in mechanical engineering at the University of Illinois at Urbana-Champaign. I continued on to my master's there and later pursued my doctorate at the University of Texas at Austin. In doing so, I knew one thing: My research had to be aimed at contributing to propulsion. In addition, my instincts were that I needed to be passionate about something if I really wanted to make a contribution.

My master's research at the University of Illinois was related to rockets, and I had the opportunity to work with a world-renowned expert in base pressure flows, Professor Helmet Korst, who was originally from Austria. Although I learned important fundamentals from him, what I learned most from him was how to be a better person. Professor Korst showed tremendous interest in me both as a future researcher and as a person. Sometimes this support came by way of a trip to Baskin Robbins in which he would build back my confidence. I had setbacks in my research (as everyone has), but I also had detractors (which most engineers do not). For instance, one male graduate student told me that I should not be there because I was taking the place of a male who would need to support his family. My instincts, and Professor Korst, told me to pursue this field despite having no women role models and despite the detractors.

After completing my master's work, I accepted a job at Lawrence Livermore National Laboratory. Two life-changing events occurred. Although the position there was very good, I had moved away from my research passion of propulsion. Second, and most important, I met my husband, Michael Alley, who gave me confidence to return to a university to get a PhD. Without his support at that time, and continued support since, I would not have had the confidence to pursue and continue an academic career. He has been my inspiration and my rock in many difficult times.

It was during my doctoral research under Professor David Bogard at the University of Texas that I learned to conduct thorough experiments and to explain my results starting at the most fundamental level. Moreover, it was at the University of Texas where I learned about the challenges related to turbine cooling from a fundamental perspective. The research I did was to evaluate how high levels of turbulence affected heat transfer, which was critical to understand for predicting how much heat transfer a turbine blade experienced. As an example, mis-predicting turbine blade temperatures by as little as 25 °F means that the blade lifespan would be reduced by a factor of two in a turbine engine. Considering the gas flow passing through the turbine is on the order of 3000 °F and the "coolant" flow temperatures passing through the blades is on the order of 1000 °F, a mis-prediction of only 25 °F is quite small. My research found that the turbulence levels on the order of 20% exiting the combustor increased convective heat transfer by as much as 20%, which led to new predictive correlations for turbine heat transfer.

At the University of Texas, I also researched film-cooling, which is one of the cooling technologies used in turbine blades and vanes, as illustrated in Fig. 2.

Film-cooling takes place as the coolant circulating inside the turbine airfoils is exhausted out of small cooling holes that are placed on the blade surface. The result of this exhaust is a film layer that further cools the external airfoil surface. Findings from my research identified relevant nondimensional parameters that scaled the effects of film-cooling. This work I had the opportunity to present at my first-ever international conference. Although Professor Bogard was unable to attend, he had the confidence to send me alone to present the work—a decision that I marvel at even today. From the lively discussions at that conference, I gathered how competitive the gas turbine research field was. In addition, I came home from the conference with an offer to do a post-doc at a reputable university in Germany. Despite having to persuade my husband to leave his teaching position at the University of Texas, my instincts told me this experience was a once in a lifetime opportunity.

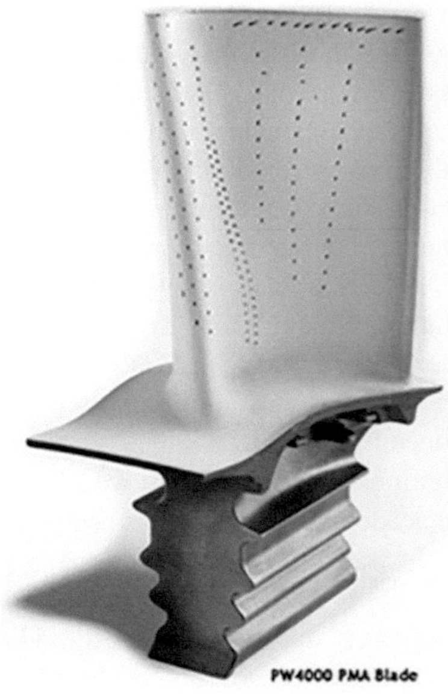

Fig. 2 Illustration of a turbine blade with film-cooling holes

Building a Research Laboratory as a New Professor

Lesson 2: Know When to Change Course Even When It Is Difficult

While I was printing out the last chapter of my dissertation at the University of Texas, Professor Bogard told me it was time to start thinking about a faculty position. Exhausted from doing the experiments, the writing, the proofreading, and everything it took to finish the degree, I let him know that an academic position would be the last job I would ever take. Besides, I was off to Germany, intent on doing more research, and expecting to return to a national lab when the post-doc was finished.

However, working at the Karlsruhe Institute of Technology made me realize the freedom that academia afforded to shape one's research directions, albeit with responsibilities. Most important, academia allowed one not only to work with vibrant, bright engineering students but also to shape those students' careers. Although I had a job waiting for me at a prestigious Department of Energy research lab after I completed the post-doc, I realized that a university position was where I needed to be.

My first academic position was at the University of Wisconsin-Madison. Both my husband and I had received good offers. As a young researcher, though, what I failed to realize was that this university lacked important scaffolding for my research area. Still, I had a good start-up package and was so proud to be a faculty member at such a prestigious university. Building my lab from the ground up was both exciting and daunting. The experience was like placing a bet—I placed all of my start-up funds on turbine heat transfer research. On the negative side, the field was competitive, and I was starting with nothing and had no industry partners, which was an essential ingredient in my field. Exhausting my small pot of precious travel money, I attended an important workshop in my field sponsored by a federal agency and involved many industry sponsors. Not only was this meeting tied to a call for research proposals, but it was also well attended by those from industry. At this meeting, one company representative told me that he was not interested in talking with me because academics did not want to work on real-world problems. A representative from another company told me that they were not interested in my ideas. Yet a third company simply ignored me. Later, after submitting my first proposal to the program and then learning that it was not funded, the program manager told me, "You did pretty good for a girl." Was this the time to change course? My instincts and my investments told me it was not.

When beginning my research program, my strongest competition naturally would be my doctoral advisor (Professor Bogard); however, he also became my strongest supporter. In those first days, we talked about how we might parse out the various regions of a turbine airfoil and how we would each stay in those swim lanes. He agreed on this arrangement, but only until I got tenure which was more than fair. After failing in my first six proposals, I thought that tenure was not in my cards.

Proposal writing is a skill, but in the end requires some luck. Most experienced academics know this point, but that does not soften the blows from rejection. Like most academics, I have my stories in which the proposal reviews seemed unfair, inconsistent, or wrong.

After the failure of my sixth proposal, my former PhD advisor, Professor Bogard, came to the rescue by supporting my case to be included in a computational fluid dynamics (CFD) competition that was being conducted by one of the major turbine manufacturers, Pratt & Whitney. This project was my first one, funded at a total of $30 K. At the time, that amount seemed like a goldmine, especially since I was considering taking out a personal loan to support my graduate students. In addition, this project was definitely a change in course given I was an experimentalist. Being part of the project was the good news. The bad news was that my competition consisted of well-respected researchers who focused solely on CFD development. While each of them had about 30 years of experience and constructed their own codes, I had about 2 months of experience and was using a commercial CFD code. My strength was that I understood both how the problem needed to be modeled and what results to expect. My graduate students and I presented this work to the community as did the others who in the competition performed state-of-the art CFD modeling [3]. My success in this competition was the start of a long-lasting research relationship with Pratt & Whitney that is present even today.

In my new laboratory at UW-Madison, my first experimental project, which was also funded by Pratt & Whitney, investigated the secondary flow structures in the endwall region of a turbine vane. The endwall region for a turbine airfoil is complex because of the boundary layer separation that occurs as the flow approaches the leading edge of the airfoil. The airfoil, specifically the first vane, turns the flow such that it is directed into the downstream blade, which causes it to spin. The separation that occurs causes a horseshoe vortex, as illustrated in Fig. 3.

These vortices result both in component inefficiencies because of entropy generation and in increased heat transfer as the vortex scrapes the endwall region. While nature-inspired engineering is now quite common, it was not so well known in 1996. At that time, though, I realized that a turbine vane had a similar shape to that of shark's dorsal fin. It is well known that sharks are able to swim at high speeds with minimal effort (little drag). The geometry of a turbine vane also looks a bit like a conning tower on a submarine, but when I tried to discuss this similarity with the US Navy, they abruptly cut off the conversation—apparently, I was treading into what was classified information at the time. Suspecting that I was onto something, I noticed that a big difference between the dorsal fin of a shark and the vane of a gas turbine engine was that the shark's fin had a nice fillet between its body and the protruding fin. It was not time to change course but to continue researching this area. Eventually, what we developed in our laboratory became the basis of what is

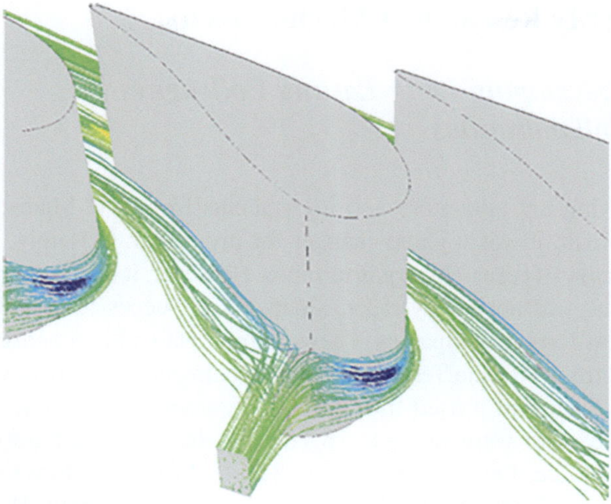

Fig. 3 Illustration of endwall secondary flow pattern (top) [8] compared with a predicted endwall flow pattern using a fillet (bottom) [4, 9]

being used in designing turbine airfoils today [5, 6]. Ironically, after the student doing the project completed his degree, he went to Pratt & Whitney and soon became my research project monitor.

After that first $30 K proposal to Pratt & Whitney was successful, I became better at writing proposals, but equally as important, I started to build my reputation in the field. My next ten proposals were funded, including an NSF CAREER award in 1995. In 1999, while still at UW-Madison, I realized what I had not seen when beginning as a faculty member there—namely, that this institution lacked important scaffolding for me to succeed as a researcher in my field. When I took the position, I had not been bothered by the university's (or, specifically, the department's) lack of commitment to my particular research area. Like many institutions, UW-Madison has a strong reputation, which they have built by supporting particular research strengths. Three years into my position, though, I was told that the college considered gas turbine heat transfer to be a dead field. Rather than return to my cubby hole and accept that my research program would always be small, I took the college's stance as a sign that I needed to change institutions, even though moving a big laboratory and graduate students was a difficult course change, particularly in mid-tenure. Although many advised me not to take this bold move, I knew that I needed to change course quickly if I wanted to continue impacting my chosen passion. My ultimate decision to move did come with strong support from my industry sponsors. The school I chose was Virginia Tech.

Expanding My Research at Virginia Tech

Lesson 3: Developing Long-Lasting Colleagues Through Collaborations

Although moving my laboratory and my students from UW-Madison to Virginia Tech was difficult, it was a game-changer for my career. Certainly, Virginia Tech had a long history of researching gas turbines. However, it also had something that, up to that point, was not on my radar: a network of successful alumni working as engineers at gas turbine companies. These alumni, who had immense pride in having graduated from Virginia Tech, wanted to remain connected with the institution. Many such alumni had passed through the classrooms and research labs of the school's faculty who were respected by the gas turbine community—professors such as John Moore, Felix Pierce, Joseph Schetz, Wing Ng, Tom Diller, William Davenport, Roger Simpson, and Walter O'Brien, the Department Head who hired me. Most important, many of those alumni made decisions about where research funding for gas turbines would go—and those alumni wanted that money to go, if possible, to Virginia Tech.

In starting over at a new university, I had the opportunity to address mistakes made when I had first set up my laboratory. Moreover, I was able to add to our lab's capabilities. Through the turbine studies that our group previously did, we had learned the importance of simulating an engine-representative velocity and temperature field upstream of the turbine vane. We also had identified the importance of the inlet conditions through CFD studies [7]. When we had presented these results to Pratt & Whitney, we had hoped that they would be willing to support our first study at Virginia Tech. They were willing, and that study was to build a non-combusting simulator upstream of our turbine vane test section in the wind tunnel that provided a relevant combustor-exit flowfield. The combustor simulator, which is illustrated in Fig. 4, was placed upstream of the corner test section that contained three large-scale turbine vanes [8].

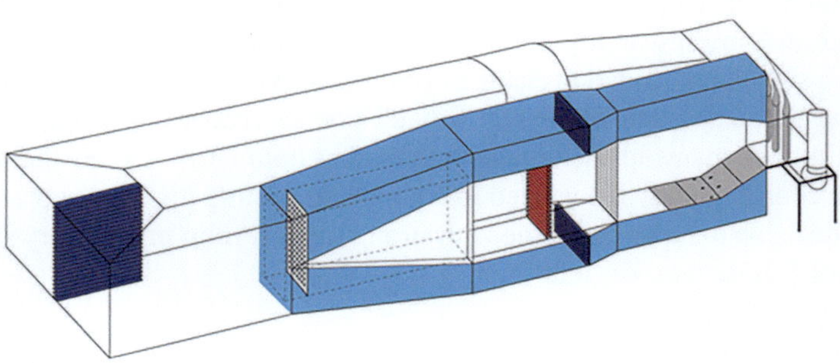

Fig. 4 Penn State's Experimental and Computational Convection Lab (ExCCL) low-speed wind tunnel with a combustor simulator upstream of the corner test section containing turbine vanes

The exiting flow and thermal fields in Fig. 5 showed that these turbine inlet conditions are dramatically different from those assumed with a uniform flowfield [9].

This combustor simulator, which was the first such facility in the open literature, became the cornerstone of many subsequent studies not only to determine relevant effects on the vane heat transfer and vane endwall flows, but also to evaluate combustor liner cooling designs as well. The facility drew international attention, particularly from the US Air Force, who decided to replicate our simulator in their Turbine Research Facility [10].

In the early years at Virginia Tech, continued research support came annually from the Department of Energy (DOE) – UTSR Program, the NSF GOALI Program, and Pratt & Whitney. Moreover, I was able to expand my industry collaborations to include Siemens and Mitsubishi Heavy Industries (now called Mitsubishi Power). Up to this point, my support and interactions with Pratt & Whitney had all been through their Military Engines Division in West Palm Beach, Florida. In 2005, United Technologies Corporation (now known as Raytheon Technologies Corporation) decided to significantly downsize their operations in Florida. Given

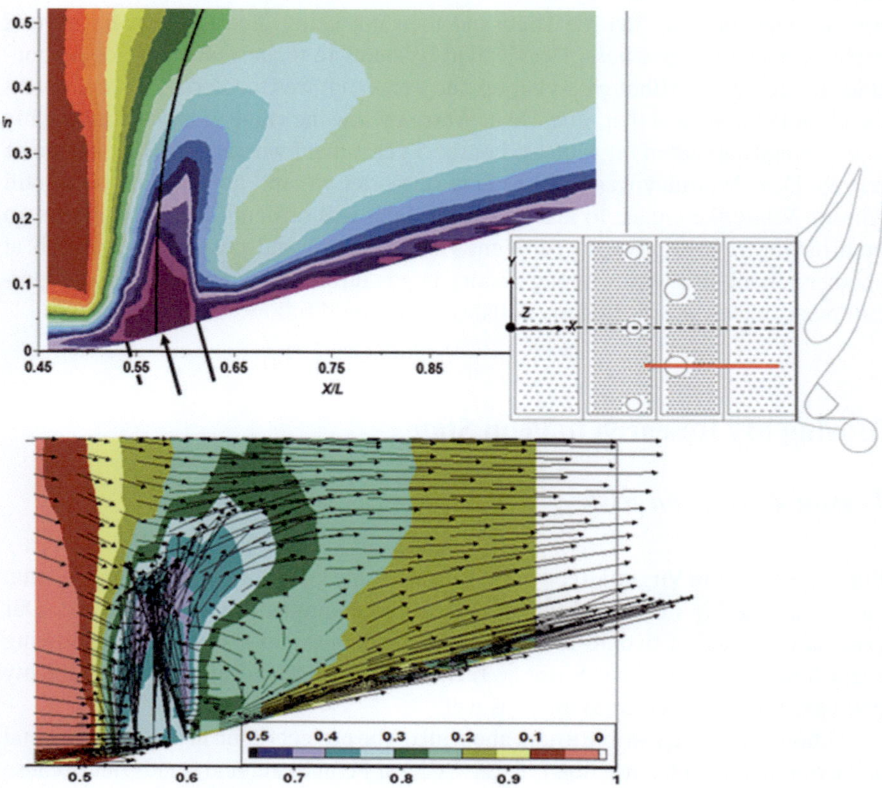

Fig. 5 Measured thermal fields (top) and flowfields (bottom) in the wind tunnel combustor simulator showing high turbulence levels

many of my collaborators were native to Florida, they elected to find other positions in Florida rather than move to East Hartford, Connecticut. While the decision to downsize in Florida was life-changing for the employees there, it also posed risks on my own research programs given that several of those employees were my primary collaborators. However, one of my first technical contacts at Pratt & Whitney, who elected not to move back to East Hartford, introduced me to numerous engineers in East Hartford. This experience taught me a great deal as a researcher in working with companies—the most important lesson being the importance of expanding my network in a company and making sure those relationships endured. In other words, it is important to work with many engineers rather than solely relying on just one, given that one person could change his or her career direction without warning.

While growing my research at Virginia Tech, I was also able to see the importance of academic administration. After being courted by the Department of Mechanical Engineering at Penn State by a faculty member who I met in the prior years and was part of my network, I applied for the position as Department Head and was selected. Much to the dismay of Pratt & Whitney, I accepted the position and consented to abandon my research program and become a full-time administrator. After announcing that decision and then not being able to sleep for several nights, I called my new boss, Dean David Wormley to request whether I could continue my research. Although giving up one's research was the norm for Department Heads at Penn State at that time, Dean Wormley saw the value of me continuing my research and supported my request. For the 5 years that I worked with him before he retired, Dean Wormley was an incredible supporter whom I am grateful to and still admire. While the choice to leave Virginia Tech was difficult, I could not pass on both having a prestigious Department Head position and continuing research at another school also with a strong history in gas turbine research. The most important piece, however, was that my collaborators would follow.

Moving My Research to Penn State

Lesson 4: Take on Risk

For the move from Virginia Tech to Penn State at the end of 2006, my research team packed up the big blue wind tunnel and all the equipment that went with it. Our caravan consisted of two 50-foot transport trucks plus and a smaller truck moving everything in the lab from Virginia Tech to Penn State. Fortunately, most of my graduate students decided to move as well.

Although it is not possible to talk about all of the projects done in our Experimental and Computational Convection Lab (ExCCL) at Penn State, one stands out because it provides an important lesson about research risks. Having access to relevant turbine geometries is extremely difficult because of their proprietary nature by the turbine companies. For decades, university researchers have used simple round film-cooling holes as a relative baseline for comparison because it was in the open literature. While round film-cooling holes do exist in turbine airfoils, so do many other complex shapes. It is well known that at high blowing ratios, the jet separates from the surface after it exits a simple round hole, whereas by adding a diffuser, such as the one in Fig. 6, high blowing ratios can be achieved that allow the jet to remain attached to the surface for better cooling.

However, there were no baseline cooling hole shapes that the research community could use for comparisons against more advanced designs. Despite my success in having proposals funded, no agency wanted to take on the project costs to support the development of a shaped film-cooling hole, but I thought it was important for our field. For that reason, I took on the financial risk of developing it through a student supported by a NASA Fellowship. More important than the financial risk was the risk of whether it would be accepted as a basis by our community. The shaped film-cooling hole we developed is shown in Fig. 6 and is referred to as the 777 cooling hole resulting from the three expansion angles each being 7 degrees. Now used by over 50 organizations (national labs, universities, and even some industries) as a basis for comparison, this cooling hole was an unqualified success given how common it appears in the literature [11]. The lesson here is that sometimes it is important to take on risk even after being told no by funding agencies.

At Penn State, research continued to grow at a steady pace until one day in 2011, when over a dinner in State College, me and one of my research staff were offered an opportunity by Pratt & Whitney to build a new "turbine lab." We quickly said yes. Later, though, we realized that we had no idea at the time what we had taken on. First, the project was enormous—only a handful of such facilities existed is the world. Second, because the project had much risk, we needed university support to gain some viable space, meaning someplace that would allow us to create noise and had the estimated 2 MW of power needed. Again, stepping up to help was Dean David Wormley, along with our Vice President for Research and our Provost. These discussions were made easier when a third research partner emerged: the Department of Energy – National Energy Technology Lab. The three-way partnership among Penn State, Department of Energy, and Pratt & Whitney was critical at the beginning of this project and continues to be important. These partners emerged

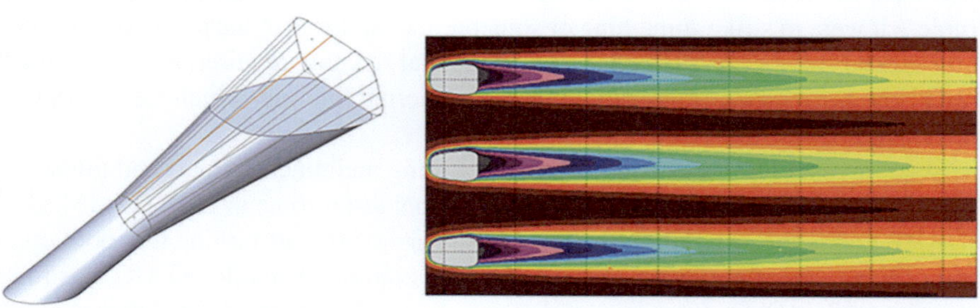

Fig. 6 A public 777 film-cooling hole geometry used by over 50 organizations and resulting cooling levels [6]

because of the long-lasting research partnerships we have had over the previous decades.

After understanding the size of what we agreed to build, I relinquished ExCCL to one of my former Virginia Tech doctoral students, Dr. Steve Lynch. Dividing the laboratories in this way was the logical choice since it would have been too risky to have Dr. Lynch (who needed to secure tenure) involved in building the new lab. Although leaving ExCCL was harder than I had anticipated and involved high risk since I could no longer fall back on doing research in a known lab, namely ExCCL, it proved to be the right decision. We were able to build the new turbine lab, and ExCCL continues to be successful under Dr. Lynch's leadership.

In developing the new turbine lab, we built a brand through a competitive naming process. The name chosen was the Steady Thermal Aero Research Turbine (START) Lab. START called for a one-stage test turbine that included a vane portion that remained stationary and a blade ring that spun at 10,000 rpm. Our vision was to develop a turbine lab in which we obtained high-quality data using actual engine hardware. Because we are at a university, we made it a priority that students gain firsthand experience in running the turbine through closely supervised efforts of permanent research staff. No other turbine lab operated in this manner and we were told by many experts that it was too risky to involve students. However, the year prior to starting our lab construction, the entire US had graduated only one doctoral student with experience in the operation of a turbine. My instincts told me that a need existed for more. Another priority for START was that the turbine had to be able to operate in a continuous manner, even though that made the facility more difficult and costlier. In other words, the facility would not be operated in a blow-down manner (lasting only a fraction of a second) as occurred at the majority of other research turbine labs in the US.

As it turned out, the seemingly straightforward project that we agreed to over dinner was actually much more complex than we could have imagined. Much of this complexity is represented in the schematic shown in Fig. 7.

Aside from the components and instruments, we needed a highly talented permanent research staff to maintain the systems, build testing and sensors, manage the various projects, and, most of all, operate in a safe manner. To give you a sense of the safety challenges, if a turbine blade, which is about half the size of a clenched fist, were to break off while spinning at 10,000 rpm, it could fly more than a mile (unless it were to strike something or someone). In addition to the safety risk, taking on four staff to operate such a lab was a financial risk that required me to also trust my instincts on supporting them all through external funding, which we have since demonstrated.

After 4 years of construction and initial testing and much support and guidance from Pratt & Whitney, we achieved an operational test turbine in 2015 [12, 13]. Our START Lab is the only lab in the US with a state-of-the-art turbine that both contains actual engine hardware and operates in a continuous mode. START has four primary goals: (1) advance novel turbine cooling technologies, (2) validate sensors in rotating environments, (3) develop additive manufacturing for turbine cooling, and (4) use the benefits of additive manufacturing for sensor integration.

Fig. 7 Schematic (top) and picture (bottom) of the START test turbine research facility at Penn State

One of our current projects is to develop the National Experimental Turbine (NExT). Support for this exciting endeavor has come from Department of Energy and four gas turbine companies: Honeywell, Pratt & Whitney, Siemens, and Solar Turbines. NExT is to serve as a testbed for US companies to advance turbines. Currently, this turbine is in the design phases, but we expect to have components in late 2021 to start testing. Other than building START, this project is, no doubt, my riskiest to date. This project requires us to design a turbine, which is already something that no other university has taken on. In addition, that turbine has to meet the needs of four turbine manufacturers—and everything within budget and in the time allotted.

Lesson 5: Trusting My Instincts That Gas Turbine Research Is Alive and Well

After working for four decades to address a host of challenges in gas turbines, I can confirm my instincts that research in gas turbines is *not* dead [14, 15]. While renewable energies are coming on strong, backup and supplemental power production will still be needed, and currently the most reliable solution for that is gas turbines. While we strive to reduce the carbon footprint, there will even be more opportunities that arise including hybrid electric propulsion where the heart of such a system will be a gas turbine. Another such opportunity is hydrogen fueled turbines that will most likely take on a very different design to maximize efficiencies for power generation applications.

Development of new gas turbine concepts is slow relative to fields such as microprocessors. However, the high degree of accuracy in gas turbines arises because of safety for both power production (imagine a power plant going out during a cold front) and for propulsion (if the engines failed, our planes would fall). The ability to cast single crystal turbine blades for development engines adds to the difficulty in which advances can be made. This level of difficulty is a principal reason that research in gas turbines is not dead. Advanced manufacturing methods and new high temperature materials such as ceramic matrix composites will lead to more efficient engines. In addition, new tools for gas turbine research are needed, such as sensors integrated into the turbine blades and machine learning for controls and maintenance.

I continue to be excited to do research in my chosen field and am grateful to the many students and long-lasting supporters I have had along the way. Without my supporters, including my father who told me that engineering was difficult, my research successes in gas turbines would not have been possible. The lessons that I learned along the way were certainly not apparent in real time, but in looking back each were so important to where I am today. I hope the audience for this article can use these lessons and achieve even higher successes leading to a fulfilling academic career in mechanical engineering.

References

1. https://science.howstuffworks.com/transport/flight/modern/question192.htm#:~:text=A%20plane%20like%20a%20Boeing,(12%20liters%20per%20kilometer) (January 1, 2021).
2. https://www.forbes.com/sites/lorenthompson/2020/12/02/pratt%2D%2Dwhitneys-geared-turbofan-is-poised-to-emerge-from-the-pandemic-in-great-shape/?sh=56ca1bb41655 (October 15, 2020).
3. C. Martin and K.A. Thole, "A CFD benchmark study: Leading edge film cooling with compound angle injection," 97-GT-297, 1997.
4. L. S. Langston, M. L. Nice, and R. M. Hooper, “Three-dimensional flow within a turbine cascade passage,” *Journal of Engineering for Power*, vol. 99, pp. 21-28, 1977.
5. G. A. Zess and K. A. Thole, “Computational design and experimental evaluation of using a leading edge fillet on a gas turbine vane,” *Journal of Turbomachinery*, vol. 124, no. 2, pp. 167-175, 2002.

6. Kvasnak, W., Zess, G. A., Thole, K. A., Soechting, O., "Apparatus and Method for Inhibiting Radial Transfer of Core Gas Flow within a Core Gas Flow Path of a Gas Turbine Engine, U.S. Patent 6 419 446, Dec. 21, 1999.
7. K. Hermanson and K. A. Thole, "Effect on non-uniform inlet conditions on endwall secondary flows," *Journal of Turbomachinery*, vol. 124, pp. 623-631, 2002.
8. M. D. Barringer, O. Richard, J. P. Walter, S. Stitzel, and K. A. Thole, K. A., "flow field simulations of a gas turbine combustor," *Journal of Turbomachinery*, vol. 124, pp. 508-516, 2002.
9. S. Vakil and K. A. Thole, "Flow and thermal field measurements in a combustor simulator relevant to a gas turbine aero-engine," *Journal of Engineering for Gas Turbines and Power*, vol. 127, pp. 257-26730, 2005.
10. M. D. Barringer, K. A. Thole, and M. D. Polanka, "Experimental evaluation of an inlet profile generator for high-pressure turbine tests," *Journal of Turbomachinery*, vol. 129, pp. 382-393, 2007.
11. R. P. Schroeder and K. A. Thole, "Adiabatic effectiveness measurements for a baseline shaped film cooling hole," GT2014-25992, 2014.
12. M. D. Barringer, A. Coward, K. Clark, K. A. Thole, J. Schmitz, J. Wagner, M. A. Alvin, P. Burke, and R. Dennis, "Development of a steady aero thermal research turbine (start) for studying secondary flow leakages and airfoil heat transfer," GT2014-25570. 2014.
13. R. A. Berdanier, I. Monge-Concepcion, B. Knisely, M. D. Barringer, K. A. Thole, E. Grover, E., "Scaling sealing effectiveness in a stator-rotor cavity for differing blade spans," *Journal of Turbomachinery,* Vol. 141 (5), pp. 051007, 2019.
14. National Academies of Sciences, Engineering, Medicine. "Commercial Aircraft Propulsion and Energy Systems Research: Reducing the Global Carbon Emissions," Washington, DC: The National Academy Press, 2016. Accessed: Jan. 1, 2021 [Online]. Available: https://sites.nationalacademies.org/DEPS/ASEB/DEPS_161178
15. National Academies of Sciences, Engineering, Medicine. "Advanced Technologies for Gas Turbines," Washington, DC: The National Academy Press, 2019. Accessed: Oct. 15, 2020 [Online]. Available: https://sites.nationalacademies.org/DEPS/ASEB/DEPS_190328.

Karen A. Thole holds the title of Distinguished Professor in the Department of Mechanical Engineering at the Pennsylvania State University. She is a Fellow of both ASME and AIAA. At Penn State, Dr. Thole is the Director of the Steady Thermal Aero Research Turbine (START) Lab, which has been recognized by two gas turbine companies as a center of excellence in turbine heat transfer and aerodynamics. She has published over 280 archival papers and advised 80 dissertations and theses with many of those students now working in the gas turbine industry. Dr. Thole chaired two recent National Academy of Engineering studies including one on low carbon aviation and the other on advancing gas turbines. For Dr. Thole's technical contributions, she has been recognized by ASME's George Westinghouse Medal and AIAA's Air Breathing Propulsion Award. As an educator, Dr. Thole has been recognized by the United States White House as a Champion of Change for STEM. She was recognized by the SWE Distinguished Engineering Educator Award, by ASME's Edwin F. Church Medal and by ABET's Felbinger Award for diversifying engineering. She holds two degrees in Mechanical Engineering from the University of Illinois at Urbana-Champaign and a Ph.D. from the University of Texas at Austin.

Nonlinear Pathways into Mechanical Engineering

B. Reeja-Jayan

I believe my perpetually immigrant family is an example of the good that can come out of immigration. I was born in Kerala, India, and brought to the Emirate of Abu Dhabi in the United Arab Emirates (UAE) when I was just a baby. My father emigrated there over a decade before in pursuit of better economic prospects. A zoologist by training, he became an accountant to earn a living. Like other wives then, my mother stayed back in Kerala until my father could find a visa for us to join him: A terribly difficult feat at that time. My brother was born in UAE. My family returned to India when I started high school and has remained there ever since. I immigrated to the US, and my brother to Singapore and later Japan.

The country of my birth, India, is the world's largest democracy; the US is the oldest. Both Singapore and Japan are democracies. UAE has a ruling royal family. These international experiences shaped my family and me in every sense: The languages I speak, the religions I follow, the values I hold, my political views, the relationships I keep, the fears I live with, the ambitions that define me, and the (nonlinear) paths I have followed in my career. This chapter focuses on the last one—the many divergent paths that led me to an academic career in engineering in the US. To my readers, I wish to share with you the joy, risks, and hardships I experienced while taking these road(s) less travelled.

Path to Engineering

In my early days of school, I was not considered a "good" student. I think I existed and went through the daily motions without delighting or annoying anybody in particular. My earliest memory of an "accomplishment" was a memory test in the

B. Reeja-Jayan (✉)
Carnegie Mellon University, Pittsburgh, PA, USA
e-mail: bjayan@andrew.cmu.edu

M. Bailey, L. Shackelford (eds.), *Women in Mechanical Engineering*, Women in Engineering and Science, https://doi.org/10.1007/978-3-030-91546-9_11

first grade. The teachers organized a large number of objects (from foods to toys) on a table, asked us to observe them, and then took us to another room where we had to list out as many items as we could remember from memory. I won that competition and received my first award. It felt good but not enough to motivate me any further in my daily learnings of math, science, language, and so on. I particularly did not like math (or maths as it was called then). As the years went by, my lowest score in this subject was even a zero. My mother tried to get me to sit down and study an hour every evening but I preferred to play with the neighborhood kids.

I loved watching cartoons on TV, though. They were in Arabic and this helped me learn the language and I did very well in Arabic classes (which were mandatory in schools in the UAE at the time). *Star Trek* (animation) reruns and Japanese sci-fi anime (all translated to run on TV in Arabic) were really the turning point which got me interested in science. These anime spanned everything from space exploration and extraterrestrials to robots and artificial intelligence. While these are all buzzwords and very real now, in the late 1980s they represented impossible dreams bordering on fantasy. I watched the 1987 movie *Robocop* so many times I knew the dialogues by heart. Research on such human–machine and other organic–inorganic hybrid systems are at the forefront of biology and materials engineering today, bringing the 8-year-old me of back then so much joy.

All the while, I continued to do poorly in school. One day after a particularly brutal parent–teacher conference in the sixth grade, my father had a serious talk with me. He told me how hard it was for him to afford the costs of sending my brother and me to school in UAE (there was a time I was almost kicked out for not making the fee payment deadline. An uncle stepped in and paid the fee). My father said he tried his best so we could have a good education and a future. He said he understood if I was not interested in school, but insisted that I persist and somehow pass the tenth grade exams. He said he is okay if I did not want to study further, as I could be married off early, which is typically what happens to girls in our family anyway.

I am not sure if it was the despair and disappointment I saw in my father or the mention of marriage, but something snapped inside me that day. I wanted to make my father proud of me, but more than that I wanted to have nothing to do with marriage. I saw all around me what that meant in our community. Women like my mother were not allowed to work after they had children. She diligently attended to our every need, spending 8 h or more in the kitchen, cooking every meal from scratch, cleaning and washing. Not to mention, I was an irresponsible child who never helped her in these chores. I have never seen my mother enjoy anything for herself. There is always a noticeable sadness in her and it makes me guilty every time we are together these days. A recent movie *The Great Indian Kitchen* does an accurate job in highlighting what is expected of married women hailing from Kerala, India; nothing fundamental has changed, even in 2021 [1]. Women in my family and the one I married into live this life.

After my sixth grade conversation with my father, I decided to take one chance to try and escape this cycle. I decided to study every day and review what was taught in each of my subjects the same day instead of my usual habit of opening my books the day before a test or exam. I thought this new resolve would last a week at the most, but that practice remains with me (in some form or the other) to this day. I ranked third among all students in the sixth grade classes in my school that year. The next year, I ranked second in seventh grade, first in eighth grade, and so on, until I ranked first among all students who took the tenth grade exams in the UAE.

Early Experiences in Engineering

My family moved back to Kerala, India, when I entered high school. I continued to do well in academics, especially in science subjects like chemistry and physics. When it was time to choose a path after graduation, I gravitated toward studying engineering as it seemed most in line with the science fiction (*Star Trek*, *Robocop*) that inspired me as a child. At the time, Kerala conducted entrance examinations to select students for engineering schools. Millions of students sit for these exams and the top 1000 or so get a free ride. The ranking determines the school and the branch of engineering you get to study. My rank was 594, and I was admitted to the University of Kerala in my hometown of Trivandrum/Thiruvananthapuram to study Electronics & Telecommunications Engineering, which is similar to what is considered Electrical & Computer Engineering (ECE) in the US. Solid state electronics was my favorite subject, which incidentally remains critical to the work I do today.

I did well in undergraduate engineering, and ranked first in the university upon graduation. My professors recommended I pursue graduate studies. Some of my peers applied to graduate schools in the US. I grew up on a healthy dose of American shows like *Star Trek*, which is why I became an engineer in the first place. An opportunity to study in the US was a dream. However, my family had fallen on financial hardships after my father left his job in UAE. I found work as a computer software programmer at a start-up to help my parents out and to support my brother's undergraduate studies. Nevertheless, I did not give up on my US dreams. I saved a little money from my salary each month toward paying for the GRE tests required to apply to the US schools. I studied on the weekends and did very well on the GREs, but my family found it inappropriate that I travel overseas without a husband. They refused to let me go alone to a country like the US (where we knew absolutely no one) unless I got married first. I knew what this meant. No husband was going let me hop on a plane and leave. Marriage tends to represent the expiration date for a woman's dreams in Kerala and most of India. In anger, I crumbled up my GRE scores and threw them into a box. I still wonder why I did not trash them. Perhaps a part of me held on to hope. The scores were valid for 5 years after all.

Two years later, I was selected (after a highly competitive process) to a scientist/engineer position at the Indian Space Research Organisation (ISRO) in the city of Bangalore/Bengaluru. ISRO surprised the world with their highly cost-effective mission to Mars in 2014 [2]. At ISRO, I got introduced to research that connects electromagnetism and materials. Spacecrafts in dangerous environments in space must be shielded to protect electronics and other subsystems from radiation and plasma. I got an opportunity to work with ultralight electromagnetic interference shielding materials based on sheets of amorphous metal alloys or metallic glasses. I currently research the fundamental origins of such electromagnetic field matter coupling behavior. This work has implications for applications from energy-efficient manufacturing to quantum electronics.

After 4 years of post-undergraduate work experience in various types of industries, I finally got my chance to pursue graduate studies. Ironically, this happened because I got married. My husband is the reason I am where I am in my career today. I had the equivalent of a tenured and permanent position at ISRO, we had bought a house in Bangalore, and were hoping to start a family. Friends and family told me it would be reckless to throw all this away and that this was the worst time to pursue my dreams. However, an offer to attend graduate school in the US (University of Texas at Austin, ECE) was not something I was about to turn down, no matter what sacrifices I had to make. On the day of my flight in August 2006, I woke up not knowing where/when I would next have a bed to sleep on. I was to fly from Bangalore, India, to Paris, France, and then Atlanta and finally into Austin, Texas. I made this journey alone. My husband, my biggest supporter, stayed behind and kept our enraged families at bay. Immigration rules prevented him from joining me, but he made it clear that I must pursue my dreams. I sold all my jewelry (wedding gifts from my parents) and used all our savings to finance my tuition and expenses for the first semester. I am grateful to The Ratan Tata Trust [3] for sponsoring my airfare to the US.

Like all international students, I, too, relied on the kindness of fellow students for accommodation, food, and generally learning the ropes. Two graduate students from India, Dr. Ashwini Gopal (ECE) and Dr. Chinmayi Krishnappa (Computer Science) remain my closest friends to this day. Their intelligence, scientific curiosity, sincerity, and accomplishments inspire me. Their support got me through very difficult times emotionally when I was lonely, separated from my family by oceans. My thrust area in ECE was called Plasma/Quantum Electronics and Optics. I was particularly interested in working at the intersection of quantum effects and nanomaterials. This was quite new in 2006, but is now a thriving scientific area. True to my nature to explore outside disciplinary boundaries, I started working in a department outside ECE. Professor Miguel Jose Yacaman in UT-Austin's Chemical Engineering department is a pioneer in atomic-scale studies of nanomaterials, particularly using transmission electron microscopy (TEM). He kindly allowed me to become a research assistant in his lab. Professor Selene Sepulveda, who was then a visiting postdoctoral researcher from Mexico, taught me everything I know about materials synthesis. I learned to synthesize and characterize nanostructures (particles, rods, wires) of materials like ZnO for broad applications in electronics, optics, and biology. I published my first peer-reviewed journal papers with Selene and the diverse group of brilliant researchers in this lab. Liquid phase materials synthesis, particularly under electromagnetic stimuli, is now at the core of my research program on energy and sustainability.

Doctoral Pursuits in Engineering

When I first saw atomic columns while imaging a gold nanoparticle specimen under the TEM, I felt very emotional. It was almost like finding a fundamental signature of God. I am a practicing Hindu, and our belief system is based on the principle of creation of matter (*Brahma*), conservation (*Vishnu*), and destruction (*Shiva*). This holy trinity is akin to the first law of Thermodynamics and connects all existence to the smallest of particles or atoms (anu). This experience with TEM had a profound impact on the rest of my academic journey. I decided I wanted to learn more about how materials form at the most fundamental level. This required learning Materials Science, which potentially meant changing majors. This is exactly what I did in 2008, when I completed my MS degree in ECE.

My future PhD adviser Professor Arumugam Manthiram offered me a position in his clean energy lab at UT-Austin. He suggested that I move disciplines to study Materials Science as part of the Texas Materials Institute (and I gladly accepted). This meant several additional years of course work and learning a whole new vocabulary (structure of materials, kinetics, phase transformation, electrochemistry, etc.). Once again, I was at a crossroads and I gladly took this fork in the road. I am so grateful I did this because Professor Manthiram's guidance continues to have a profound impact on me as a scientist and educator. In 2019, Professor Goodenough was one of three researchers awarded the Nobel Prize in Chemistry for inventing the lithium-ion battery that revolutionized portable electronics and is now set to save our planet by powering electric, pollution-free automobiles [4]. Professor Manthiram gave the Nobel lecture on the request of Professor Goodenough, who is now 98. The contributions made by both these extraordinary human beings to energy storage research is profound. For a nervous PhD student from India, seeing Professor Manthiram's passion for his work, scientific rigor and diligence, meticulous organization of his day-to-day activities, infallible work ethic (he used to be in his office every day around 6 am), and passion for teaching and mentoring inspired the way in which I advise my students now.

Professor Manthiram believed in me from the very beginning. So much so that as a first-year PhD student he put me in charge of setting up an energy-harnessing (solar photovoltaic) research thrust in his group. Step 1 of my PhD thesis was to find a broom; clean out an old furnace lab; and carve out space for a workbench, glovebox, and a solar simulator. Together with a dedicated team of undergraduate and MS students I was able to get this project going. We eventually discovered a process for low-temperature synthesis of ceramic oxide thin films using microwaves, which found use in the electron transport layer of flexible polymer–ceramic hybrid solar cells [5]. Along with materials for solar cells, I also developed coatings to passivate the reactive electrode–electrolyte interfaces inside lithium-ion batteries. NASA funded these fast-charging batteries, for use in spacesuits worn by astronauts during extravehicular activities [6, 7].

My husband had joined me in the US by this time and just as things were starting to feel well again, I suddenly got very sick. It began as stomach problems, which got worse after I got a stomach bug during a 2011 trip to India to renew my F1 student visa. One night it got so bad I had to go to the ER with severe vomiting and diarrhea preventing me from even keeping down water. Doctors could not find anything wrong and sent me home to rest. As the weeks went by, I felt better but started shedding weight, I lost 40 pounds in a span of months. When tests revealed nothing wrong with me, I was sent to a dietician who suggested I might be bulimic. Even family members and several acquaintances also thought I was doing this to myself. It was a dark and miserable time and I threw myself into my research work to escape it all. I read scores of books on all sorts of topics. I also read a lot about the interaction between microwaves and matter, and how these effects can help synthesize new materials with interesting properties. Years later, this idea would become the scientific nucleus for my own lab. I have this time of sickness to thank for it. If I had not fallen ill, I would not have had the time to pause, reflect, and think.

Sometime in 2012 my weight started dipping again and reached as low as 85 lb. as shown in Fig. 1. Fearing cardiac damage, I was subjected to a battery of tests to figure out what was wrong. Celiac disease was suggested (based on my gastrointestinal symptoms). Celiac is an autoimmune condition in which the ingestion of food containing the protein gluten (found in wheat, barley, rye, etc.) triggers the immune system to attack the gastrointestinal tract. Over the years this attack damages the ability of the intestinal villi to absorb nutrients from food, resulting in undiagnosed patients literally wasting away (as I was) [8]. However, testing blood markers for Celiac disease was still new and insurance would not approve the highly sensitive

Fig. 1 Sudden and excessive weight loss is a symptom of undiagnosed Celiac disease. Coupled with the stresses of graduate school, my weight dropped to dangerously low levels and I was at a very low point in my life in 2011–2012

Celiac test suggested by my doctor. I was not a Caucasian and it is widely (and wrongly) presumed that Celiac disease does not affect people of color. Months later I got the test conducted, and tested positive for Celiac disease in the summer of 2012. I continue to suffer from Celiac consequences even now. I joke that I have visited ERs in all the major tourist places in the US. I even went to the one in Napa Valley during a vacation. The reason for all these ER admits is that it is almost impossible to maintain a gluten-free diet in our modern world. Gluten is the second most widely used material in the food industry (after sugar). I was particularly sensitive to even trace levels, which made it very hard for my intestines to heal.

When I showed up for my PhD defense in May 2012, many members of my dissertation committee did not recognize me because of how frail I had become. They were extremely concerned. But I soldiered on and passed. I stayed back in UT-Austin as a Postdoctoral Research Fellow in Professor Manthiram's lab. I am grateful for this time, which gave me space to heal and plan for my future. In academics, taking time off the perpetual treadmill is usually frowned upon. Once again, this time was critical to the next step in my career.

My husband and I wanted to start a family but decided to wait due to my ongoing health issues and uncertainties with our immigration status (neither of us had a green card or permanent resident status yet). Disapproval from family and my own awareness of the fertility time limit facing me as a woman were constant sources of worry. Years later as an early career faculty, I would be faced with these questions again: "*Is this the right time to have a baby? What happens if I wait too long?*" This is a dilemma faced by all women pursuing growth in their careers.

MIT Bound

In 2013, I started applying to jobs in industry (becoming a faculty member was the farthest idea from my mind then. I simply believed I was not good enough for such a job). After a disastrous industry job interview, I decided to also apply for postdoctoral research positions at universities. I had taken up a part-time position as the Associate Technical Editor for the Materials Research Society (MRS)'s flagship journal *MRS Bulletin*. *MRS* is the largest international society for materials scientists. I edited journal articles in the *Bulletin* for their technical content. While editing one such article, I learned about the work done by Professor Karen Gleason's lab at MIT Chemical Engineering on paper-based photovoltaics, sensors, and other devices. It was a long shot but I was so excited that I curbed my fear and wrote to her. I immediately got a reply asking if we could meet. My husband bought me air tickets to fly to Boston and meet Professor Gleason. Years later, Professor Gleason said she liked my "chutzpah" for just showing up in her office. That—along with my background in lithium-ion battery work, polymers, and devices—made her offer me a position in her lab. For me, this opportunity to train at MIT was unbelievable and as far away from the dreams of the shy immigrant girl who consistently felt she was not good enough to be an engineer.

In Chemical Engineering, I learned a whole new discipline yet again, this time about reactor design, transport phenomena, and process engineering. These topics helped form the foundation of the ceramic manufacturing thrust in my lab. Professor Gleason was the first female professor with whom I have worked. She is my role model. She taught me to be confident, to believe in the quality of my work, to stop second-guessing myself. Her scientific prowess (she discovered the vapor-phase polymerization process using chemical vapor deposition (CVD) and developed it into a commercial technology now used internationally); her engineering acumen and all-around optimism and confidence is so infectious. As the first woman to get tenure in MIT Chemical Engineering, Professor Gleason has experienced her share of challenges, and she works tirelessly to mentor and support her students and postdocs. She is especially sensitive to the challenges faced by her female trainees. She once told me that every time I come to her with data, I spend the first 5 min berating myself for the things I did not do, before finally getting to the data from the experiment I actually ran. As women, we tend to assume intellectual modesty is the "proper" thing to do, but it can have consequences to succeed in the self-promoting world of academia. I still suffer from imposter syndrome [9]; it never goes away. But over time I learned to consciously tame it. Working with Professor Gleason was instrumental to this change in me.

While my PhD focused on ceramic and polymer–ceramic hybrids, my postdoctoral work at MIT went deeper into polymers, particularly CVD polymerization which allows us to precisely engineer complex surfaces with a variety of functionalities. I focused on the applications of these conformal coatings in 3D microbatteries, a way of miniaturizing energy storage. These days we talk a lot about scaling up batteries for electric vehicles, but going the other way to scale down batteries is still hard.

I collaborated across departments and even institutes on campus like the Institute for Soldier Nanotechnology [10]. A frequent hangout for me was with Dr. Andrea Mershin's team at the Center for Bits and Atoms [11], a wonderous place where engineering intersects with space exploration, movie making (*Star Trek* director J. J. Abrams hangs out in the building), building artificial life and intelligence, and other extraordinary pursuits. These experiences helped me dream big, I mean *really* big, when it became time to contemplate my next steps.

Professor Gleason was clear that I should apply for tenure-track jobs at R1 universities. She connected me to other mentors at MIT like Professor Hadley Sikes, who coached me through the application, interview, seminar, chalk–talk marathon of around 10 faculty job interviews. A challenge for me was managing Celiac disease. Long days on the road and eating out at restaurants were dangerous for me due to chances of inadvertent gluten exposure. Almost all my hosts accommodated my requests and made every attempt, but sometimes things just went wrong. I gave one job talk while being very sick (my faculty host kindly pointed out that I must rest before the next event on my agenda). One of the ideas I was pitching as part of my independent research was to engineer polymer-based sensors to detect gluten. This personally inspired research goal was highly appreciated in the interviews and this hobby of a project eventually became a funded research thrust in my lab, graduating my second PhD student, Dr. Phil Smith.

Starting My Own Lab @CMU

In 2015, when I started at Carnegie Mellon University (CMU) as a faculty in Mechanical Engineering, I was a lot older than other starting faculty in my cohort, mainly on account of the extra time I spent to get there. I could feel the weight of all these experiences. Indeed, the Dean of the college told me during our first meeting that my degrees and diverse experiences made me a candidate that stood out in the search that year. An academic career is stressful, period. There is no way to sugar-coat this. My autoimmune condition is triggered by gluten, but my physicians told me that stress accelerates these responses.

However, I also got to do some extraordinarily fun things. I had heard about work being done in game-based learning. This, along with my own early interest in video games (I played *Atari* games a lot), inspired me to develop a course where I used the game *Minecraft* to teach materials science to mechanical engineering students. Minecraft is a game where you can start with simple cube (blocks) and build anything from a Tesla Gigafactory to a space station. Students in my course can browse through rooms that teach them about Bravais lattices, crystal structures, defects, etc., as shown in Fig. 2, as well as build their own projects dealing with materials processing (plastic recycling, brick-making factory). Experiences from my first job as a computer programmer helped me embark on this project, which I believe can one day replace or augment traditional pedagogical methods. This project [12] was made possible by the Wimmer and Struminger Fellowships, the 2018 National Science Foundation CAREER award, and most importantly by my student Michael Oden working with a dedicated team of CMU undergraduate students.

An overarching goal of research in my lab (J-Lab) [13] is to use electric and electromagnetic fields for engineering energy-efficient processes to synthesize materials. When we succeed, our work can impact the way we produce ceramic materials that find applications in energy devices, sensors, electronics in areas as diverse as environment monitoring, transportation, aerospace, telecommunications, and healthcare [14]. Currently, producing these materials requires extremely high temperatures, making these processes highly energy intensive. We engineer reactor systems that allow us to use synchrotron X-rays to monitor the growth of materials, under electromagnetic fields (microwaves). Our highly multidisciplinary research merges together insights in electromagnetics, engineering physics, and materials chemistry. Events along my personal journey were often trigger points that helped me to navigate the map of these diverse disciplines and form connections that became invaluable to my career. My first PhD student, Dr. Nathan Nakamura (now a National Research Council (NRC) Postdoctoral fellow at National Institute of Standards and Technology (NIST)) collected the very first experimental evidence that microwaves change the atomic-scale structure of metal oxides. He took a

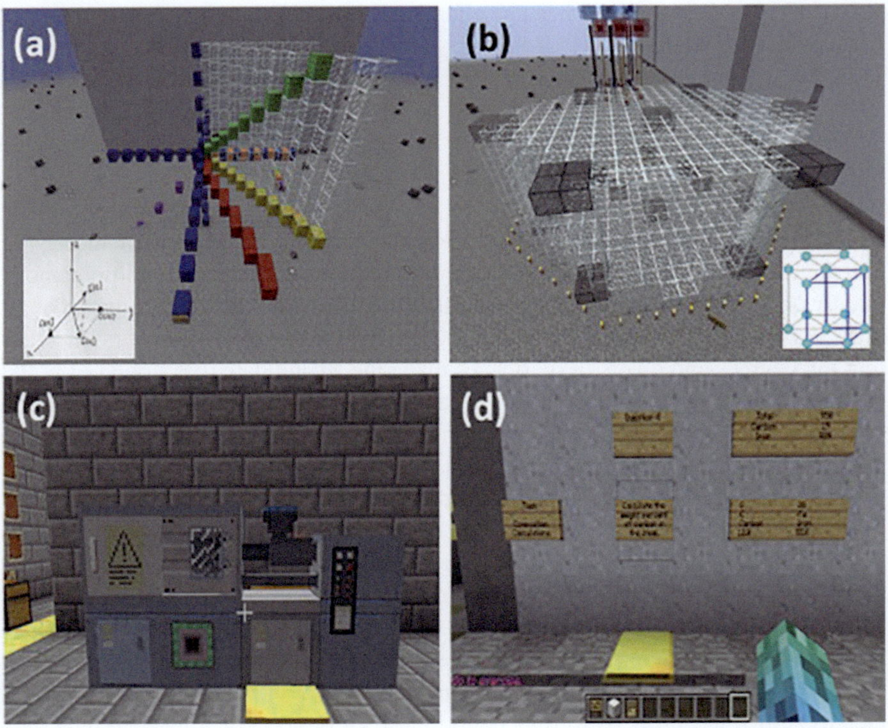

Fig. 2 Video games and science fiction inspired me to become a scientist and an engineer. Teaching engineering using games like Minecraft is a dream come true to me. This picture shows examples of various learning modules set up in my course to teach processing–structure–property relationships in materials

chance on me as a new faculty with a strange idea. My PhD student, Dr. Laisuo Su, was able to apply our techniques to develop faster charging lithium-ion batteries. He is now a Postdoctoral fellow in Professor Manthiram's lab. Laisuo starting his post-doctoral work in my old lab was a proud and emotional moment for me. I am grateful to all the undergraduate and graduate students (not enough space to name them all here) who continue to push hard so that our lab can pursue and realize seemingly impossible feats.

In 2019, my husband and I had a son, as shown in Fig. 3. Mine was a complicated pregnancy with many ER visits, ending in an emergency C-section, and our baby spending time in the NICU. It was a difficult time but we are grateful and lucky to have him. Coming back to work a few months after was hard. My parents flew from India to help me and I am ever so grateful for their support when I was so vulnerable. Then COVID-19 hit and my parents could not travel back home. Cooped up indoors during this strange time brought back old arguments and resentments but I am glad we got this time together. We agreed to respect each other's choices and perspectives, and to be thankful that we have each other to love.

Fig. 3 Our first child and my best friend is a Hungarian Puli Tiberius (Tubby) who is named after the father of Captain James T. Kirk (Captain of the Starship Enterprise in *Star Trek*). This was the moment where my husband Aji introduced our newborn baby to Tubby

Science Fiction Dreams

In true journal fashion, Fig. 4 summarizes all the forks in the road that I took to become a faculty member in a top Mechanical Engineering program. I don't even hold a Mechanical Engineering degree! Could I have planned such a nonlinear trajectory for myself? I believe not. Did I have mentors? Yes, definitely. Was it luck? Yes, partly. But it was also a philosophy. Did it always work? Definitely not. At each of these transitions in my life, the decision-making had nuances and sometimes I took one fork, followed it, failed, and went back to the road to take the other fork. Along the way, also came the challenges of immigrant life, family pressures, and health struggles. Failure is especially hard when you come from years of imposter syndrome. But you can push back through hard work, determination, continuous improvement, and never giving up. This philosophy that defines my journey is the one piece of me I wish to leave with my students.

In all our endeavors, it is important to keep dreaming. For me the magic comes from science fiction. Through our electromagnetic field–assisted synthesis work, we may one day make a *Star Trek*–style replicator that can assemble any material and tailor its properties to our liking. I am so thrilled to be working with colleagues who are pushing the boundaries of human–machine interfaces (something that was fantasy when I watched *Robocop* in 1987). Materials play a key role in these advancements. I am so excited that a part of my dreams are coming true already: Through

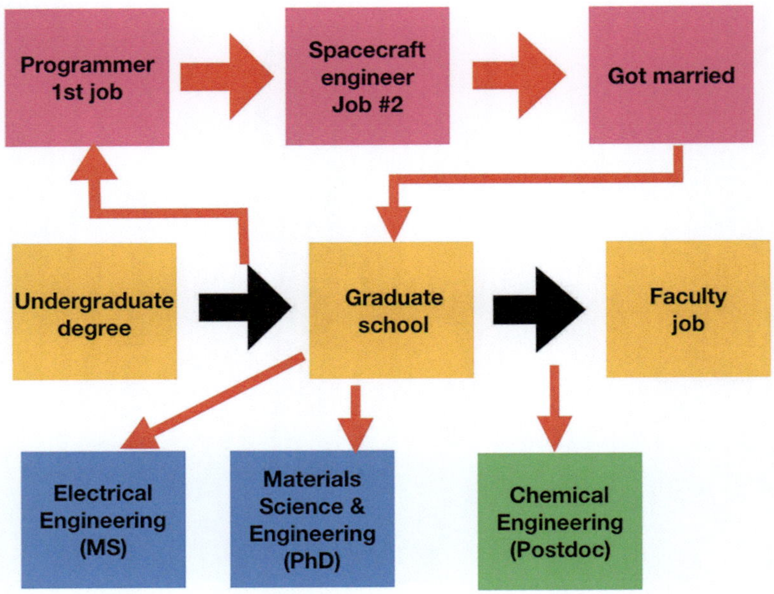

Fig. 4 The black arrows and yellow boxes represent what is typically considered a traditional path to a faculty position in Mechanical Engineering. The red arrows and other color boxes represent my nonlinear path

our collaborations on *Terminator*-style liquid metal–based stretchable batteries [15] and using machine learning for autonomous discovery in science and engineering [16].

When this pandemic ends, I hope we can all start to dream again and search for, find, and journey down strange new roads we previously might not have considered. These *roads not taken* may make all the difference in our new post COVID world. As the Robert Frost poem goes [17]:

I shall be telling this with a sigh.
Somewhere ages and ages hence:
Two roads diverged in a wood, and I—.
I took the one less traveled by,
And that has made all the difference.

References

1. G. Pandey. "The Great Indian Kitchen: Serving an unsavoury tale of sexism in home." BBC.com. https://www.bbc.com/news/world-asia-india-55919305 (accessed July. 26, 2021).
2. J. Amos. "Why India's Mars mission is so cheap – and thrilling." BBC.com. https://www.bbc.com/news/science-environment-29341850 (accessed July. 26, 2021).
3. Website provides details about the Sir Ratan Tata Travel Grants. https://ngosindia.com/sir--ratan-tata-travel-grants/ (accessed July. 26, 2021).

4. Website provides recording of the lecture of Professor John Goodenough's 2019 Nobel prize in Chemistry. https://www.nobelprize.org/prizes/chemistry/2019/goodenough/lecture/ (accessed July. 26, 2021).
5. B. Reeja-Jayan, K. L. Harrison, K. Yang, Chih-Liang Wang, A. Yilmaz, and A. Manthiram, "Microwave-assisted Low Temperature Thin Film Growth in Solution", *Sci. Rep. 2*, vol. 1003. 2012, https://doi.org/10.1038/srep01003.
6. J. Liu, Q. Wang, B. Reeja-Jayan, and A. Manthiram, "Carbon-coated high capacity layered $Li[Li_{0.2}Mn_{0.54}Ni_{0.13}Co_{0.13}]O_2$ cathodes", *Electrochemistry Communications*, vol. 12, pp. 750-753, 2010.
7. J. Liu, B. Reeja-Jayan, and A. Manthiram, "Conductive surface modification with aluminum of high capacity layered $Li[Li_{0.2}Mn_{0.54}Ni_{0.13}Co_{0.13}]O_2$ cathodes", *J. of Physical Chemistry C*, vol. 114, pp. 9528-9533, 2010.
8. Website provides details about the autoimmune condition Celiac Disease https://celiac.org/about-celiac-disease/what-is-celiac-disease/ (accessed July. 26, 2021).
9. T. Crowe and S. Slocum, "How to stop imposter syndrome from sabotaging your career," in *Women in Mechanical Engineering - Energy and the Environment. Women in Engineering Series*, M. Bailey and L. Shackelford, Eds., Cham, Switzerland: Springer Nature Switzerland AG, 2022.
10. Website for MIT's Institute for Soldier Nanotechnology https://isn.mit.edu (accessed July. 26, 2021).
11. Website for MIT Media Lab's Center for Bits and Atoms https://www.media.mit.edu/graduate--program/center-for-bits-and-atoms/ (accessed July. 26, 2021).
12. Website provides details and videos about teaching engineering using games http://jayanlab.com/teaching/ (accessed July. 26, 2021).
13. The lab website for Professor Jayan http://jayanlab.com (accessed July. 26, 2021).
14. B. Reeja-Jayan and J. Luo "Far-from-equilibrium effects of electric and electromagnetic fields in ceramics synthesis and processing," *MRS Bulletin*, vol. 46, no. 26, 2021.
15. D. Liu, L. Su, J. Liao, B. Reeja-Jayan, and C. Majidi, "Rechargeable Soft-Matter EGaIn-MnO_2 battery for stretchable electronics," *Advanced Energy Materials*, vol. 9, no. 46, Dec. 2019, https://doi.org/10.1002/aenm.201902798.
16. Website for workshop co-organized by Professor Jayan on using artificial intelligence for scientific discovery https://autonomous-discovery.lbl.gov (accessed July. 26, 2021).
17. Frost, Robert, Louis Untermeyer, and Robert Frost. *The Road Not Taken: A Selection of Robert Frost's Poems*. New York: H. Holt and Co, 1991.

B. Reeja-Jayan is an Associate Professor in Mechanical Engineering at Carnegie Mellon University. She also holds courtesy appointments in Materials Science and Engineering, Chemical Engineering, and Electrical & Computer Engineering departments. She is also a Faculty Fellow of the Scott Institute for Energy Innovation and Dean's Early Career Fellow in Engineering. Her multidisciplinary lab explores ways by which electromagnetic fields can synthesize materials hitherto unavailable to conventional synthesis routes. These low temperature processed materials directly grow on flexible, lightweight substrates, enabling structurally integrated energy and sensing. Dr. Jayan is a strong believer in game based learning methodologies that she uses extensively in her undergraduate and graduate engineering courses. Dr. Jayan is a recipient of the 2018 National Science Foundation CAREER Award, 2017 Army Research Office (ARO) Young Investigator Award, 2016 Air Force Office of Scientific Research (AFOSR) Young Investigator Award, the George Tallman Ladd Research Award, the Donald L. and Rhonda Struminger Faculty Fellowship, the Berkman Faculty Development Fund, and Pittsburgh Magazine's 40 Under 40 Award. Her research is also funded by the Department of Energy (DOE), Defense Advanced Research Project Agency (DARPA), and by private sponsors.

Originally, from Kerala, India, Dr. Jayan now lives in Pittsburgh, PA with her husband Aji, her toddler Neil, and her puppy Tubby.

My Journey from Fixing the Lawnmower to a Career in Fuels and Combustion

Elana Chapman

Sometimes we look back to determine if there was a single event that might have set us on our course in life, especially an event or experience connected to our personal interests and eventually our careers. As I made an intentional and mindful effort in pondering this question, I realized that there was not one single event that set me on my course, but perhaps a few events that guided and directed me to where I now find myself in my career. As I thought about it, I recalled my youth when I used to hang out in the garage with my Dad while he fixed the lawnmower and I curiously wondered how it worked. I remember those who buffered me and boosted my growth when I was stuck trying to figure out the next challenging "lawnmower" question—parents, educators, friends, family, sorority sisters, professors, and advisors who grounded me to "keep it simple" like the lawnmower. I have been a Senior Fuels/Biofuels Engineer for General Motors for the last 12 years. I am aspiring to become a Society of Automotive Engineers' Fellow through my technical accomplishments, leadership, and service, and, potentially, someday honored to be an invited member of the National Academy of Science and/or Engineering. In this chapter, I would like to give a historical perspective on those instances that impacted my course in life and talk about how I went from fixing the lawnmower to my career in fuels and combustion. In these instances, there are lessons and key learnings about intellectual curiosity, lifelong learning, and pursuing and rising to new challenges for young women interested in pursuing careers in science and engineering, college graduates thinking about going to grad school, PhD students working on their research, and educators and advocates interested in increasing women in engineering and science, technology, engineering and mathematics (STEM).

It was Christmas time between the Fall and Spring sophomore semesters while I was attending the University of Dayton studying Mechanical Engineering. I was exhausted and struggling with school. I just thought that there was some other degree that I should pursue, and I had decided to change my major. But before I did

E. Chapman (✉)
General Motors, Detroit, MI, USA

M. Bailey, L. Shackelford (eds.), *Women in Mechanical Engineering*, Women in Engineering and Science, https://doi.org/10.1007/978-3-030-91546-9_12

that, I had to let my mom know what I was planning to do. I was not looking forward to the conversation, but I knew that I had to talk to her about it. Even though I knew that I could figure out how to fix the lawnmower, I was doubting that I could finish the coursework to get the degree. However, my mom knew that I was giving up on myself. As we talked through the entire night, she heard all the reasons that I had to give up and convinced me that I should not give in to my defeated feelings. But let me start from the beginning.

There were lots of things that interested me as an inquisitive young learner. I can recall some of the struggles of being at a new grade school, trying to fit in with the cliques, and just trying to keep up with the homework. Yes, I was one of those students who took an interest in learning, but also worried that I might forget my homework at school, or which book I needed to bring home. I found myself bringing home perhaps more than I needed each night, but I was confident that I had all my materials and references with me to complete my tasks. My memories are quite clear on those instances which had the most profound effects on me. During my fourth grade year, I recall an outside group coming to the classroom with a special project and presentation. While it escapes me who the group was, I was most impacted by how those adults in that group helped us to be successful, as well as by the materials that I was given to build my own digital timer circuit. In fact, I still have these parts to this day. I still recall the adults circulating in the room to make sure we were able to get the circuit built and confirm with each of us that it was working. That was perhaps my first introduction to electrical circuits, and to problem-solving. Looking back on this now, groups that bring simple projects like this to the classroom actually encourage and inspire young children to be inquisitive, interest them in how things work and foster continued curiosity. Throughout my life, trips to *Radio Shack* for those electronic projects and parts would be part of my continued curiosity.

My mother supported my curiosity and, during this time, she specifically looked for toys that would interest me; I played with such things as *Lego* blocks, *Erector* sets, *Matchbox* cars, and a children's tool kit for woodworking. And given my insatiable curiosity, I soon tired of these and searched for something even more interesting. I found it in the garage at my Dad's workbench with the many tools and garden equipment on hand. While the tools were interesting, they were used to make something or fix something. And the things that needed to be fixed were the garden equipment, especially the lawn mowers. We had two riding lawnmowers and two push mowers to cover the large yard we had to mow each week. Usually my dad did all the maintenance, but when he was not around my mom asked if I could take a look. I was enchanted and of course, I had to first figure out how they worked. My mom was good at suggesting and then showing me the owner's manuals. It was good practice to look at diagrams of how the parts fit together and the direction they should turn, and then look back at the parts on the actual machine. Although I was not always successful at fixing them, I was learning about them. Even more so, I was also learning that I wanted to learn more.

I thought the most effective way for me to "tinker" would be for my parents to buy a crashed car at the junkyard for me to work on. Looking back now, I realize why that never happened; my parents knew this would have created a nice "mouse house" while living out in the country at the farm. I was left to settle for opening the hood of my mom's car and learning what I could from observation, and tinkering with and comparing to the lawnmowers. Actually, it was a bit more than tinkering. When the mower's engine would not start or was hard to start, it was a problem to solve. I had to focus my mind on how it was supposed to work and think through what might be causing it not to start. The first thing was to check if the engine had the proper fluids and if they were getting to the engine. Of course, when we tried to start it over and over again it was deemed "flooded" with fuel and fuel vapor, and then we needed to let it set for a while and try again later. I wouldn't understand then why that seemed to work, but it just did. It was learning by trial and error, yet not understanding how to explain it.

I carried that questioning curiosity throughout grade school, which especially showed up with the subjects I liked to learn about the most—math and science. I was placed in the advanced classes in grade school, with the hope that I would advance into the honors classes in High School. Although I chose to go into the Honors Sciences in High School, I was hesitant to take Honors Math. It was not until after my Freshman year, and being bored relearning Algebra that, when offered to springboard into the Honors Math track, I immediately took the offer and studied Honors Geometry and Trigonometry the summer of my Freshman year. (Thanks to the Chaminade Julienne High School Honors Math teacher Mrs. Martha Duffy for her encouragement and lessons that summer to springboard a select group of students forward.) I was now in both the Honors Math and Science tracks for my High School career. During those years, besides studying what I could from my texts and the library, my research also included both *Popular Science* and *Popular Mechanics*. They were certainly not the most technical literature, but they did peak my interests and allowed me to view aspects of science that I had not considered. I am thrilled to say that I finished third in my class and was given many awards including, to my surprise, the award for highest achievements in Honors Chemistry.

While reading those *Popular Science* and *Popular Mechanics* magazines, I began considering what might be next for me. I recall thinking that working on a project like hip replacements might be really interesting and an engineering degree might be the avenue to do that. I dreamed about going to Massachusetts Institute of Technology (MIT), however, knowing that it was quite far away and quite expensive, I chose the University of Dayton (UD), much closer to home and with scholarships, it was affordable. I decided on entering Mechanical Engineering, as it was a degree that would be a "Jill of All Trades," meaning that I would not focus specifically on electrical, chemical, or materials engineering, but would have knowledge and a skill set of all. I was glad to know that the curriculum had some classes related to the focus area of energy and internal combustion engines as well.

College was extremely tough, and I eventually found my cohort group who supported me throughout my Mechanical Engineering degree and through college in general. I am grateful for the group of women who helped me in a class when I was struggling, and I'm grateful that I could return the favor and help them through classes when they needed it. I recall how much our group of four was there for each other, especially when studying for a Mechanical Vibrations test, and meeting at midnight to review the materials. I was relieved to be able to count on my tribe for support throughout this time. (From this UD Mechanical Engineering tribe, every one of us went on to graduate school in engineering, with three of us going on for and receiving PhDs. This was and continues to be a very unusual thing!)

In addition, that tribe expanded into a bigger support system at UD. One of our fellow woman engineers pulled a larger group together to consider starting a social sorority on campus. Since my smaller tribe was all in, we decided to embark on the new adventure. Founding a social sorority on campus was a little harder than we imagined. Despite concerns from the university with the nature of the group and the potential liability, the drive behind the group of women engineers was such that a small group went to the College of Engineering and the University Vice Presidents to make their case. The group was not going to take "no" for an answer. Through their perseverance, this group became the University of Dayton Gamma Chapter of Phi Sigma Rho sorority. In that process, many of us came to understand the liability concerns, and committed ourselves to evolving the sorority to a point of financial stability. During this time at UD, this was not just the fun group of women student engineers to hang out with, but a community of women who could best understand the struggles of working hard to become an engineer. It is through this tribe that we were able to be successful and make it to the finish line to complete our degrees. And it was not until the end that we found out how impactful this group was to our success (Fig. 1).

Along my journey, there are many individuals who shaped my direction and, perhaps, changed the course of my life. The first was during my time at the University of Dayton. Dr. Kevin Hallinan, professor of Mechanical Engineering, invited me to support him with some of his research activities. While it was only a small amount of work that I did for him, he taught me about many aspects of research. At one point, he suggested I consider going to graduate school. "Who me?"—I was thinking inside my head. Most days I felt like I was barely surviving and succeeding in my classwork. However, he must have seen something in me then that I did not recognize. Perhaps it was the abstract and paper I wrote from my cooperative/research experience for ASME which won the prize for Best Student Paper at the regional meeting, and he saw that I was capable of doing research, drawing conclusions, and had a skill with writing. Or was it the senior design lab project which Dr. Hallinan helped me with measuring flame speeds of fuels by video frames in a Lexan Tube? I didn't pursue graduate school right after graduation and followed the

Fig. 1 Current logo of Phi Sigma Rho Sorority for Women in Engineering and Engineering Technology

rest of the graduating seniors into the job market. It wasn't until a few years later when I became somewhat disinterested in what I was doing, and more interested in learning about engines and fuels, that I realized something about myself that perhaps Dr. Hallinan saw in me. I realized that I thirst to learn, and I needed to be in a position where I am continually learning. Those are good skills to have when you are a researcher. It was then that I decided to start looking at grad schools and applying to programs.

During my Senior year at the University of Dayton, I was offered and took a position at Delphi Delco Electronics, which was a components provider for General Motors (GM). Rather than go directly to graduate school, I needed to find a job so that I could take care of my needs. Also, I was not ready at the time to go back to school, and I was not sure I was capable of achieving an advanced degree. I started as an Electronics Packaging Engineer in the group that made engine controls for both GM and many of the International GM and non-GM groups. Through this position, I learned that I was also interested in the work of the others on the team which included the Systems Engineering role, the objective of which was to architect the hardware and software functions, as well as the Software Engineering role which programmed the controller to operate the engine. It was one of those "how does that work and what does that do" moments for me. I was very interested in their kind of work since they interfaced with the engines and vehicles that the equipment was designed for, and I wanted to have the same "hands-on learning."

My first approach to gaining additional education to make the transition to these roles was to get a bachelor's degree in Electrical Engineering Technology at Indiana University–Purdue University—Indianapolis (IUPUI) in hopes that I could become a Systems Engineers. However, taking the classes while working full time proved to be too difficult. But the seed that was planted in me to go to grad school grew faster, taller, and stronger. My interest in "hands-on learning" working on a project involving an engine and a vehicle also grew. So, I investigated programs at universities in which I could work on engines and controls, with a preference for the project to move the design work from an engine dyno laboratory onto a concept vehicle demonstration. It wasn't until I decided upon the combustion research area that I began to explore schools and faculty advisors.

Perhaps it was my fortune at the time that all the schools I applied to accepted me. After a review of my decision matrix, I chose to go to Penn State University and work for Dr. Andre Boehman, Assistant Professor of Fuel Science, and father of Dr. Louis "Lou" Boehman, Professor at the University of Dayton, who taught me about internal combustion engines and was instrumental in putting me on my initial trajectory. I feel fortunate to say I am perhaps the one and only student who was taught by both father and son. The project at Penn State involved a diesel engine and a bus demonstration project. The graduate degree was in Fuel Science, and a program that interested me involved fuel utilization in an engine, with the ability to take courses in Mechanical Engineering and Combustion, as well as Fuel Science. I wouldn't realize then that this was a special program and that the fuel would play a major role in my research work while the engine was just the tool used to design the system and collect the data. I was right where I wanted to be (Fig. 2).

Fig. 2 DME Shuttle Bus Project team. From left to right: Dr. Andre Boehman, Dr. Elana Chapman, Dr. Louis "Lou" Boehman, and Jennifer Stefanik. (Photo credit: Andre Boehman)

My project in grad school involved a diesel engine, and a special fuel called dimethyl ether (DME). I was not thinking so much about the fuel when I chose the school and the degree program, but I was very interested in just working on an engine and in the lab. Working with my colleague on the project, our plan was to run the engine with the DME blended in the diesel fuel (diesel–DME blend) to enable the fuel system operation similar to diesel fuel operation, but to demonstrate the potential for environmental improvements (particulate matter and NOx reduction) with this fuel combination versus a normal diesel fuel operation. I can recall one day working really hard to just get the engine started. Everything seemed to be set up correctly, but nothing happened. It was frustrating and a point at which I thought, "I can't do this. I don't know what to do." However, I thought back to the days I was working on that lawnmower at home. I did not know then what was wrong, but I focused on the simple aspects of the operation: It needs fuel and air to start, then it needs an ignition source. "I CAN do this." Remembering that helped to keep me focused on the simple aspects of how the engine should operate, and what was needed to start it. Eventually, I was checking wiring harness voltage sources and sensors, and finally traced the issue to the throttle position sensor. It was a huge achievement just to get the engine running. But it was also a confidence booster to learn that I was ready to tackle the next part of the research.

Our first project regarding diesel–DME blends, blossomed to an evaluation of which materials in the fuel system could withstand the blend and which would degrade. To begin, we contacted technical experts and learned all we could about DME and potential issues of the fuel with the elastomer interactions (swelling of the elastomers, degradation of the materials, loss of function, and fuel leaking from the engine). We eventually determined that certain seals and O-rings had to be replaced, and only specific tubes could be used in the fuel system. Next, we had to understand if the fuel properties would potentially compromise the fuel system, especially the fuel injectors. Again, another project blossomed from this issue, which was addressed through the testing of lubricity additives mixed with the fuel blends. Lubricity additives in the fuel reduce the amount of wear that occurs in the fuel injector. The most challenging fuel property to master was the phase-change behavior of the DME. The fuel itself is a gas at standard temperature and pressure (20 °C and 1 atmosphere), or the normal conditions in a room. The fuel system had to be developed so that it could keep both the diesel and the DME blended in a liquid state as it passed through the engine and fuel system. Through trial and error, we purchased a large enough heat exchange bath to remove the heat that was injected into the fuel through the engine head, thus the fuel would remain a liquid in the closed fuel system. We also needed to ensure that we had enough pressure in the system to keep the fuel in the liquid state as well as to push the fuel into the fuel system lines for the testing. So, we needed to not only develop the loop for the fuel system, but also the tank and blending approach for preparing the fuel prior to testing. All of this was done to prove the fuel system operation and integrity, as well as demonstrate the improvement in emissions from the vehicle concept. Once this was accomplished, it was time to move this to the vehicle demonstration [1] (Figs. 3 and 4).

The fuel system would be replicated on a university bus which was only operated for short time periods. The bus fuel system was similar to our lab version, except for a few things that needed to be adapted for use on the vehicle. The cooling system developed for the laboratory was adapted for use in the vehicle, and the fill up and daily close down procedures were modified as well. At the end of every day, the fuel tank would go through a "blow down" procedure to release pressure in the lines. This was necessary since the closed fuel system would continue to increase in heat after the engine and cooling system was turned off. That increase in heat would build pressure in the system, which was unsafe for those working around it, and also the pressure would be so high that the special fuel pump and cooling system would not be able to overcome the vapor pressure in the system below the threshold for the DME to change phases from gas to liquid. To resolve this issue, we installed a short off-gas tube on the bus, well above the bus top, for the daily close down procedure which we called the fuel system blow down. We did not want to release the DME at the ground level due to a potential fuel ignition issue with the mixing of the gaseous fuel and air at the ground. So, it was released perhaps 20 ft above the ground, and 5 ft above the bus. As an interesting aside, we did not have a concern with the potential toxic release of the DME into the air as we knew that DME not only degrades in the atmosphere in 3 days but is also a very common gaseous chemical used in the

Fig. 3 Photo of the Navistar T444E Turbodiesel Engine in the 450 hp. engine test cell prior to conversion to DME–diesel blends

Fig. 4 Navistar T444E showing conversion to DME–diesel operation

delivery of hair spray. Therefore, it was relatively safe to use and safe for the environment [2].

For my PhD work, I continued my study of DME at Penn State, but with another fuel system concept application, and with a light duty diesel engine. In this work, I focused on using a combination of the DME as a gas mixed with the intake air and running the engine with a normal diesel liquid fuel injection. I studied specifically the direct impact of this combination on the particulate emissions, NOx, and other criteria pollutants [3]. For this concept, the intake air system had an upstream heater on the inlet side so that the DME and air combination could be heated, and the fuel easily ignited in the engine. This kind of fuel system made delivery of the DME a bit easier, but more difficult to keep track of the fuel and air in each combustion cycle [4]. To monitor the fuel and other hydrocarbon species in the exhaust, a special gaseous gas chromatograph (GC) was set up and used to collect the data. Separately, I was wondering if there could be some increase in the efficiency of the engine due to the DME assisting with a more consistent autoignition of the diesel fuel. Studies were carried out with both diesel fuel and with a 20% blend of biodiesel. There were clearly reductions in the particulate emissions with this concept. However, the other emissions increased, due to the lack of sufficiently controlling the ignition and complete combustion of the DME and perhaps some internal fuel mixing that led to complications with the NOx. I completed explaining the research results using Chemkin, a reaction chemistry simulation software, (https://www.ansys.com/products/fluids/ansys-chemkin-pro) to model the fuel radical species in the DME ignition process. The work and process involved in explaining my research was not easy, but I kept my mind on the lawnmower to ground me each time I was struggling (Figs. 5 and 6).

In addition to the work I did for my Fuel Science PhD degree, I was also involved with some very interesting Department of Energy/university student sponsored programs. Future Truck, Challenge X, and Eco Car are a few of the Department of Energy (DOE) student competitions. These student programs were a great opportunity for me as a graduate student to work with and teach undergraduate students, and to diversify my education by working on other energy and vehicle related projects. For all of these student-based hybrid electric vehicle projects, the objective was to highlight advanced hybrid electric technologies, potentially resulting in high efficiency and lower emissions. As the graduate student team lead for the diesel engine and aftertreatment package, I was given the task of developing the aftertreatment approach which included inviting suppliers to participate with us in the demonstration of their technology. I had the opportunity to learn a lot about Selective Catalytic Reduction (SCR) (also known as urea-SCR because the urea is the reductant) which is used to reduce and remove hydrocarbons and NOx from diesel exhaust emissions. Also, I worked with a separate supplier to bring forward a Diesel Particulate Filter (DPF) sized for the engine to remove the diesel exhaust particulates. Each student-sponsored program involved the choice of fuel for improved overall CO_2 reduction (net carbon CO_2 through the fuel and powertrain selection), or as we also call it, a renewable fuel. For the project we only had one choice for the diesel engine which was biodiesel blended as a B20 (20% biodiesel blended with

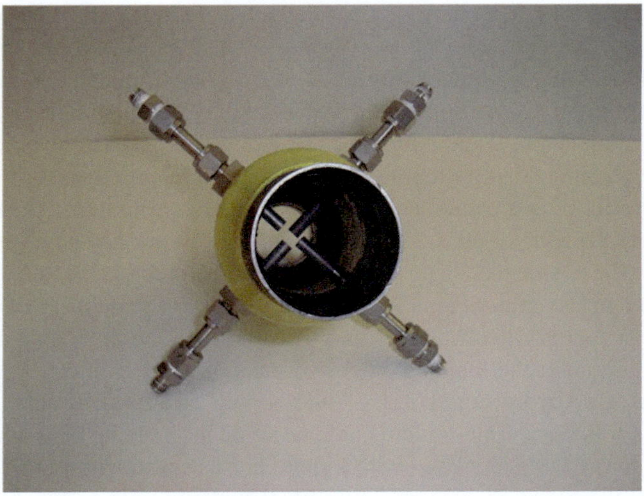

Fig. 5 Custom intake air manifold aspiration system used for fueling the engine, connected in front of an intake air heating system [4]

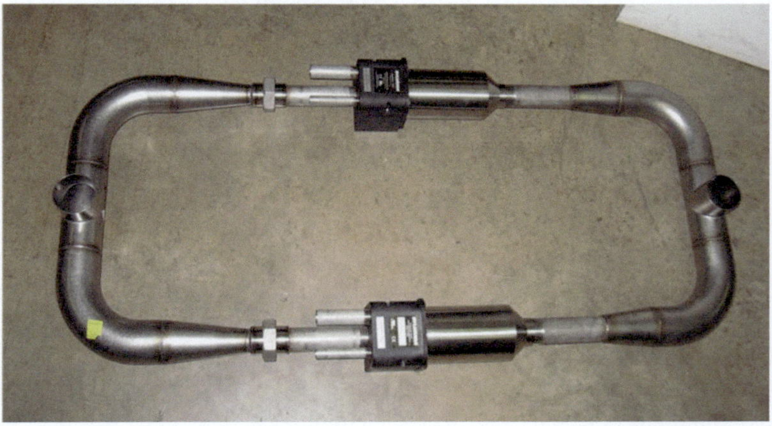

Fig. 6 Layout of the intake air heating system prior to installation on the engine stand [4]

diesel). With my support of the project, our team was able to finish in fifth place out of 20 teams that year. Most important to me was that the vehicle ran through the entire emissions testing and the system was able to function correctly [5].

During my PhD research, I was funded by a National Science Foundation Grade K–12 program. This involved using the technical aspects of the DOE student competition based on advanced transportation research to educate both teachers and students in the classrooms in central Pennsylvania. This included developing K–12 content for the classrooms, providing a website to document the K–12 lessons, and preparing updates for the K–12 students about the DOE student vehicle competitions. The college students involved in the NSF program and the vehicles competitions were pursuing a certificate in the DOE program called GATE: Graduate Automotive Technology Education. This program focused on Advanced Storage Systems technologies for transportation, including batteries, fuel cells, flywheels, and ultracapacitors. I was fortunate to take classes in all these areas, developing a breadth of knowledge in these technologies and earning a second Master's Degree in Mechanical Engineering.

I finished my PhD work at Penn State and accepted a position at General Motors (GM) in the Advanced Diesel group. I was not in the Advanced Diesel group very long before I was sought after to fill a role as a Senior Fuels/Biofuels Engineering in the Fuels Group at GM. My work in this group involves understanding all the basic fuel properties of gasoline and ethanol fuel blends, and how those combust in engines to produce improvements in efficiency and reductions in vehicle emissions. This includes research fuels used for engine studies, and market fuels that are from retail stations.

Since 2010, I have been working on this issue called Stochastic Preignition (SPI). Stochastic is defined as something that occurs randomly and without an explainable and understood pattern. SPI is a phenomenon in an engine cylinder which causes the early ignition of the gasoline (spark ignition) fuel and air prior to the spark discharge. This abnormal engine combustion cycle event causes many issues with the normal operation, reducing fuel efficiency, degrading emissions, and potentially causing damage. It has been very difficult to find not only the cause of this but the potential solutions as well. As a typical researcher might do, I looked to the literature to determine if "we have been here before" [6]. Our literature review on the subject provided the technical community a comprehensive review and propelled us forward with an idea of where to start and how to work the issue. Because it is stochastic, it takes a significant number of engine cycles (almost a million engine cycles per test which takes about 16 hours to collect) to get enough events to occur to be able to differentiate the causes. Additionally, those of us working on this particular issue have become experts in statistical analysis and how to renormalize data. (In data collection, we understand that a data set is "normally distributed," which means it has a Gaussian distribution with a peak and a tail on either side. When the data set is non-normal, there are methods that can be used to renormalize the data so that the traditional statistical methods can be used to draw conclusions from the data.) The quest to understand this issue has led me to understand the impact of the engine oil and oil additives, the engine fuel property relationship, and engine operating conditions that increase and decrease the propensity of this kind of event. Some improvements have been made to engines and oils, but continued research in this area is ongoing due to the need to reduce overall costs, improve fuel economy loss and emissions, and to allow for continued downsized boosted turbo direct injection engines as a technology option.

In 2014, I decided to set myself the goal to become a Society of Automotive Engineers (SAE) Technical Fellow. As part of the nomination process, I needed a significant number of reference letters. Now, I needed to be intentional in what I was asking for from my mentors. One of the amazing skills I gained from being a PhD researcher is being humble with the ability to be open to learning and to coaching. With this goal in mind, it became clearer to me where I needed help and how to ask for it.

Along my journey, I've learned some things about myself that guided my choices and I wanted to end with these insights. Most important is that I keep myself grounded by remembering the lawnmower and that I am resilient and can succeed in whatever I decide to do next.

Key Learnings About Myself

- I love to learn, so I needed a career where I would be continually learning.
- Be mindful of yourself and what stimulates your passion—for me that was continually learning about fuels, combustion, and engines.
- Use positive self-talk to get you through those tough situations—You CAN do it!
- Believe in yourself. With every new challenge, there is a new opportunity to learn.
- When it is hard to believe in yourself, turn to your tribe to build you up and keep going!
- Be mindful of and trust your instincts.
- Be humble and be open to coaching. Ask for coaching and ask for help.
- Build a network of colleagues, advisors and mentors.

Key Things That Supported Me—From Parents, Faculty, and Others—In My Eventual Career

- When my dad was not around to fix the lawnmower, my mother asked me to see what I could do. Listen to your mom!
- Using the opportunity of the research at my cooperative experience, writing up the work in an abstract and paper, and presenting at a conference during my undergraduate degree.
- Having a professor plant a seed in me that I might consider graduate school.
- Having a great graduate degree research project, a great advisor, and the tools to succeed.
- My tribe of colleagues during my undergraduate degree who were always cheering me on.
- People in my life who were building me up and believing in me until I could sustain my own confidence and belief in my abilities.
- DOE- and NSF-funded programs that develop technical talent in K–12 students, teachers in the K–12 classroom, and students at universities.

What Helps Me Now

- Believe in yourself. With every new challenge, there is a new opportunity to learn.
- Exude confidence.
- Take a pause to consider where you are at, what's missing, and then what's next.

References

1. E. Chapman, A. Boehman, P. Tijm, and F. Waller, "Emission Characteristics of a Navistar 7.3L Turbodiesel Fueled with Blends of Dimethyl Ether and Diesel Fuel," SAE Technical Paper 2001-01-3626, 2001, https://doi.org/10.4271/2001-01-3626.
2. J. Eirich, E. Chapman, H. Glunt, D. Klinikowski, A. Boehman, J. Hansel, and E. Heydorn, "Development of a Dimethyl Ether (DME)-Fueled Shuttle Bus," SAE Technical Paper 2003-01-0756, 2003, https://doi.org/10.4271/2003-01-0756.
3. E. Chapman, A. Boehman, "Pilot ignited premixed combustion of dimethyl ether in a turbodiesel engine", *Fuel Processing Technology*, vol. 89, no. 12,pp. 1262–1271, 2008, https://doi.org/10.1016/j.fuproc.2008.08.010.
4. E. Chapman, "NOx Reduction Strategies for Compression Ignition Engines", Ph.D. dissertation, Dept. Energy & Geo-environmental Eng., Penn State Univ., University Park, PA, USA, 2008. [Online]. Available: https://etda.libraries.psu.edu/files/final_submissions/1797
5. E. Chapman, S. Pflumm, E. Kung, R. Acharya, J. Saxon, B. Feldman, B. Herrold, K. Wilson, P. Safabakhsh, M. Shirk, J. Caserta, A. Boehman, D. Haworth, H. Koga, T. Tadros, D. Maglast, and L. Blackman, "Penn State Future Truck Hybrid Electric Vehicle: Light-Duty Diesel Exhaust Emission Control System to Meet ULEV Emissions Standard," SAE Technical Paper 2005-01-3877, 2005, https://doi.org/10.4271/2005-01-3877.
6. E. Chapman, and V. Costanzo, "A Literature Review of Abnormal Ignition by Fuel and Lubricant Derivatives," *SAE Int. J. Engines*, vol. 9, no. 1, pp. 107–142, 2016, https://doi.org/10.4271/2015-01-1869.

Dr. Elana Chapman is a recognized expert in the area of fuel science and combustion. Currently a Senior Fuels/Biofuels Engineer at General Motors (GM), she has spent over 13 years supporting the Advanced Engineering and Production Combustion groups with test fuels, and performed other projects related to fuels and engine combustion to meet future vehicle emissions and performance standards. Dr. Chapman serves as GM's Executive Committee member and Co-Chair on the Coordinating Research Council—Advanced Vehicle Fuels and Lubricants Committee, and member of the Performance Committee. She serves as the Chair of the Society of Automotive Engineers (SAE) Fuels and Lubricants Committee and has authored over 30 technical papers. She earned a B.S. in Mechanical Engineering from the University of Dayton and an M.S. in Fuel Science, M.S. in Mechanical Engineering, Ph.D. in Fuel Science, and a Department of Energy (DOE) Advanced Energy Storage Technologies Certificate from Pennsylvania State University.

As a longtime member of SAE, Dr. Chapman received a 2018 Forest R. McFarland Award, which recognizes individuals for their sustained outstanding contributions to SAE. In 2020, she received the Rodica Baranescu Award for Technical and Leadership Excellence, celebrating the successes of women in the engineering profession, and recognizing their leadership and technical contributions in the automotive sector. Dr. Chapman has been heavily involved in supporting future women in STEM through the Phi Sigma Rho Sorority and Foundation, the GM-Penn State Recruiting Team, and on the Penn State's Dean of Engineering Diversity and Inclusion Advisory Board.

Do Pipeline Engineers Want to Pollute the Environment?

Claudia Soriano Vazquez

When I graduated with a BSc in Mechatronics Engineering in Mexico, little did I know that my professional career would take me to Canada to develop their energy industry. The purpose of this chapter is twofold. First, I want students to know that engineering is a rewarding profession for those of us who stay curious, and that the possibilities to shape their careers depend on how nimble they are to challenge themselves and welcome opportunities. Second, I want to inform the general public that pipelines are an intrinsic part of our life and share how engineers like me use the best of our collective expertise to supply the energy that society needs while protecting people and the environment.

Why Study Engineering?

Engineering is still a male-dominated profession. In my career, there have been many times where I have been the only woman in the room, but through the years I have seen a significant change in the workforce as more and more women choose exciting careers in engineering, to the point where I have also been part of all-female engineering teams.

My interest in studying engineering was thanks to early exposure to it. My dad was a communications engineer who worked at the Mexican national telephone company. In his career, he was involved in many milestones that transformed people's lives, from replacing vintage rotary telephones to modern phones with buttons and voicemail, the advent of the Internet and cell phones, and implementing city-wide surveillance systems for vehicle traffic control. He instilled my interest in engineering by bringing home magazines that highlighted technological advances. My favourite one was a Concorde special edition with a two-page picture of the

C. S. Vazquez (✉)
Enbridge, Calgary, AB, Canada

M. Bailey, L. Shackelford (eds.), *Women in Mechanical Engineering*, Women in Engineering and Science, https://doi.org/10.1007/978-3-030-91546-9_13

supersonic plane being inspected by an aeronautics female engineer. This image made me think how cool engineers are!

But what finally defined my choice was a scholarship to study at a French high school, where I took mechanical and electronic courses in grades 11 and 12. I remember having lots of fun learning basic programming, loading the program into a machine and seeing how the machine would follow my commands. So when my time came to choose an undergraduate program, I wanted to learn how to create solutions to problems in everyday life using the combined knowledge of mechanical and electronics engineering. A novel program called Mechatronics Engineering was offered at the Universidad Nacional Autonoma de Mexico and I knew this is what I wanted to study.

A Young Engineer in Oil and Gas

When I graduated from university, I was interested in industrial automation, so I accepted a rotational trainee position at a company recognized as one of the world's industry automation leaders. In my first year, I worked on short assignments in several business units in Mexico City, and at the end of it, I was offered a permanent position as a Project Engineer in the Oil and Gas Department. I confess that the oil and gas industry is not one that I was particularly interested in, mostly because I did not know much about it, except that it is a main contributor to the economy in North America. My boss-to-be explained to me that I could participate in automation projects in this sector, so with all the excitement of a young engineer eager to learn everything, I gladly accepted the position. This is how my journey in the energy industry started.

Shortly after, I had many opportunities to visit oil and gas facilities in the country. Wearing personal protective equipment such as coveralls, steel-toe boots, helmet, and safety glasses was always a requirement and quickly became a normal part of my job. My earliest visit was to a refinery where we were to commission a gas turbine. When we arrived, I remember feeling appalled because my presence was not acknowledged by any of the all-male workers who addressed all their conversation only to the senior engineer I was with. As it was my first site experience, I did not know how to handle the uncomfortable situation so I said nothing. Later, I was shocked when I realized that there were no female washrooms near my worksite, while the men's washrooms were just a few steps away. Seriously? No women's washrooms around in 2010? Certainly, these anecdotes were uninviting for someone new to the industry, but I was not going to let those minor incidents get in my way of gaining valuable engineering experience. If anything, they only made my skin a bit thicker.

For the next 3 years, I worked on a variety of projects. The most memorable one for me, due to its magnitude, was the turnkey construction project of a combined cycle plant for a mine. In a nutshell, the project would allow this client to produce its own energy, and we were in charge of building the plant from the ground up. I was managing the project coordination, which was important to facilitate work on site, but I was still interested in understanding the nuts and bolts of the plant. However, I noticed that the engineers in charge of designing the plant and giving technical recommendations were based in the USA and Germany. I wanted to become an engineer like them, capable of answering detailed technical questions and I realized that I would need to prepare further. So, I looked for the best engineering graduate programs in the world and I found a dream graduate student and research assistant opportunity in Vancouver, Canada, in the Department of Materials Engineering at the University of British Columbia.

Corrosion Is Not 'Bad', Corrosion Is 'Natural'

My first interaction with materials engineering was during my undergraduate years. It blew my mind to learn that the atoms of metallic materials are arranged in crystalline structures and that these structures can change with processing, which results in different mechanical properties. Material engineering students also learn about the stress–strain curve, which describes the mechanical behaviour of materials when an external load is applied to them. The explanation of this curve can be found elsewhere [1], but for this chapter, suffice to say that most metals can be stretched like an elastic band to a certain limit, called the yield point. If the external load is removed before the material reaches the yield point, then metals return to their original shape (formally known as Hooke's Law). However, if the metal is stretched beyond the yield point, the material will be permanently deformed. Furthermore, if the external load continues to be applied, there will be a point where the material reaches its maximum strength, known as ultimate tensile strength, and then will inevitably break.

These basic concepts are fundamental to designing engineered structures because any design begins with material selection and stress–strain calculations to ensure the structure will have the ability to withstand the required service conditions. This is valid whether we refer to planes, ships, bridges, pipelines, or other engineering marvels.

But what happens with materials over time? I learned more about this in my graduate years when I was engaged in corrosion research [2]. Before being educated in corrosion science, I used to associate corrosion with red rust on steel and to characterize it as 'bad'. However, I have learned that corrosion is simply a 'natural process'. This was important to understand because it is the foundation of my current role as an integrity engineer, where we aim to protect ageing infrastructure against corrosion. But let me elaborate.

Very few metals exist in nature as we know them; gold is one of them. Most metals are found in nature in the form of oxides on rocks. To obtain these metals, intensive extraction and processing are required. Such processes place the material in a higher energy level than it had in its natural form. But just as a ball at a high altitude tends to drop to the ground when it is released, materials at a high energy level 'want' to return to a lower energy state. This tendency is what produces corrosion, which is nothing else than the metal naturally returning to its original oxide form.

As engineers, we see corrosion as detrimental because it reduces the design life of engineered products. The good news is that corrosion can be managed. How? Let's first take a peek at the corrosion process.

Corrosion is an electrochemical process that requires four elements in order to take place as shown in Fig. 1 [3]:

1. Anode is the component that corrodes, it loses cations (ions with net positive charge) and electrons, which results in loss of material.
2. Cathode is the component that receives the electrons lost by the anode.
3. Ionic path is the electrolyte that facilitates the transfer of ions.

   Water is the best example of the ionic path, but it is not the only one. Other examples are blood that surrounds metallic implants in the human body [4] and concrete that covers steel reinforcements in a bridge [5].
4. Electronic path is the path through which the electrons lost by the anode travel to the cathode. This electronic path is provided by metals themselves, because they are excellent conductors.

Corrosion can be stopped by removing one of these elements. For example, a coating adequately adhered to the substrate material will be a barrier between the metal and the ionic path in the environment and hence prevent corrosion from happening. But if the coating gets damaged, the four corrosion elements may be at play again.

After gaining more knowledge in materials engineering and corrosion, I was ready to go back to work in industry. This time, I found an opportunity as a Pipeline Engineer in Calgary, Alberta. I was ecstatic about my first professional job in Canada. As I had to relocate, when I shared my good news and bid farewell to my

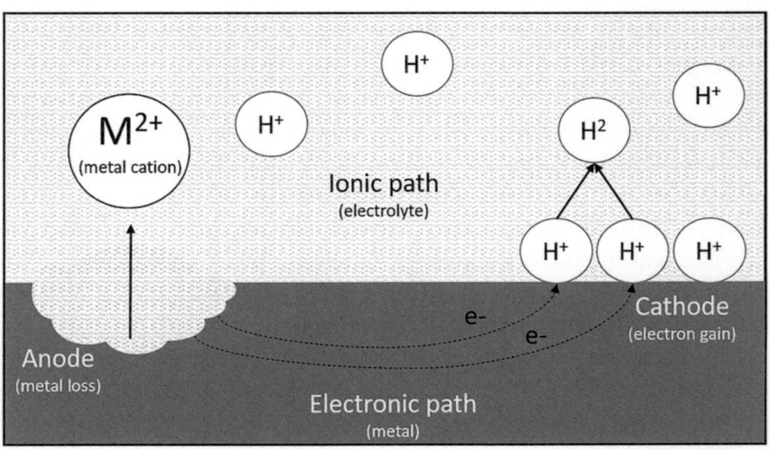

Fig. 1 Four elements in the corrosion process

friends, I was shocked when one of them said: “A pipeline engineer? So, do you want to pollute the environment?”

“What? Where does the association between being a pipeline engineer and polluting the environment come from?”

I hope that through sharing some personal stories and learning from almost 7 years working in this industry, I am able to address that pipeline engineers care about the environment and do everything in their hands to protect it.

Delivering Energy

Pipelines are the arteries that deliver the energy that our daily life demands. Consider that every time we wear manufactured clothes, cook our food, turn the lights on, work on a computer, make a cell phone call, make our daily commute, get health services at the hospital, take a flight, etc., in every single moment of our lives, our quality of life depends on energy. Traditionally, oil and gas have been the primary sources to generate this energy. Even now, despite the rise of renewables, fossil fuels continue to represent the main share in the energy mix to meet global energy demand. In 2018, they accounted for more than 60% of total world gross electricity production [6] and it is expected that the global oil demand will continue to grow at least for the next decade before flattening out.

However, these fuels need to be transported from the remote sites where they are produced to the locations where they will be further processed and then consumed. Pipelines are the safest and most reliable method to transport oil and gas [7].

While there are various types of pipelines, the remainder of this chapter focuses on transmission pipelines, which are superhighways that travel hundreds to thousands of miles to carry the natural gas or crude oil from the sites where they are produced to refineries.

The transmission pipeline infrastructure in North America consists of more than 117,000 km (72,700 mi) in Canada [8], 620,000 km (385,250 mi) in the USA [9] and 27,000 km (16,777 mi) in Mexico [10, 11], according to public records of government agencies in each country. If these pipelines were aligned one after the other, their length would be almost enough to get to the Moon and back. Even if we do not see them, pipeline networks are ubiquitous. The reason why we do not see them is because, generally, they are buried. But pipeline networks are as geographically spread out as telecommunication and electricity networks.

Pipelines are stringently regulated throughout their life cycle. No pipeline can be built without the regulators' approval, and in order to obtain it, a series of strict conditions must be met. These conditions include landowner and indigenous consultations, environmental and archaeological studies, and technical requirements [12, 13], to name a few. But in recent years, operators need to also consider acceptance from society, from which there is increasing scrutiny and concern. My friend who asked if I wanted to pollute the environment was only the first person of many who have questioned what I do. I think it is a fair question and people have the right to be informed, so I invite you to keep reading.

Winter Pipeline Construction

A few years ago, when I was still fairly new to the pipeline industry, I had the great opportunity to participate in the construction project of a natural gas pipeline. With the ever-decreasing number of approved transmission pipeline projects, this was an experience that I feel lucky and proud to count under my belt.

The segment we built was a short one (about 30 km or 18 miles) but it had many of the technical challenges that one can imagine. This pipeline was to be constructed in a mountainous region, crossing sensitive ecological areas, farmers' fields, several creeks, an important river, and a highway. To make things even more complicated, this was a winter construction project in northern Canada. For those of you unfamiliar with Canadian winters, they are bitter. Temperatures on any given day at that latitude can vary between −15 °C (5 °F) and – 25 °C (−13 °F) and some days during that project, they dropped to less than −30 °C (−22 °F). With the humidity in the region, it feels even colder. There is no way to describe this cold to someone who has not experienced it, but instead of feeling cold, the skin feels pain, it aches, and if you are not careful, exposure of the skin over a prolonged time can cause frostbite. On top of that, with the constant snowfalls, the pipeline corridor was buried under at least a foot of snow.

There I was, the most junior member of the team, the only female engineer and the only non-Canadian participating in my first pipeline construction project. After having discussed the job duties with my supervisor, I was eager to travel to that remote location and spend a few months rotating every couple of weeks in and out of a town with a population of 2000 people to be part of this project and learn as much as possible. Only, there was a problem. This town was a 3-h drive from the closest airport and, being from Mexico, I had never driven in wintery conditions. Being surrounded by colleagues who consider snowy conditions as a normal part of life, I felt slightly apprehensive to speak up about my fear of winter driving. Nonetheless, I was not going to let this fear step in my way of gaining valuable site experience. Furthermore, I had been constantly reminded ever since I started working in the pipeline industry that 'safety is first'. So, I spoke with my supervisor about my concern with my lack of experience with winter driving skills and felt relieved when he showed empathy and understanding. He immediately registered me in a winter driving course and developed a communications plan with the team on site. Once I finished the course, I gained confidence about driving myself to the site. But also, I felt protected by the communications plan: I was to inform the site office when I was about to start driving from the airport and they would be expecting to see me within 3 h. If for any reason I did not show up or failed to communicate with them, they would send someone to look for me in case I had driven into a ditch in an area where there was no cell phone reception. Luckily, this plan never had to be implemented because I always made it safe and sound to the project site and back to the airport.

Every workday would kick off at 6 a.m. with the construction meeting where all inspectors would report to the construction manager. Lacking any site experience, I thought that people started to gather at 6 a.m. and perhaps have some informal conversations before going to the construction corridor because it was too early and too cold. So, the first day, I arrived at 6 a.m. to the office trailer site and I was surprised because all the parking spots were already taken! It took me a while to drive to the far end of the parking lot, make myself a parking spot on a pile of fresh snow and then walk through the thick layer of snow to the trailer where the meeting was held. Needless to say that I made a big entrance. I arrived about 20 min late and when I opened the door, I felt all eyes on me. There must have been 30 or more older men in the room. I sat at the nearest chair I found, at the far back. The purpose of the daily construction meetings is for every person in the room to report their progress of the previous day and the challenges for the present day. The chief inspector did the roll call, but he missed asking me if I had anything to say, maybe he did not see me. But a voice at the other side of the room pointed out his omission. The chief inspector excused himself for it and gave me a chance to speak. As it was my very first day on site, I did not have anything to report, really, so I simply introduced myself to the crowd. The voice that spoke up for me belonged to the other woman in the room that I had not seen before. She was older than me and much more experienced in fieldwork; she was a safety inspector. Her gesture was extremely powerful to me because she showed me that even if I had nothing to say and I was new, I had the right to be acknowledged. I deeply appreciated it.

After a few days arriving earlier each time, I realized that 5:30 a.m. was the best time to start the day to have the chance to discuss face to face with inspectors any questions before they headed to the pipeline corridor to continue with construction activities. The workdays were 12-h journeys that started at 6 a.m. sharp from Monday to Saturday. Despite the long work hours, time on site evaporated between the many challenges that arose.

My duty as a field engineer was to resolve any design or engineering questions from the construction crews or inspectors. This was a huge responsibility! Although there were other engineers on site, I was the only one from the engineering department designated to be the first line of response to an assortment of queries, all important and urgent. But I was not alone, similar to the winter driving experience, I could always reach out to senior design engineers and subject matter experts in the head office to discuss the issue and provide direction to the field. By addressing those questions, I grew a bit wiser every day.

Sometimes the questions could be fairly easy to answer, such as what are the applicable construction drawings and procedures for a given section of the line. Other times, the questions could be very specific, such as what is the voltage range that should be used for *jeeping* Fusion Bonded Epoxy coating. *Jeeping?* Is this English? It turns out, that the pipe comes in short sections that are coated at the plant by the pipe manufacturer. Each joint is welded on site to the next joint and so on and welds are non-destructively inspected to ensure their good quality. Before the longer pipe section is laid down in the trench that has previously been excavated for it, the coating is checked to ensure that it has not been damaged by scratches produced during construction because any damage can remove the protectiveness of the coating. This coating inspection procedure is called *jeeping* [14]. If *holidays*, which are minute discontinuities in the coating and not visible by the bare eye, are found beyond the acceptance criteria, the coating needs to be repaired. The objective is that the welding and coating conditions are optimum before burying the pipeline.

Other issues were more difficult to resolve, even for the seasoned construction manager and chief inspector on site, and required a multidisciplinary approach. For example, the crossing of a river. In a similar way to the rest of the pipeline corridor, the depth of cover, which is the thickness of the soil layer between the top of the pipe and the soil surface, must meet certain requirements to protect the pipeline from damage by external factors. One impressive trenchless method that exists for installing pipelines beneath obstructions, such as river crossings, is Horizontal Directional Drilling [15]. This process entails boring a path for the pipeline by launching a directional drill from the entry point on one side of the river to the exit point on the other side, without disturbing the ground surface. Once the bore is completed, the pipeline segment is pulled through. With this method, the pipeline crosses several meters underneath the river and is protected from threats such as water erosion, or being hit by rocks or debris carried by the river during flash floods. This way, the pipeline is protected and therefore the river itself is protected from a potential release of the products transported by the pipeline.

Towards the end of the construction project, when the entire pipeline has been built, buried and ready to start operation, a pressure test is conducted. In this test, the pipeline is filled up with a fluid, typically water, and the pressure is gradually increased to beyond the intended operating pressure of the line by a safety factor but less than the burst pressure that is calculated based on the strain–stress principles discussed earlier. The purpose of this test is to identify any possible imperfection remaining on the line that could pose a threat in the future. If there were any manufacturing or construction defects, the pressure test would uncover them. If the pressure test is successful, the pipeline operator sends all the required information to the regulator and once approved, the pipeline can begin operating. The pipeline we built is one segment of a natural gas system that delivers energy to the Vancouver metropolitan area and the United States.

By the time the construction project was completed, I did not feel like a foreigner anymore, but rather an important member of the core project team. Being a field engineer was a phenomenal learning experience and one that marked my future approach with field personnel. Now that I work mostly in the office, when I receive a call from them, I consider it of high importance because I know that timely responses are of the essence.

How Are Pipelines Inspected?

After the pipeline construction ended, I moved to a new role as a pipeline integrity engineer, where I have developed for over 5 years. Pipeline integrity is a branch of engineering that focuses on identifying and managing threats to the pipeline for the operating life of the asset, as explained herein [16]. In simple terms, pipeline integrity engineers are like pipeline doctors because analogously as the human body requires medical attention for diverse reasons throughout a person's lifetime, pipeline infrastructure also needs to be inspected and maintained to continue operating safely for decades.

Pipeline integrity engineers are well aware that a pipeline incident, such as a leak or rupture, could have catastrophic consequences for people and the environment. Our goal is to prevent any incident from happening. Given the diversity of pipeline threats, integrity engineers must be specialized, just as doctors are. Some integrity engineers specialize in managing external forces of geotechnical or hydro-technical nature; others focus on providing adequate cathodic protection to the pipeline, which is an additional layer of protection against corrosion; and others, like me, focus on guaranteeing that the condition of the steel the pipeline is made of is appropriate for continued service. The remainder of this section focuses on the latter type of integrity engineering.

Pipelines are fully inspected at selected frequencies, which usually span a few years, with in-line inspection (ILI) tools. As their name suggests, all in-line inspection tools travel inside the pipeline and cannot be seen from the ground surface because pipelines are generally buried. ILI tools are cylindrical-shaped machines that carry a multitude of sensors to collect vast amounts of data about the condition of the steel pipe. ILI tools detect, identify and size imperfections and provide their specific dimensions (length, width, and depth) and exact position (clock position, relative and absolute distances, and coordinates).

Due to the varied nature of possible material imperfections, ILI tools are also specialized in detecting and identifying specific type of imperfections, as listed in Table 1. ILI tools are powerful instruments because they provide information at every inch over the hundreds of miles of the pipeline. Integrity engineers use the data of the steel imperfections reported by the tools to assess if any imperfections are of concern. To do this, we perform calculations to determine the burst pressure at the corrosion points, the strain on the pipeline at deformations and predict the time to failure of any imperfection to determine which ones need to be repaired.

All imperfections that have the potential to initiate a failure are scheduled to be repaired promptly. In these cases, the pipeline is excavated, the coating is removed, the pipeline is sandblasted and inspected with non-destructive examination methods such as magnetic particle inspection, radiography, ultrasonic, etc. After a thorough inspection is completed and sufficient data exists, the appropriate repair method is selected. Once the repair is completed, the pipe is recoated and reburied.

On a day-to-day basis, I assess every inch of the line and it is my responsibility to make educated and informed decisions that ensure the integrity of the pipeline

Table 1 ILI tools and pipeline threats

ILI tool technology	Description	Threat	What causes it?	Why is this a threat?
Magnetic flux leakage (MFL)	Primarily designed to identify metal loss in the pipeline Most of the time, metal loss is due to corrosion but in some instances it may also be caused by scratches from machinery MFL tools can discriminate whether the imperfection is on the external or the internal surface of the pipeline	Corrosion (internal and external)	*External corrosion* may be caused by damaged or disbonded coating that no longer isolates the steel from the environment or it may be an indication that the cathodic protection system is not being effective *Internal corrosion* may be a symptom of water being transported in the fluid or *microbiologically influenced corrosion*, which results from the metabolic processes of bacteria that form colonies adhered to the pipeline	A loss of metal translates into a reduction of the thickness of the pipe wall. Therefore, the maximum pressure that a given pipeline can withstand is reduced in the areas where corrosion exists
Caliper	Primarily designed to identify geometric deformations	Deformations (dents, buckles, wrinkles, ovalities)	*Dents* may be caused by equipment that accidentally hits the pipe, or sometimes by rocks that were buried under the pipeline and were not seen at the time of construction but overtime entered in contact with the pipe and dented it *Buckles* are caused by combinations of loads *Wrinkles* may be a symptom of ground movement that applies large forces on the pipe	Deformations might introduce additional strain, pushing the pipe properties closer to the point of failure [13]

(continued)

Table 1 (continued)

ILI tool technology	Description	Threat	What causes it?	Why is this a threat?
Inertial mapping units (IMU)	IMU tools provide positioning (coordinates) and altitude along the pipeline trajectory	Pipeline displacement	Ground movement	Pipeline displacement from its original centreline may introduce high strain and potentially rupture the line
Electromagnetic acoustic transducer (EMAT)	Designed to identify cracks	*Cracks* are brittle linear indications with sharp tips and no volumetric loss of steel	Various mechanisms (e.g. environment, fatigue, mill defects)	*Cracks* could initiate brittle fractures
Hard spot (HS)	Designed to identify hard spots	*Hard spots* are areas of higher hardness compared to the rest of the pipe and they may reside in the pipeline from the time it was manufactured	Localized quenching of the steel plate during the hot rolling process. Quenching is a rapid drop in temperature produced by droplets of water splashed on the hot steel	*Hard spots* are not a threat by themselves, but they make the pipe susceptible to *hydrogen stress cracking* if they are subjected to sufficiently high stress levels in the presence of atomic hydrogen [18]

and, hence, guarantee the safety of the public and the environment. For this reason, I believe it is a profession to be proud of.

Future Opportunities

As the millennial that I am, I acknowledge the challenge posed by climate change and I aim to be part of the transition to a net-zero energy future. As a pipeline engineer who has been confronted by people questioning if I want to pollute the environment, my answer is no. Engineers who work in the pipeline industry do not want to pollute the environment. On the contrary, we are committed to using the best of our collective expertise to safely deliver energy using the best methods available.

Looking into the future, I am convinced that pipelines will continue to transport energy for decades to come and they will contribute towards a decarbonized future [17]. How? Over the past year, various governments have launched initiatives to advance the use of hydrogen as a fuel that has the advantage of producing energy without CO_2 emissions. In alignment with this vision, Canada has already piloted pipeline projects that transport blended hydrogen [18, 19], while Europe is proposing a continent-wide pipeline network that will supply hydrogen in the next two decades [20, 21].

To finalize, I hope that this chapter has planted a seed of curiosity about the various aspects of pipeline engineering and reassured the public about the careful work done to safely operate them. Lastly, I wish that the energy sector, which is the least gender-diverse with only 22% of women in the global workforce [22], will gain more female engineers that choose to take the lead to generate and deliver the energy of the future.

References

1. W. D. J. Callister and D. G. Rethwisch, *Materials Science and Engineering: An Introduction*, 8th ed. 2010.
2. C. Soriano and A. Alfantazi, "Corrosion behavior of galvanized steel due to typical soil organics," *Constr. Build. Mater.*, vol. 102, pp. 904–912, Jan. 2016, https://doi.org/10.1016/j.conbuildmat.2015.11.009.
3. D. A. Jones, *Principles and prevention of corrosion*, 2nd ed. Upper Saddle River, NJ: Prentice Hall, 1996.
4. S. Karimi and A. Alfantazi, "Ion release and surface oxide composition of AISI 316L, Co-28Cr-6Mo, and Ti-6Al-4V alloys immersed in human serum albumin solutions," *Mater. Sci. Eng.*, vol. 40C, pp. 435–444, 2014.
5. B. S. Berke, A. A. Sagues, and R. G. Powers, "Long term corrosion performance of soil reinforcement in mechanically stabilized earth walls," in *Corrosion 2008, March 16, 2008 – March 20*, 2008, pp. 83191–83199.
6. IEA, "Electricity Information: Overview," International Energy Agency, Paris, France, 2020. [Online]. Available: https://www.iea.org/reports/electricity-information-overview.
7. Canadian Energy Pipeline Association. "Pipeline Performance". pr20cepa.com. https://pr20.cepa.com/pipeline-performance/. (accessed Jan. 20, 2021).
8. Government of Canada. "Pipelines Across Canada". Canada.ca. https://www.nrcan.gc.ca/our-natural-resources/energy-sources-distribution/clean-fossil-fuels/pipelines/pipelines-across-canada/18856 (accessed Dec. 12, 2020).
9. U.S. Department of Transportation. "Pipeline Mileage and Facilities." phmsa.dot.gov. https://www.phmsa.dot.gov/data-and-statistics/pipeline/pipeline-mileage-and-facilities (accessed Dec. 12, 2020).
10. Petroleos Mexicanos. "Descripción General de los Sistemas". pemex.com https://www.pemex.com/nuestro-negocio/logistica/ductos/Paginas/default.aspx (accessed Dec. 13, 2020).
11. Centro Nacional de Control de Gas Natural. "Sistema de Transporte y Almacenamiento Nacional Integrado de Gas Natural". cenagas.gob.mx https://boletin-gestor.cenagas.gob.mx/ (accessed Dec. 13, 2020).
12. CFR, "Code of Federal Regulations Title 49 Part 192." U.S. Government Publishing Office, 2019, [Online]. Available: https://www.ecfr.gov/cgi-bin/text-idx?SID=5be5f3c8026687564c41086e1a74dbaf&mc=true&node=pt49.3.192&rgn=div5.
13. *Oil & Gas Pipeline Systems*, CSA Z662:19, CSA Group, Ontario, Canada, 2019.
14. *Discontinuity (Holiday) Testing of New Protective Coatings on Conductive Substrates*, Standard Practice 0188, NACE, 2006.

15. J. D. Hair & Associates, "Installation of Pipelines by Horizontal Directional Drilling Engineering Design Guide," Pipeline Research Council International, PR-277-144507-E01, Sep. 23, 2015.
16. *Managing System Integrity of Gas Pipelines,* B31.8S, ASME, 2018.
17. Enbridge, "Net Zero By 2050 Pathways to reducing our emissions." Enbridge Inc., Calgary, 2020.
18. The Canadian Press, "Enbridge Gas to blend hydrogen with natural gas for consumers in Markham, Ont.," thestar.com. https://www.thestar.com/business/2020/11/18/enbridge-gas-to-blend-hydrogen-with-natural-gas-for-consumers-in-markham-ont.html (accessed Nov. 18, 2020).
19. T. Seskus. "Hydrogen-injected natural gas to heat homes in Alberta city next year." cbc.ca. https://www.cbc.ca/news/canada/calgary/alberta-hydrogen-home-heating-1.5657736 (accessed Oct. 21, 2020).
20. European Commission, "A hydrogen strategy for a climate-neutral Europe," Brussels, 2020. [Online]. Available: https://knowledge4policy.ec.europa.eu/publication/communication-com2020301-hydrogen-strategy-climate-neutral-europe_en.
21. A. Wang, K. van der Leun, D. Peters, and M. Buseman, "European Hydrogen Backbone," 2020. [Online]. Available: https://gasforclimate2050.eu/sdm_downloads/european-hydrogen-backbone/.
22. IEA, "Energy and gender A critical issue in energy sector employment and energy access," 2019. [Online]. Available: https://www.iea.org/topics/energy-and-gender.

Claudia Soriano Vazquez is a Senior Pipeline Integrity Engineer at Enbridge and Vice Chair of the ASME Southern Alberta Section. Claudia is a Professional Engineer, registered in the province of Alberta. For over a decade, she has worked in energy, pipeline, and mining projects in Mexico, the USA, and Canada through her various roles at Siemens, Spectra Energy, Golder Associates, and Enbridge.

Claudia graduated with honors with B.Sc. in Mechatronics Engineering at the Universidad Nacional Autonoma de Mexico in 2009 and with the Master of Applied Sciences in Materials Engineering at the University of British Columbia in 2015. She is fluent in Spanish, English, and French.

Claudia is also an advocate for STEM education. She is a mentor in the "Women Leaders in STEAM" program from the US-Mexico Leaders Network, where she shares her passion for engineering with high school girls.

In her spare time, she enjoys hiking in the rocky mountains.

Intelligent Control to Reduce Vehicle Energy Consumption and Greenhouse Gas Emissions

Bo Chen

Introduction

The transportation sector is a large energy consumer and one of the largest contributors to the US greenhouse gas (GHG) emissions. Energy use for transportation was about 26% of the total US energy consumption in 2020 [1]. In addition, transportation accounted for the largest portion (29%) of the total US GHG emissions in 2019 [2]. To reduce the energy consumption and emissions of the transportation sector, the market for electric vehicles (EVs) and hybrid electric vehicles (HEVs) is growing at an unprecedented pace. The electrified transportation presents two areas of research trends: (1) building efficient EVs/HEVs and EV charging infrastructure to form an EV ecosystem for all-electric future; (2) exploring advanced technologies using vehicle-to-vehicle (V2V) and vehicle-to-infrastructure (V2I) communications to further reduce energy consumption of connected and automated vehicles (CAV). My recent research in automotive controls aligns with these two emerging areas.

Prior to joining Michigan Tech in 2007, I was a postdoctoral associate in the department of Mechanical and Aerospace Engineering at the University of California, Davis. My postdoctoral research primarily lies in intelligent agent systems. An intelligent agent is an autonomous computational entity that is able to perceive, reason, and initiate activities in its environment. It is intelligent and adaptive, usually programmed with artificial intelligence approaches. It can communicate with other agents, work cooperatively, and take autonomous actions in order to achieve its design goals. I developed a mobile agent system called Mobile-C [3]. Mobile-C supports mobile agents by integrating an embeddable C/C++ interpreter into an agent system platform as a mobile agent execution engine. Mobile agents are

B. Chen (✉)
Department of Mechanical Engineering – Engineering Mechanics,
Department of Electrical and Computer Engineering, Michigan Technological University, Houghton, MI, USA
e-mail: bochen@mtu.edu

M. Bailey, L. Shackelford (eds.), *Women in Mechanical Engineering*, Women in Engineering and Science, https://doi.org/10.1007/978-3-030-91546-9_14

intelligent agents that can move from one computer to another in a network, while being executed in multiple computers. Mobile-C provides communication services to support agent communication and migration through message passing. With mobile agent systems, a control network is able to adopt newly developed control algorithms and make adjustments in response to operational or task changes. To apply Mobile-C in intelligent transportation systems, a comprehensive literature review has been conducted and resulted in a survey paper [4] (Best Survey Paper Award) that has been cited over 680 times. Mobile-C has been applied for distributed traffic detection and management to enhance the flexibility and reduce raw data transmission [5]. After joining Michigan Tech, I received research funding from the National Science Foundation to explore fundamental scientific issues that could potentially lead to adaptive sensing and monitoring based on agent technology and immune-inspired pattern recognition methods. My research group established a mobile agent–based monitoring paradigm to overcome the major limitations of a wireless monitoring network, such as the adaptability and communication bandwidth. A mobile agent–based monitoring network was developed by integrating Mobile-C with high-computational power-sensing hardware [6]. The mobile agent–based monitoring network moves detection algorithms instead of raw sensor data, which significantly reduces the amount of data transferred. Multi-objective optimization algorithms were developed to optimally control the generation and distribution of mobile monitoring agents [7]. The agent distribution was controlled to increase the detection probability and extend the network lifetime. The immune-based pattern recognition algorithms for damage detection, classification, and emergent pattern recognition have also been developed [8, 9]. The developed pattern recognition algorithms mimic immune recognition mechanisms that possess adaptation, evolution, and immune learning capabilities. The damage patterns are represented by feature vectors that are extracted from structure's dynamic measurements. The training process is designed based on the clonal selection principle of the immune system, which allows the classifier to improve the quality of representative feature vectors based on input data. One paper based on this research work received a Best Paper Award at 2010 IEEE/ASME International Conference on Mechatronic and Embedded Systems and Applications.

Shortly after joining Michigan Tech, I realized that it is important to expand my research areas so I can submit proposals to more funding agencies. I also noticed that research collaboration is common in large research projects. To write collaborative research proposals, I actively searched the potential collaborators who have synergy with my research expertise. An opportunity arose in a project in which I collaborated with colleagues at Michigan Tech to develop engine controllers for Nostrum Energy LLC. The research area is very different from my prior research in intelligent agent systems. Through the collaboration, I not only learned required knowledge in internal combustion engines, but also gained research experience in conducting industrial research projects. This collaboration also helped me to build long-lasting research collaborations, which has resulted in multiple large research projects funded by the Department of Energy and industrial partners. These projects include an Interdisciplinary Program for Education and Outreach in Transportation

Electrification; MTU subcontract for Ford DOE Project: Advanced Gasoline Turbocharged Direct Injection Engine Development; NEXTCAR: Connected and Automated Control for Vehicle Dynamics and Powertrain Operation on a Light-Duty Multi-Mode Hybrid Electric Vehicle; Energy Optimization of Light and Heavy-Duty Vehicle Cohorts of Mixed Connectivity, Automation and Propulsion System Capabilities via Meshed V2V–V2I and Expanded Data Sharing; and ARPA-E NEXTCAR Phase II – L4/L5 CAV Enabled Energy Reduction, where L4/L5 stands for level 4/level 5 of vehicle autonomy. My experience has shown that research collaboration is important for junior faculty members to gain research skills necessary to conduct large research projects, expand professional network, and build long-lasting research collaborators. Through these research projects, my research in automotive control has also expanded from small-scale to large-scale automotive control systems, such as the development of engine controller, HEV powertrain supervisory controller, predictive vehicle dynamics and powertrain control, and connected vehicle control for intelligent transportation systems.

In addition to collaboration with colleagues at MTU, I also seek external research collaborations to build long-term partnerships. During my sabbatical leave from 2014 to 2015, I worked with the EV–Smart Grid Interoperability Center at DOE Argonne National Laboratory (ANL) in Chicago. The ANL EV Smart Grid Interoperability Center supports global harmonization of standards and technology for the EV–grid interface, as well as charging interoperability to ensure future electric vehicles and charging stations worldwide work together seamlessly. The work experience at ANL expanded my research area into EV–smart grid integration and provided me opportunities to work with other national laboratories and international organizations, such as the Joint Research Centre of the European Commission.

In the past 14 years, my research group has developed various control algorithms for internal combustion (IC) engines, energy management strategies of hybrid electric vehicles, predictive control systems for connected and automated vehicles, and advanced control strategies for EV grid integration. My research projects have been supported by grants and contracts totaling over ten million dollars from government funding agencies, Argonne National Laboratory, and industry. The research has resulted in over 100 refereed publications in premier archival journals and conference papers and a number of Best Paper Awards/Best Student Paper Awards. My research and educational achievements have been recognized by the American Society of Mechanical Engineers (ASME). I have been elected a Fellow of ASME in 2020.

This chapter introduces my research and educational efforts in the automotive control area to reduce vehicle energy consumption and greenhouse gas emissions. The rest of the chapter is organized as follows. Section "Research on developing intelligent control systems for advancing vehicle electrification" introduces my research work on control development for HEVs, connected and automated vehicles, and electric vehicle–grid integration. Section "Educational effort to prepare students for transportation electrification" presents my educational effort for transportation electrification. Section "Conclusions" concludes the chapter.

Research on Developing Intelligent Control Systems for Advancing Vehicle Electrification

This section introduces two of my research areas: vehicle control and electric vehicle–grid integration. The scope of vehicle control ranges from powertrain component control such as IC engine control, HEV control, connected and automated vehicle control, and energy optimization of vehicle cohorts. The electric vehicle–grid integration research studies the impact of EV charging on grid and develops optimal control strategies to mitigate this impact.

Advanced Vehicle/Powertrain Control and Predictive Control of Connected and Automated Vehicles

Over the past decade, my research group has been collaborating with Ford Motor Company and Nostrum Energy LLC to develop advanced control strategies for IC engines. The group has developed stochastic knock detection and control strategies for a Ford EcoBoost engine. The developed algorithms have been integrated with a production engine controller for the real-time knock detection and control on an engine test cell [10]. A recent work in combustion engine control is a model-based control system for cycle-by-cycle control of a gasoline turbocharged direct injection spark-ignition engine using an economic nonlinear model predictive controller (E-NMPC) [11, 12]. The E-NMPC engine control system is designed to meet driver-requested torque output, minimize fuel consumption, and reduce NOx emissions. These control objectives are achieved by controlling throttle position, spark timing, intake and exhaust valve phasing, and wastegate position with the consideration of engine operating constraints, including both physical limitations of actuators and thresholds of abnormal combustion metrics comprising high variation of indicated mean effective pressure and combustion knock.

With the experience gained from engine control projects, I have learned working principles of engines and vehicles, which enables me to expand my research scope to HEV control. In the hybrid electric vehicle control area, my group has studied various optimal control strategies for HEV energy management, such as Equivalent Consumption Management Strategy (ECMS), adaptive ECMS, dynamic programming, and model predictive control. The group has received three Best Student Paper Awards for HEV modeling and control. One example of this area of work is the development of a predicative HEV energy management strategy with the consideration of both fuel economy and quantified lithium-ion battery aging [13]. In this work, the battery aging rate is quantified based on an electrochemical lithium-ion battery model, which is able to provide quantified aging characteristics for the increase of internal resistance and decrease of battery capacity. The battery aging factor is quantified by changing factors of the battery, including state of charge (SOC), charging rate and internal temperature of the cell. The energy management system is developed using a nonlinear model predictive control (NMPC) method to find an optimized control sequence over the prediction horizon, which minimizes engine fuel consumption and improves battery aging. The NMPC energy management considering battery aging is compared to an NMPC that does not consider the battery aging. It is found that with the optimized weighting factor selection, the NMPC with the consideration of battery aging has better battery aging performance and similar fuel economy performance compared to the NMPC without the consideration of battery aging. In addition to HEV powertrain energy management, my group has also studied the driving pattern recognition using features such as average cycle speed, acceleration, percentage time of low/medium/high speed, etc. to identify driving scenarios such as urban driving or highway driving [14]. The impact of aggressive driving on HEVs has also been investigated with a focus of powertrain energy flows, the energy consumption of individual powertrain components, their operating regions, and the energy losses of these components.

My recent research on vehicle control has shifted from isolated and reactive control to connected and predictive control. Vehicles on road currently operate in isolation and rely on a human driver to provide high-level dynamic control of the vehicle. The vehicle controllers lack the ability of prediction and automatic adaptation to the changes of traffic and road conditions. With the advancement of V2V and V2I technologies, more and more real-time information regarding traffic and transportation systems will be available to vehicles. As the control thrust leader of the MTU NEXTCAR project funded by DOE ARPA-E, my group has developed a model predictive control system to reduce energy consumption of a multimode plug-in hybrid electric vehicle (PHEV) by leveraging future traffic and road information obtained through V2V and V2I communications. This information is incorporated with vehicle dynamics and constraints for making the control decisions on vehicle-operating mode and powertrain energy management. The model predictive control system is also designed to enable connected and automated vehicle applications such as eco-approach and departure at signalized intersections, platooning, and cooperative adaptive cruise control.

The predictive control system is a multilevel control system as shown in Fig. 1. The predictive control system receives velocity bounds and road grade for a prediction horizon from a cloud computing center. The velocity bounds define a range of velocities within which a connected vehicle can operate. These bounds are generated by a traffic simulator considering real-time traffic condition and safety. The control system outputs can be displayed on a human–machine interface (HMI) to provide real-time feedback to the driver. The top level of the predictive control system is a novel algorithm that uses velocity bounds and powertrain information to generate an optimal velocity trajectory over the prediction horizon [15, 16]. The objectives of the velocity trajectory generation algorithm are to reduce dynamic losses, required tractive force, and to complete trip distance with a given travel time. When applied to a GM Volt-2, the generated velocity trajectory saves fuel compared to baseline energy consumption for a real-world drive cycle for both charge sustaining and charge depleting operations. The charge sustaining mode of a PHEV utilizes a combination of engine and motor power to maintain the SOC at a specified level,

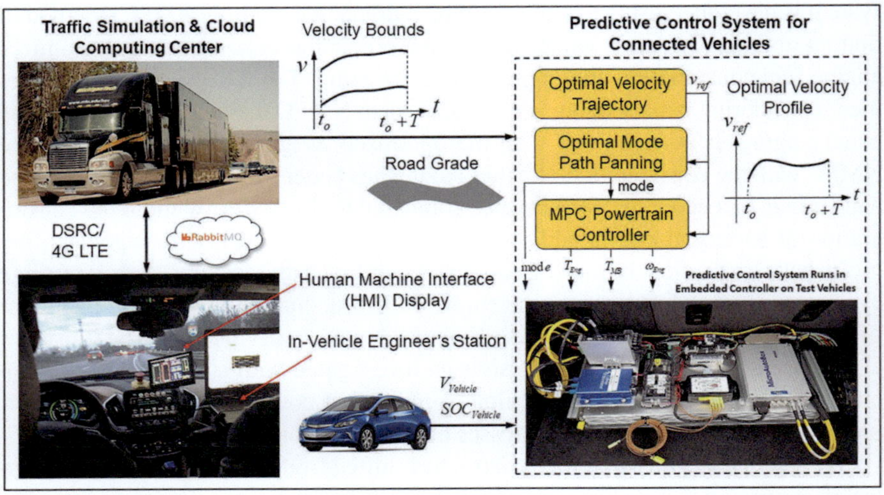

Fig. 1 Overview of MTU NEXTCAR predictive control system for connected vehicles

while the charge-depleting mode operates the vehicle solely on the battery energy. The simulation results show the energy savings ranging from 1.36% to 9.16% for charge sustaining case and 6.91% to 9.63% for charge depleting case. The baseline energy consumption is obtained using logged vehicle velocity profiles for the same drive cycle. The middle level of the predictive control system is the optimal mode selection for the drive unit using a discrete optimal mode path planning (OMPP) algorithm [17]. GM Volt-2 has five models: 1-EV mode, 2-EV mode, low extended range mode, fixed ratio extended range mode, and high extended range mode. The optimal mode at each time instant within a given prediction horizon is selected to minimize the engine fuel consumption, the deviation of the actual SOC from the reference SOC, and the energy required for mode shifts. The bottom-level of the control system is an NMPC power-split controller [18]. With the vehicle velocity trajectory and the drive unit mode selected by the upper levels of the control system, the NMPC power-split controller makes torque-split decisions among two electric motors and one combustion engine such that fuel consumption is minimized while battery SOC and vehicle velocity targets are met. The OMPP algorithm is integrated with NMPC power-split controller in order to create an integrated predictive powertrain controller (IPPC). The IPPC has been extensively tested in simulation across multiple real-world driving cycles where energy savings have been demonstrated. Simulation testing reveals that the IPPC can provide a 4–10% energy savings in standard drive cycles and a 3–7% energy savings over nonstandard, real-world drive cycles. The IPPC has also been deployed and tested in real time on test vehicles equipped with rapid prototyping embedded controllers. Real-time in-vehicle testing shows that the IPPC provides an energy savings of 4–6% over baseline vehicle control while achieving computational turnaround times suitable for real-time control. This work has demonstrated the feasibility of utilizing CAV technologies and predictive controls to reduce the energy consumption of connected vehicles.

My research in vehicle electrification to reduce energy consumption and GHG emissions continues. Currently, I am participating in two recently funded research projects by the Department of Energy. One project, "Energy Optimization of Light and Heavy-Duty Vehicle Cohorts of Mixed Connectivity, Automation and Propulsion System Capabilities via Meshed V2V–V2I and Expanded Data Sharing" aims to reduce energy consumption through expanded V2V–V2I communication and the use of cloud computing and multi-agent optimization. Energy optimization for connected and automated vehicles has been focused for a single vehicle or a fleet of identical vehicles. This MTU project, in partnership with AVL Powertrain Engineering Inc., Borg Warner Inc., Traffic Technology Services Inc., American Center for Mobility and Navistar, will investigate the energy optimization for mixed vehicle cohorts consisting of light and heavy-duty vehicles with various levels of connectivity, driving automation, and propulsion systems.

The second project is MTU NEXTCAR Phase II. The shift to fully autonomous transport is moving forward fast. Based on NEXTCAR Phase I technologies, MTU NEXTCAR Phase II is developing connected and automated vehicle technologies to demonstrate energy saving on level 4/level 5 autonomous vehicles. Partnership with GM and FCA, technologies for level 4 automated vehicles will be added to Chevrolet Bolt EV, 48 V mild HEV RAM1500, and Chrysler Pacifica PHEV. The level 4 vehicle test fleet of the project provides representative propulsion systems: EV to HEV and compact to full-sized truck. The diversity of vehicle and propulsion systems will provide an understanding of the energy reduction potential for level 4 autonomous vehicles with respect to mass, size, and propulsion system.

Electric Vehicle and Smart Grid Integration

Accelerating EV adoption requires more public charging stations nationwide in addition to charging capability at home and at work. The evolution of electrified transportation represents a potentially large growth in electrical load that may impact electrical power grids by contributing to peak loads and changing the loading patterns of electricity distribution equipment if vehicle charging is not properly coordinated and integrated as part of a larger electricity system. I started to do research in this area when I was on sabbatical leave at Argonne National Laboratory from 2014 to 2015. Since the sabbatical leave, my group has been working with ANL EV–Smart Grid Interoperability Center to develop advanced control strategies for effective EV and grid integration. A Hardware-in-the-Loop (HIL) framework, which consists of a real-time power grid simulator Opal-RT, various types of EV charging stations (AC/DC/extreme fast charger), solar panels, and building loads, is developed to validate vehicle-to-grid integration (VGI) communication and control algorithms at different power levels and scales. The HIL system is able to read real-time power generation/consumption measurements, simulate charging impact on power grid, and communicate with charging controllers via cloud communication for the coordination of EV charging with renewable energy generation and building

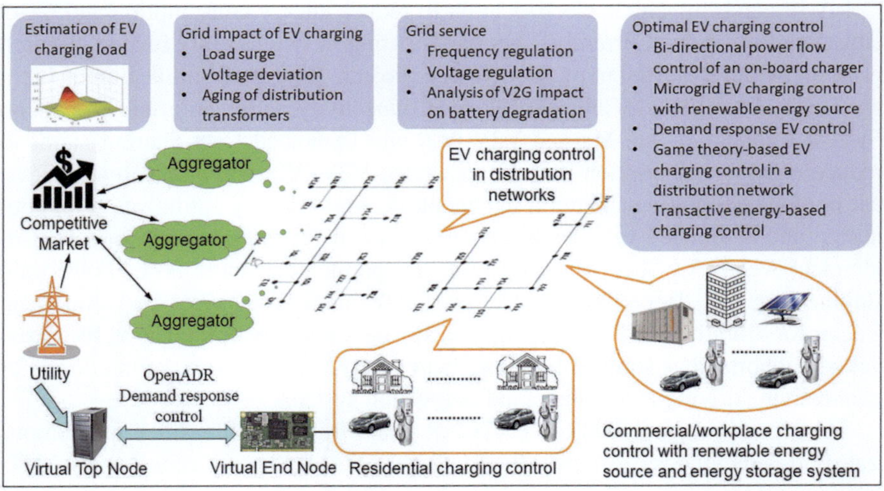

Fig. 2 EV–grid integration research in the Laboratory of Intelligent Mechatronics and Embedded Systems at Michigan Tech

loads to mitigate charging impact. Currently, the work is focused on translating charging behavior of ANL Smart Energy Building to a distributed network model, integrating utility grid, and testing the impact of EV charging on grid stability with EVs at scale.

In the EV–grid integration area, my group has investigated various aspects as shown in Fig. 2, including the estimation of EV charging load, the impact of EV charging on distribution grid, optimal EV charging control strategies to mitigate this impact, and using EV battery to provide grid services such as frequency regulation and voltage regulation [19]. The optimal EV charging control strategies have been developed for different levels and scales, including bidirectional power flow control of an onboard charger [20], plug-in electric vehicle (PEV) charging control in microgrids with renewable energy sources [21] (Best Student Paper Award in 2017), integrating utility demand response control signals for charging control, game theory approach for the management of PEV-charging activities in a distribution network [22], and transactive energy–based charging control [23] (Student Paper Award in 2019), which enables the integration of a market in charging control to achieve a socially optimal solution. Paper [24] studies the impact of aggregated residential and PEV charging load to several aspects of a grid, including load surge, voltage deviation, and the aging of distribution transformers. An optimal charging control method is proposed with the consideration of utility demand response control signals (time-of-use and direct load control) to mitigate the impact. Paper [25] presents a distributed control strategy to solve the power management problem for large-scale PEVs. Paper [26] further considers the use of EV batteries to put the energy back to the grid for regulating grid frequency. This grid service is called vehicle-to-grid (V2G) integration. Unlike stationary energy storage systems, using EVs to provide grid service is difficult because the EVs are movable and the available battery energy (the state of charge of the batteries) depends on individual travel plans. A distributed control scheme is developed that is able to make real-time charging/discharging control decisions based on the current grid frequency deviation, real-time electricity price, and individual EV charging requirements. The control algorithm is able to handle the dynamics of available EVs, the location of EVs, and the grid conditions.

The successful research collaboration with National Laboratories has made me an expert in EV–grid integration. I have contributed to a DOE EV–Smart Grid Integration Requirements Study conducted by multiple National Laboratories to define the engineering requirements for smart grid operations to enable vehicle-to-grid integration in the US. I identified key VGI factors and suggested research opportunities in various aspects of VGI, including system architecture, communication standards, sensing, control, grid management, and cyber security. I have also been invited to attend technical meetings on the cybersecurity of electric vehicle–charging infrastructure organized by several offices at DOE and NIST to identify the gaps and vulnerabilities in this threat space. I have also participated in a series of DOE-organized conversation topic calls on forecasting approaches that address the growing adoption of electric vehicles.

Educational Effort to Prepare Students for Transportation Electrification

I have actively participated in curriculum development and contributed to an Interdisciplinary Program for Education and Outreach in Transportation Electrification funded by DOE. The primary objective of this project is the development of an interdisciplinary curriculum that can lead to a Master of Engineering degree, and graduate and undergraduate certificates in Advanced Electric-Drive Vehicles. I developed a graduate-level course, "Distributed Embedded Control Systems," for this project. The course is co-listed in both Mechanical Engineering–Engineering Mechanics (ME–EM) and Electrical and Computer Engineering (ECE) departments. The course aims at developing an understanding of the model-based embedded control systems design and provides students with hands-on experience that is critical for the control development for hybrid electric vehicles. Several teaching labs and a final project have also been designed allowing students to develop control strategies, using an industrial standard rapid control prototype system. The course has been very well received by graduate students in both ME–EM and ECE departments at Michigan Technological University. This course has been selected as one of the courses of a graduate certificate in Advanced Electric-Drive Vehicles and a graduate certificate in Automotive Systems and Controls.

Requested by graduate students both from ME–EM and ECE departments, I have developed 20 project-based courses related to Vehicle Electrification, including Modeling and Hardware-in-the-Loop Test of Hybrid Vehicle Powertrain Systems, Model Predictive Control for Hybrid Electric Vehicle Powertrain Systems,

CAN Communication for dSPACE HEV HIL Simulation System, Optimization and Scheduling of PEV Charging/Discharging, and Path Planning and Tracking Control for Automatic Parking Assist Systems. Project-based courses offer opportunities for graduate students to learn by actively engaging in real-world engineering projects that they are interested in. Through project-based courses, students develop critical thinking, creativity, and communication skills. Feedback from students reflects that this learning experience is very helpful and valued by the companies that they have interviewed with.

I advise graduate students in both ME–EM and ECE departments. Most of my graduated students work in the automotive industry after graduation. These students are well prepared for the control development for electrified vehicles. They have very good academic records which are evidenced by multiple awards received by them, including four Best Student Paper Awards, one Best Paper Award, one Student Paper Award, one SAE Oral Presentation Award, four Michigan Tech Outstanding Graduate Student Teaching Awards, one Michigan Tech Finishing Fellowship Award, and four Outstanding Scholarship Awards in the Mechanical Engineering Department.

I have contributed to education outreach for undergraduate students. My NSF project "REU Site: Research in Advanced Propulsion and Fuel Technology for Sustainable Transportation," provided the opportunity for 30 highly talented undergraduate students from the US universities to participate in interdisciplinary research projects related to advanced propulsion control and sustainable transportation. REU students spent 40 h per week on research activities over a 10-week period in the summer. REU students also attended professional meetings, seminars, and in-depth information sessions related to applying to graduate school and obtaining funding. These opportunities increased students' interest in conducting research, problem-solving skills, ability to effectively collaborate as part of a team, and improved communication skills. I also mentored several undergraduate students in research projects and summer undergraduate research fellowship programs.

Conclusions

In this chapter, I have presented the research trends of transportation electrification and my professional journey at Michigan Tech. Several emerging research areas in electrified transportation and smart mobility have been discussed, including vehicle electrification, connected and automated vehicles enabled by V2V and V2I communications, EV-charging infrastructure, and EV–grid integration. For smart mobility, vehicle control has shifted from isolated and reactive control to connected and predictive control to allow vehicles automatically adapting to the changes of traffic and road conditions.

My professional career at Michigan Tech spans more than a decade from an Assistant Professor to a Full Professor. I feel that establishing an externally funded research program is important, however, this may be challenging for junior faculty members. From my experience, research collaboration helps to expand research areas, gain research skills for conducting large research projects, learn graduate student mentoring, expand professional networks, and build long-lasting research collaborators.

References

1. "Energy use for transportation." eia.gov. https://www.eia.gov/energyexplained/use-of-energy/transportation.php (accessed July 3, 2021).
2. "Fast facts on transportation greenhouse gas emissions." epa.gov. https://www.epa.gov/green-vehicles/fast-facts-transportation-greenhouse-gas-emissions (accessed July 3, 2021).
3. B. Chen, H. H. Cheng, and J. Palen, "Mobile-C: a mobile agent platform for mobile C/C++ agents," *Software-Practice & Experience,* vol. 36, pp. 1711–1733, 2006.
4. B. Chen and H. H. Cheng, "A review of the applications of agent technology in traffic and transportation systems," *IEEE Transaction on Intelligent Transportation Systems,* vol. 11, pp. 485–497, 2010.
5. B. Chen, H. H. Cheng, and J. Palen, "Integrating mobile agent technology with multi-agent systems for distributed traffic detection and management systems," *Transportation Research Part C: Emerging Technologies,* vol. 17, pp. 1–10, 2009.
6. B. Chen and W. Liu, "Mobile agent computing paradigm for building a flexible structural health monitoring sensor network," *Journal of Computer-Aided Civil and Infrastructure Engineering,* vol. 25, pp. 504–516, 2010.
7. W. Liu and B. Chen, "Optimal control of mobile monitoring agents in immune-inspired wireless monitoring networks," *Journal of Network and Computer Applications,* vol. 34, pp. 1818–1826, 2011.
8. B. Chen and C. Zang, "A hybrid immune model for unsupervised structural damage pattern recognition," *Expert Systems with Applications,* vol. 38, pp. 1650–1658, 2011.
9. B. Chen and C. Zang, "Emergent damage pattern recognition using immune network theory," *Smart Structures and System,* vol. 8, pp. 69–92, 2011.
10. W. Luo, B. Chen, J. Naber, and C. Glugla, "Stochastic knock detection, control, software integration, and evaluation on a V6 spark-ignition engine under steady-state operation," presented at the SAE 2014 World Congress, Detroit, MI, 2014.
11. X. Wang, "A study of model-based control strategy for a gasoline turbocharged direct injection spark ignited engine," Ph.D. dissertation, Mechanical Engineering – Engineering Mechanics, Michigan Technological University, Houghton, MI, USA, 2020.
12. X. Wang, A. Khameneian, P. Dice, B. Chen, M. Shahbakhti, J. D. Naber, *et al.*, "Control-oriented model-based burn duration and ignition timing prediction with recursive-least-square adaptation for closed-loop combustion phasing control of spark-ignition engines," presented at the ASME 2019 Dynamic Systems and Control Conference, Park City, Utah, 2019.
13. M. Cheng and B. Chen, "Nonlinear model predictive control of a power-split hybrid electric vehicle with consideration of battery aging," *Journal of Dynamic Systems, Measurement and Control,* vol. 141, 2019.
14. L. Feng, W. Liu, and B. Chen, "Driving pattern recognition for adaptive hybrid vehicle control," *SAE International Journal of Alternative Powertrains,* vol. 1, pp. 169–179, 2012.
15. P. K. Bhat, J. Oncken, R. Yadav, B. Chen, M. Shahbakhti, and D. Robinette, "Generation of optimal velocity trajectory for real-time predictive control of a multi-mode PHEV," presented at the 2019 IEEE 90th Vehicular Technology Conference: VTC2019-Fall, Honolulu, Hawaii, USA, 2019.
16. B. Barik, P. K. Bhat, J. Oncken, B. Chen, J. Orlando, and D. Robinette, "Optimal velocity prediction for fuel economy improvement of connected vehicles," *IET Intelligent Transport Systems,* vol. 12, pp. 1329 – 1335, 2018.

17. J. Oncken, K. Sachdeva, H. Wang, and B. Chen, "Integrated predictive powertrain control for a multi-mode plug-in hybrid electric vehicle," *IEEE/ASME Transactions on Mechatronics,* vol. 26, pp. 1248–1259, 2021.
18. J. Oncken and B. Chen, "Real-time model predictive powertrain control for a connected plug-in hybrid electric vehicle," *IEEE Transactions on Vehicular Technology,* vol. 69, pp. 8420–8432, 2020.
19. C. Cao, Z. Wu, and B. Chen, "Electric vehicle – Grid integration with voltage regulation in radial distribution networks," *Energies,* vol. 30, 2020.
20. L. Wang, C. Cao, and B. Chen, "Grid-tied single-phase bi-directional PEV charging/discharging control," *SAE Int. J. Passeng. Cars – Electron. Electr. Syst.* , vol. 9, pp. 275–285, 2016.
21. C. Cao, L. Wang, B. Chen, J. D. Harper, T. P. Bohn, D. S. Dobrzynski, *et al.*, "Real-time modeling to enable hardware-in-the-loop simulation of plug-in electric vehicle-grid interaction," presented at the 2017 ASME/IEEE International Conference on Mechatronic and Embedded Systems and Applications (MESA2017), Cleveland, Ohio, 2017.
22. C. Cao and B. Chen, "Generalized Nash equilibrium problem based electric vehicle charging management in distribution networks," *International Journal of Energy Research,* vol. 42, pp. 4584–4596, 2018.
23. Z. Wu, C. Cao, and B. Chen, "Transactive energy based approach for large-scale plug-in electric vehicle charging control," presented at the 2019 IEEE PES Asia-Pacific Power and Energy Engineering Conference (APPEEC), Macao, China, 2019.
24. C. Cao, L. Wang, and B. Chen, "Mitigation of the impact of high plug-in electric vehicle penetration on residential distribution grid using smart charging strategies," *Energies,* vol. 9, 2016.
25. L. Wang and B. Chen, "Distributed control for large-scale plug-in electric vehicle charging with a consensus algorithm," *International Journal of Electrical Power and Energy Systems,* vol. 109, pp. 369–383, 2019.
26. L. Wang and B. Chen, "Dual-level consensus-based frequency regulation using vehicle-to-grid service," *Electric Power Systems Research,* vol. 167, pp. 261–276, 2019.

Dr. Bo Chen is a Professor with the Department of Mechanical Engineering—Engineering Mechanics and the Department of Electrical and Computer Engineering at Michigan Technological University. She is the Director of the Intelligent Mechatronics and Embedded Systems Laboratory focused on the investigation of advanced controls, optimization, and artificial intelligence for connected and automated vehicles, electric vehicle–grid integration, and smart mobility. Sponsors of her research include National Science Foundation, US Department of Energy, Argonne National Laboratory, Ford Motor Company, Nexteer Automotive, Nostrum Energy, and Michigan Space Grant Consortium. Dr. Chen has authored or co-authored more than 100 peer-reviewed journal/conference papers. She has received one Best Paper Award from IEEE/ASME International Conference on Mechatronic and Embedded Systems and Applications, one Top Cited Article Award from the journal of *Computers & Graphics*, and one Best Survey Paper Award from *IEEE Transactions on Intelligent Transportation Systems*. She was also the co-recipient of four Best Student Paper Awards and one Best Paper Award at IEEE/ASME conferences with her graduate students.

Dr. Chen is an ASME Fellow. She has served as the Chair of the Technical Committee on Mechatronic and Embedded Systems and Applications in the ASME Design Engineering Division and the Chair of the Technical Committee on Mechatronics and Embedded Systems in the IEEE Intelligent Transportation Systems Society. She was the General Chair of 2013 ASME/IEEE International Conference on Mechatronic and Embedded Systems and Applications. She was an Associate Editor of the *IEEE Transactions on Intelligent Transportation Systems* from 2012 to 2019.

Silicon Solar Photovoltaics: Slow Ascent to Exponential Growth

Santosh K. Kurinec

Introduction

Semiconductors such as silicon (Si) and gallium arsenide (GaAs) are materials that generate free carriers and include some organics that generate excitons (which are dissociated to form free carriers) when exposed to photons with energies exceeding their optical bandgaps. The principle of photovoltaics (PV) is that the photogenerated excess carriers get collected by the junction built in electric field and extracted at the contacts, providing useful power output. PV devices and modules made from crystalline silicon currently dominate the market. In a continued quest for lowering their cost, many efforts are being pursued to involve the use of alternative materials and multi-junctions.

The US National Renewable Energy Laboratory (NREL) maintains a plot of compiled values of highest confirmed conversion efficiencies of research cells, from 1976 to the present for a range of photovoltaic technologies [1]. This chart highlights cell efficiency results within different families of semiconductors: (1) multijunction cells, (2) single-junction gallium arsenide cells, (3) crystalline silicon cells, (4) thin film technologies, and (5) emerging photovoltaics. The graph sums up the historic quest of the solar industry to improve the conversion efficiencies in all PV technologies.

The first silicon p–n junction solar cell was fabricated in 1954 at Bell Laboratories [2]. It drew limited attention as it was perceived as a feeble power source. However, the solar cells successfully powered the first Soviet *Sputnik* satellite launched in 1954, triggering the space race. In 1958, the American satellite *Vanguard 1* entered orbit with six silicon solar cells, which generated about 1 watt power in total [3, 4]. In comparison, the power produced by a typical rooftop solar

S. K. Kurinec (✉)
Department of Electrical and Microelectronic Engineering, Rochester Institute of Technology (RIT), Rochester, NY, USA
e-mail: skkemc@rit.edu

M. Bailey, L. Shackelford (eds.), *Women in Mechanical Engineering*, Women in Engineering and Science, https://doi.org/10.1007/978-3-030-91546-9_15

PV system 63 years later is a thousand times greater [5]. Why did it take so long? In contrast, the invention of the first transistor in 1947 took just 11 years to lead to the creation of the first integrated circuit by Texas Instruments, Fairchild, and Intel. The limited momentum of the PV industry was due to the absence of a few dominant players like IBM, Intel, and Motorola, resulting in PV remaining in research laboratories [6]. Most PV interests in the US were for space applications. Growing interest in reevaluating terrestrial applications emerged with the release of small commercial modules by Sharp, Philips, and Solar Power in the early 1970s. The energy crisis and subsequent oil embargoes stimulated the rapid development of PV technology in the mid-1970s, with the first modern modules fabricated in 1976. In 1978, President Carter founded the Solar Energy Research Institute (SERI) in Golden, Colorado, and later in 1991, President George H. W. Bush elevated SERI to a national member of the Department of Energy (DOE)—National Renewable Energy Laboratory (NREL).

After dramatic achievements in module designs supported by the US Government programs, declining funding and interests shifted the key research and development to other countries in Europe, Japan, and Australia. Martin Green, PV pioneer and world-leading specialist in crystalline silicon solar cells, founded in University of New South Wales, the largest and best-known university-based photovoltaic research group in the world in 2003 [5, 7]. Remarkable progress was made with cells on high-quality monocrystalline, multicrystalline (Generation I), and thin film amorphous silicon, while III–V GaAs-based single junction and multi-junction solar cells were investigated for space-based applications. Generation II, III, and IV photovoltaics include thin film, multi-junction, multiband, hot carrier, and perovskite-based solar cells and all these technologies are promising, showing record efficiencies. For terrestrial applications, cost, environmental stability, and long lifetime with minimum degradation (>25 years) are the key factors. The US DOE SunShot initiative was launched in 2011 with a goal of cost reduction of utility-scale solar to approximately \$1 per watt or \$0.06 per kilowatt-hour. The program is now at 90% of its goal and recently expanded its target to \$0.03 per kilowatt-hour by 2030 [8]. This chapter provides a comprehensive overview of silicon photovoltaics and my experiential journey encountering its reluctant ascent to a dominating technology.

Brief Historical Lookback

My interest in Physics sparked when my science teacher in high school tenth grade taught about the peaceful use of nuclear energy using controlled nuclear fission for power generation. I decided to pursue a BS and MS in Physics at the University of Delhi. I interned at the nuclear research reactor of Bhabha Atomic Research Center, India, during a summer of my Master's program. After graduating, I joined the National Physical Laboratory (NPL), New Delhi, India, for my PhD research. India was exploring the development of magnetic materials for booming electrical, electronics, and automotive industries, using indigenous raw materials.

Dust in Rust

Iron oxide is the main raw material used in magnetic ceramics like ferrites. Silicon dioxide is a common impurity in indigenous iron oxide. My PhD research was focused on understanding the effect of silicon impurity on the magnetic properties of Mn–Zn ferrites. The Indian Institute of Technology in Delhi had acquired its first scanning electron microscope (SEM) from Cambridge Instruments (Cambridge Stereoscan S4–10 SEM). I employed it to examine microstructures by imaging secondary electrons, electron dispersive diffraction (EDX), and electron beam induced current (EBIC) modes. It was found through detailed microstructural examinations that a certain level of silicon is soluble in the ferrite and, in fact, improved the magnetic properties. When it exceeds solid solubility, it segregates at grain boundaries, degrading the magnetic properties such as permeability and loss factor. Figure 1 shows SEM and EDX image of a Mn–Zn ferrite sample showing Si segregation at the grain boundaries [9]. At the time I finished my PhD, the energy crisis was being felt worldwide, directing a focus on renewables. Major research institutions drew attention to photovoltaics.

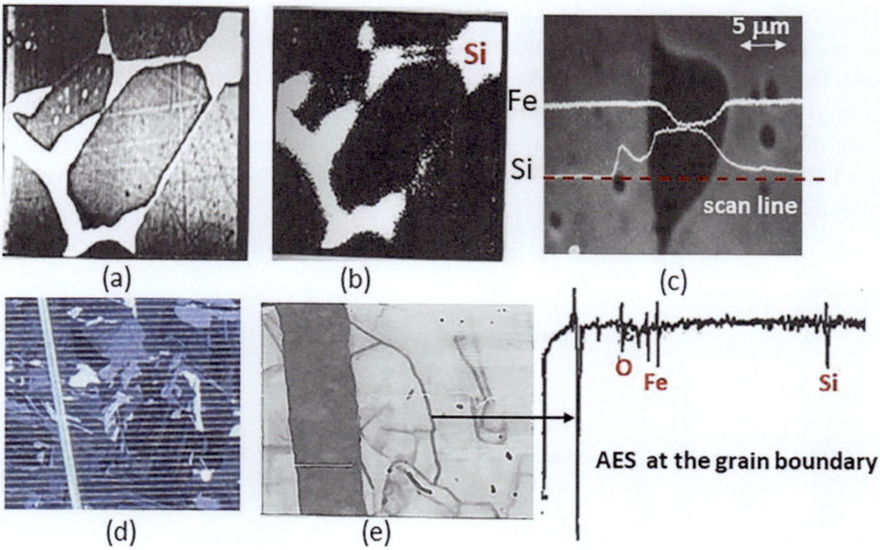

Fig. 1 SEM (**a**) and EDX map and line scans (**b**, **c**) of Mn–Zn ferrite grain showing Si segregation at the grain boundaries; (**d**) mc-Si solar cell examined at a grain boundary using Auger Electron Spectroscopy (AES) showing Fe segregation at the Si grain boundary

Rust in Dust

The Department of Materials at NPL set up a complete process for purification of metallurgical grade silicon obtained from a regional steel plant. It consisted of reacting silicon with HCl to get liquid trichlorosilane (TCS), which was fractionally distilled to purify to solar grade level. After purification, TCS was cracked in a reactor to get polycrystalline silicon (polysilicon) rods. On the industrial scale, it is known as the Siemen's process and is the flagship process for silicon purification developed in the early 1980s. The polysilicon rods were zone-refined for compaction and we obtained large grain polysilicon rods (Fig. 2). Then, wafers were cut, lapped, and polished. The wafers were *n*-type as the silicon obtained was rich in phosphorus. We had in-house doping techniques using solid paper and liquid spin-on dopant sources. Solar cells were fabricated with Al/Ag and Ti/Al/Ag metallization yielding 10% efficiency [10]. Detailed analyses revealed significant iron impurity in silicon that affects the PV quality of the material. My research in PV began with investigating the effect of iron impurity in multi-crystalline silicon solar cells.

TAET Program

The University of Florida and the US Agency for International Development established a training program in Alternative Energy Technologies (TAET) in the 1980s. The TAET program provided training in both the technical principles and socioeconomic aspects of the selection and implementation of renewables. In 1982, I was selected as a participant from India among 35 participants from around the world. I carried out a project on investigating the nature of grain boundaries and their influence on the photovoltaic properties of polysilicon. This work attracted me to join the University of Florida as a Postdoctoral Research Associate to further my PV research.

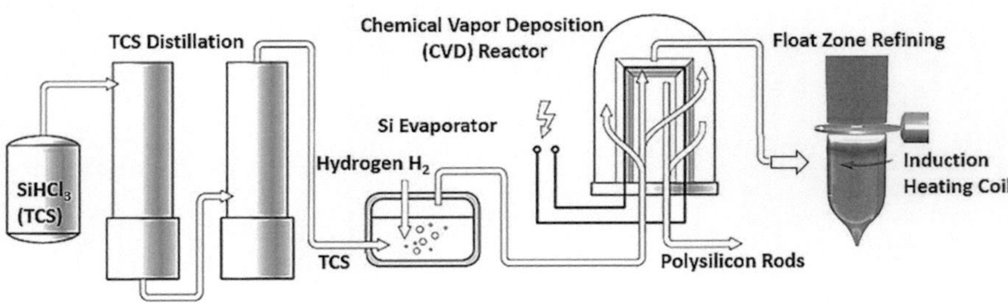

Fig. 2 In-house developed silicon purification process followed by float zone refining, used at NPL to fabricate mc-Si solar cells of 10% efficiency

Photovoltaics

The solar spectrum is modeled as a black body radiation emitted by the Sun at its surface temperature of 5800 K according to Planck's law. The reference Air Mass 1.5 (AM1.5) spectrum describes solar insolation on a terrestrial horizontal surface at a solar zenith angle of 48.19° and it is specified as 1 kW/m^2. In a single junction diode solar cell, incident photons with energy equal or higher than the bandgap of the semiconductor create electron hole pairs, which are collected by the junction electric field. This gives rise to radiation-generated current, shifting the diode current–voltage curve to power generation mode. Figure 3 shows a baseline cell fabrication process, electrical equivalent circuit, and current–voltage characteristics of a solar cell. The power conversion efficiency is defined as the ratio of the maximum power generated to the incoming incident power.

Shockley and Queisser obtained the theoretical limit to the conversion efficiency through detailed balance limit of p–n junction solar cells on the basis of absorption and reemission processes [11]. AM1.5 solar spectrum with distinct dips due to molecular absorption in Earth's atmosphere is shown in Fig. 4a. Photons with energies below the band gap (E_g) are not absorbed, whereas the energy of photons with energy higher than the band gap is not fully converted to electrical energy and is dissipated as the thermalization loss. The inset in Fig. 4a depicts the electronic band structure with the separation of the quasi-Fermi levels determining the open-circuit voltage V_{oc}. Theoretical Shockley–Queisser detailed-balance efficiency limit as a function of band gap (black line) is shown in Fig. 4b. The record efficiencies for different materials are plotted for the corresponding band gaps. An optimum bandgap (~1.4 eV) gives the highest theoretical limit of the efficiency as 33%. NWO-I Institute for Atomic and Molecular Physics, Netherlands, constantly updates this efficiency chart [12].

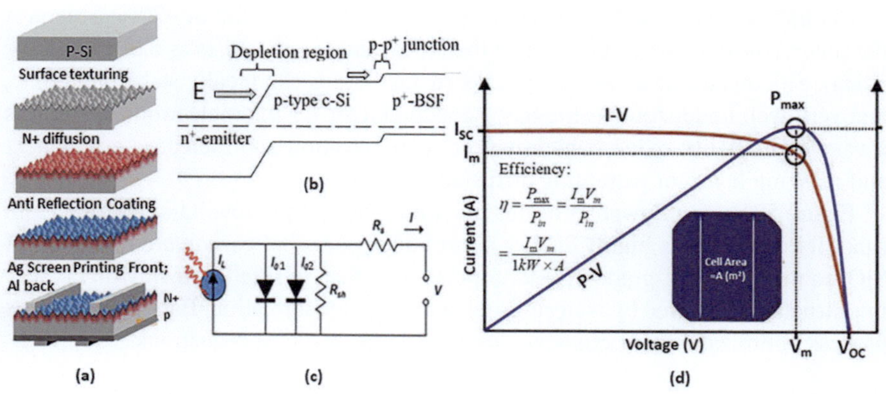

Fig. 3 (**a**) Process flow for a basic n + p solar cell; (**b**) equilibrium band diagram of n+pp.+; (**c**) a two diode equivalent circuit of a solar cell; (**d**) typical current–voltage characteristics of a solar cell defining the key parameters: open circuit voltage (V_{oc}), short circuit current (I_{sc}), maximum power (P_{max}), and power conversion efficiency

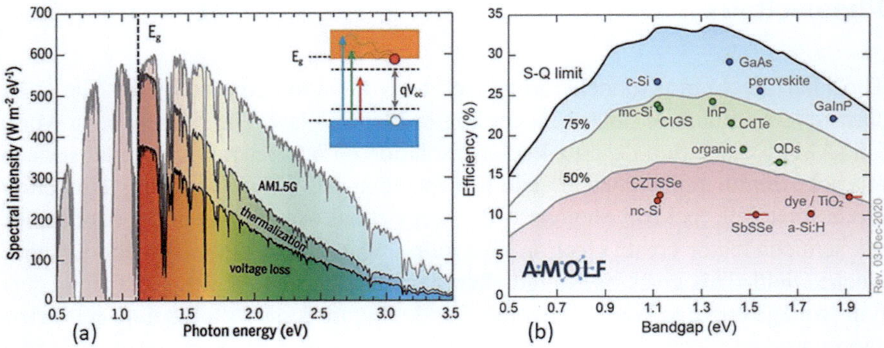

Fig. 4 (**a**) AM1.5 solar spectrum and (**b**) fundamentals solar cell efficiency limits [12]. (Reprinted with permission)

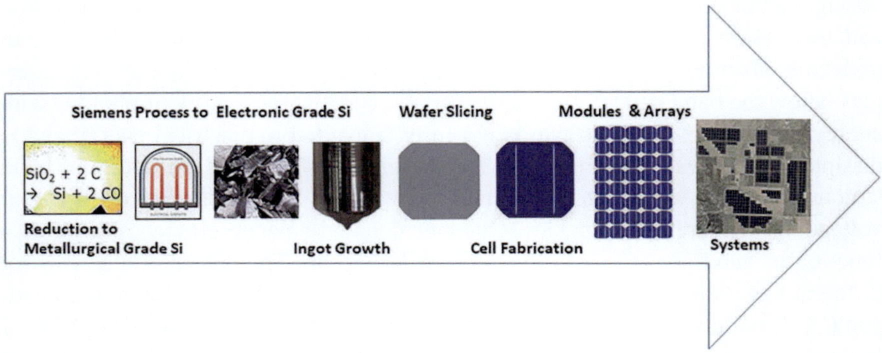

Fig. 5 Process flow chart for fabricating monocrystalline silicon solar PV systems

Crystalline silicon with the bandgap of 1.12 eV has this limit of ~30%. Silicon is the second most abundant element on the Earth, and mostly exists as oxide (SiO_2, or silica). Silicon has the diamond crystal structure and is chemically stable, nontoxic, and very well understood as the base semiconductor for microelectronics. It forms a very high-quality oxide, which is used as passivation and dielectric component and masking layer for patterning diffusion of dopants.

Figure 5 shows a flowchart of manufacturing silicon photovoltaics starting from sand. It begins with a highly energy-consuming process of carbon arc reduction of SiO_2 to metallurgical-grade Si (~98% pure). It is then purified, grown with crystalline structure, followed by wafering, cell and module fabrication. Each step requires process optimizations for efficiency improvements and cost reductions.

Innovations in Si PV

Over the last several decades, academics, scientific laboratories, and niche industries worldwide have remained diligently committed in achieving higher efficiencies across cell, module, and system level. These accomplishments are summarized in the following subsections.

Cell Level

The simplest form of a solar cell is a p–n junction connected with grid top metal and blanket back metal contacts as shown in Fig. 4a. In order to improve the efficiency, approaches to reduce reflection losses, optimization of diffusion, improved minority carrier lifetimes, and reduction of parasitic resistances were necessary. These improvements required novel cell structures. A series of high-efficiency crystalline silicon solar cell structures have emerged, which include passivated emitter rear cell (PERC), passivated emitter, rear locally diffused cell (PERL), interdigitated back contact cell (IBC), heterojunction with intrinsic thin-layer cell (HIT), heterojunction solar cells with interdigitated back contacts (HBC), bifacial cells, and TOPCon solar cells [13, 14]. When deployed on a conventional solar farm, bifacial cells absorb direct incoming light, while also taking advantage of ground reflection, which can contribute up to additional 30% power generation [15]. PERC cells with another addition of a special nano-coating layer, known as Q cells, can capture previously unused sunlight back into the cell where it can be converted into solar electricity [16]. It enhances cell efficiency further. Table 1 gives the schematics of these structures and their best known efficiencies to date.

The first efficient silicon solar cell was made on a n-type substrate. The selection of p-type as the substrate was primarily due to the development of Al-BSF process that creates back surface field (BSF) to mitigate minority carrier recombinations at the back contact. On a per-watt basis, passivated emitter and rear totally diffused, silicon heterojunction (SHJ or HIT), and interdigitated back contact cells currently cost more than standard aluminum back surface field (Al-BSF) and PERC cells owing to smaller production scales and use of n-type wafers [14]. However, if demand for high-efficiency cell architectures grows, these advanced cell technologies may gain market share and their cost may decline due to benefits from economies of scale. Cells with higher efficiencies could reduce per-watt balance-of-module and balance-of-system costs [17].

Table 1 Record reported efficiencies in single junction silicon solar cells (as of 2020)

Solar Cell Type	Cross-section Schematic	η	Solar Cell Type	Cross-section Schematic	η
I. Passivated emitter rear cell (PERC)- Mono • 'Workhorse' of the PV industry today **PERC- Multi**	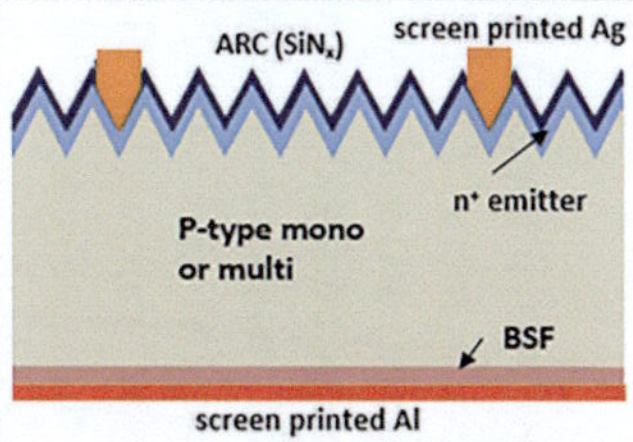	24 22	**IV. Interdigitated back contact cell (IBC)** • Both electrodes at rear side, no shading • High demands on material quality. Carriers must travel through complete wafer. Excellent passivation required: low front surface recombination velocity • Challenges:defining and separating both polarities on same side	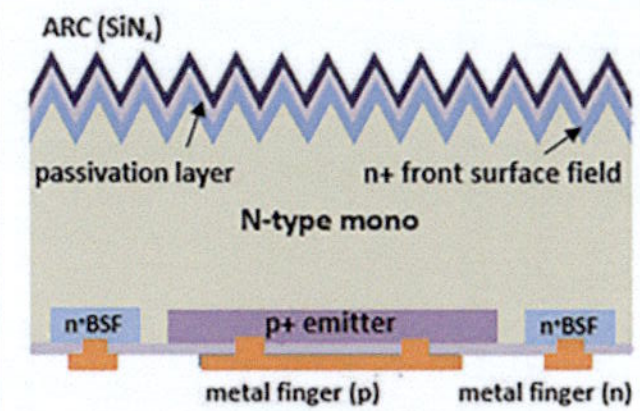	22.5
II. Bifacial PERL • Increased power generation due to additional rear side illumination. • Can be made PERC as well	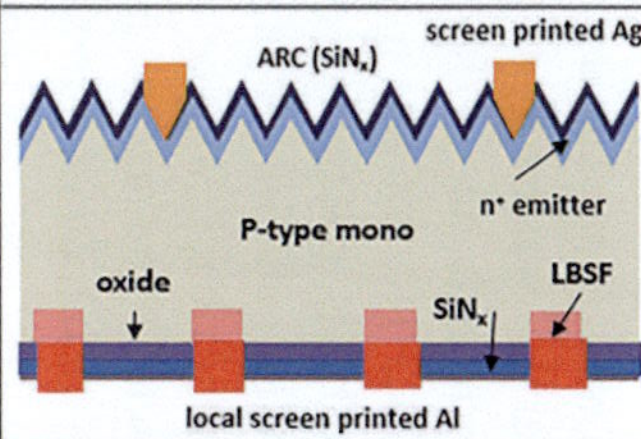	22	**V. Heterojunction solar cells with interdigitated back contacts (HBC)**	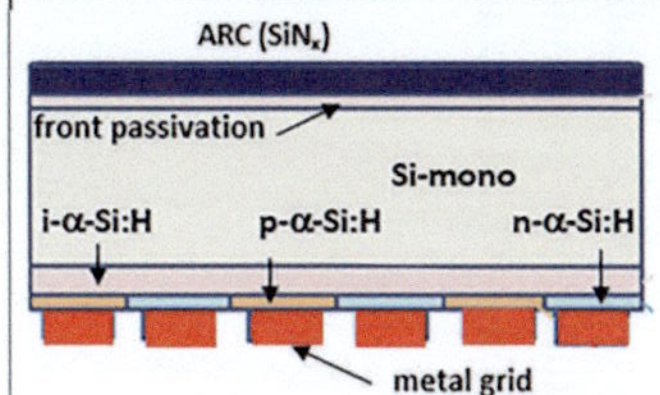	26.7
III. Heterojunction with intrinsic thin-layer cell (HIT) • Cells make use of passivating contacts based on a layer stack of intrinsic and doped amorphous silicon	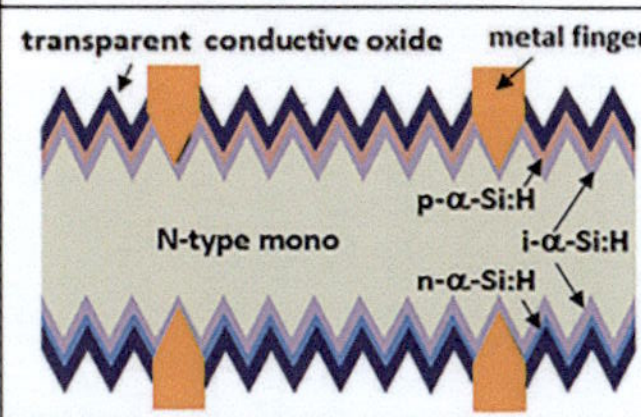	25.1	**VI. TOPCon solar cells** • Carrier selective, thin SiOx tunnel oxide to passivate rear contact • Heavily doped pseudo-crystalline Si transport layer • No rear side patterning required • Low contact resistance, high FF • One-dimensional flow of charge carriers	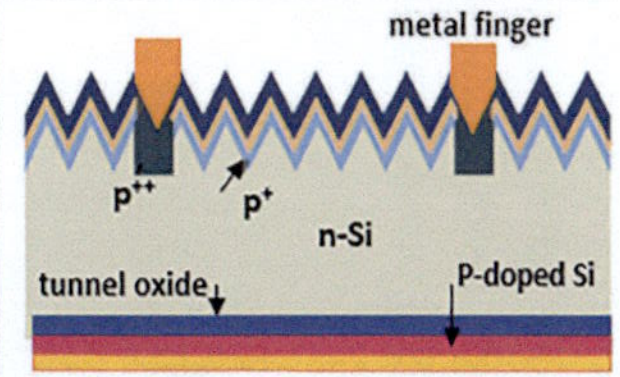	25.7

Substrate Level

Solar-grade polysilicon typically has purity levels of 6N (99.9999% pure) to 8N (99.999999%) and it is used to make solar cells; 9N polysilicon may also be used in some premium solar cells. Electronic-grade polysilicon has higher purity levels of 9N to 11 N for producing silicon wafers for integrated circuits. The Siemens process, mentioned above, is the most widely used method for polysilicon production. Polysilicon sits on top of the photovoltaic supply chain. One megawatt (MW) of photovoltaic power requires 7 tons, or 7000 kg, of polysilicon material [18].

Polysilicon is cast into monocrystalline (mono-Si) or multicrystalline (mc-Si) silicon. Mono-Si is grown using the Czochralski (CZ) crystal growth process. At the beginning of the CZ process, the polysilicon is melted in a cylindrically shaped crucible. After the feed material is completely molten, a seed crystal with a diameter of typically a few millimeters is dipped from top into the free melt surface forming a melt meniscus at the contact interface between seed and melt. Then, the seed is slowly withdrawn from the melt (often under rotation) and the melt crystallizes at the interface by forming a new crystal portion. During the further growth process, the shape of the crystal, especially the diameter, is controlled by carefully adjusting the heating power, the pulling rate, and the rotation rate of the crystal [19]. Silicon ingots are cylindrical giving circular wafers, which require a larger surface area of a solar panel compared to square-shaped. Multicrystalline silicon (mc-Si) is produced by melting Si and directionally solidifying into bricks. The mc-Si process yields rectangular wafers, which can be packed densely on a panel. The production of mc-Si is cost efficient compared to the mono-Si CZ process and therefore has prevailed as a major technology for the solar panels. However, the defects present in the mc-Si because of the production process lead to less-efficient solar cells.

Thinner wafers are preferred to lower the cost and the weight of the panels. While the technical limit of slurry-based wire saws lies at a wire thickness of 100 μm, diamond wire can be as thin as 60 μm and thus reduce the kerf loss significantly [20]. The cylindrical ingot is first cut along its length on four sides to make its shape closer to a square in cross section, known as the pseudo squares. Pseudo square diameter is the diameter of a square within a circle. Wafer sizes have increased over the last 7 years from M0 (156/205 mm, flat length/diagonal length) to M2 (156.75/210 mm), and to M6 (166/223 mm). Larger wafer size offers balance of systems cost reductions. For installation, a 72-M6-cell module weighs around 30 kg, which is near the limits of manual installation. Considering the increased power output, including potential bifacial gains, the M6 wafer will increase the operational current to around 13 amps, which is the limitation of current string inverters. Modules using M6 wafers are also compatible with centralized inverters [21].

In August 2019, Zhonghuan Semiconductor unveiled its latest wafer product—the new M12 size—featuring a much bigger size of 210 mm in length and 295 mm on the diagonal [22]. A 60-cell PV module with this enlarged wafer would easily boost module power output above 600 W. M12 is targeting next-generation PV cell technologies such as interdigitated back contact technology. Figure 6 shows a picture of a wafer puller capable of pulling M10–M12 ingots developed by Linton Crystal Technologies [23], a Rochester, New York–based company. This puller employs the Magnetic Confined Czochralski method that lowers oxygen contamination in the crystal.

Most PV modules are fabricated using p-type silicon substrate, which is highly susceptible to light-induced degradation (LID) when exposed to sunlight. Boron-doped p-type silicon thus requires additional processing steps to mitigate this degradation [24]. An alternative method for the production of stable lifetime material is to dope silicon with a different Group III element, such as aluminum, gallium, or indium. Gallium is being explored as the dominant dopant for p-type silicon solar cells [25]. However, it has much lower segregation coefficient ($k = 0.008$), which means that gallium has a much stronger thermodynamic tendency to stay in the melt rather than be incorporated into the solid silicon crystal. One way to overcome this is to constantly replenish the melt with gallium during the crystal growth to achieve uniform resistivity (1–2 Ohm.cm) along the crystal. Another method being considered is the Continuous Czochralski method, which does not require a large crucible as Si is fed while the ingot grows [26].

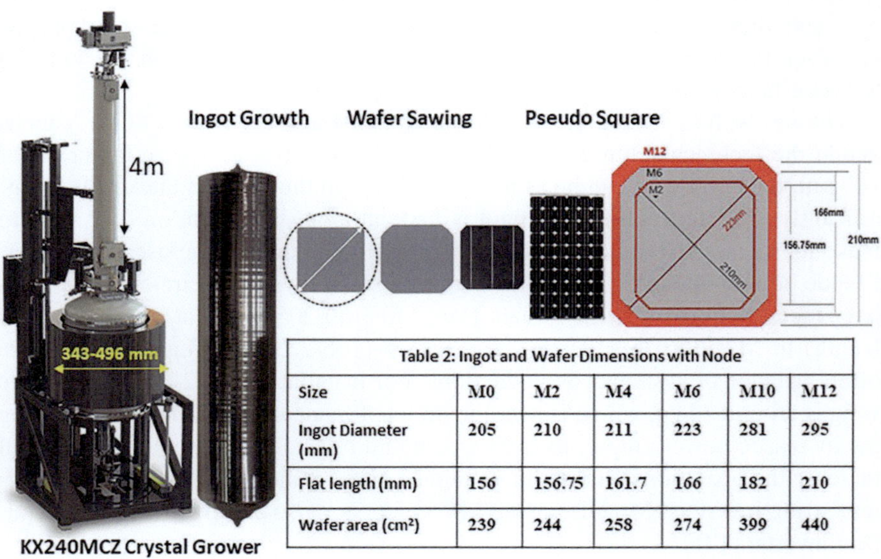

Table 2: Ingot and Wafer Dimensions with Node

Size	M0	M2	M4	M6	M10	M12
Ingot Diameter (mm)	205	210	211	223	281	295
Flat length (mm)	156	156.75	161.7	166	182	210
Wafer area (cm²)	239	244	258	274	399	440

Fig. 6 Czochralski (CZ) crystal puller (Courtesy: Linton Crystal Technologies) designed to grow M10–M12 ingots. Circular wafers are shaped into pseudo squares with dimensions shown in Table II for each generation

Modules

Solar cells are connected in series and parallel to construct modules for required voltage, current, and power output. Series connection adds voltages and parallel connections add currents of individual identical cells. Typically, PV modules are fabricated by electrically connecting 36 to 72 solar cells together in a sealed, weather-proof packaging and are the fundamental building block of a PV system [27].

Mismatch occurs when some cells degrade in their performance due to shading effects or other degradation causes. Series strings are more prone to shading effects and therefore bypass diodes are used to allow bypassing current through the damaged cell. Solar cells transport current using the thin metal ribbons that connect them to neighboring wires and cells, which leads to some energy lost. By cutting solar cells in half, the current generated from each cell is halved resulting in lower interconnect resistive losses. Two parallel half-cut cell strings replace one full cell series string. Half-cut cell modules increase module power by ~1.5% due to reduced electrical losses in cell connectors. Parallel substrings allow the module to save up to 50% of the string's power under partial shading conditions [28].

Another simple and accessible way to reduce resistance losses in solar cells is to add more busbars. Adding more busbars reduces the gap between them, which shortens the finger length. The forthcoming trends are—increasing the number of busbars, while maintaining the same shading factor, and switching to multi-busbars using round wires (9–15 wires) instead of flat ribbons as interconnections [29]. The round-shaped busbar, a wire indeed, increases the light scattering effect towards the cell surface for higher cell absorption, resulting in increased power generation.

Most bulk silicon PV modules consist of a transparent top surface, an encapsulant, a rear layer, and a frame around the outer edge [27]. The main attributes of glass used are transmission, mechanical strength, and specific weight. The front surface of a Si PV module must have a high transmission in the wavelength range of 350–1200 nm. In addition, the reflection from the front surface should be low. Currently, 3 mm–thick glass is the predominant cover material for PV modules accounting for 10–25% of the total cost. The cover glasses can also provide enhanced ultraviolet protection of polymeric PV module components, potentially increasing module service lifetimes [30]. The properties of PV module materials are of great importance to ensure optimal light capture and module lifetime as well as ultimately reducing the cost. Traditional opaque-backsheeted panels are monofacial. Bifacial modules expose both the front and backside of the solar cells. When bifacial modules are installed on a highly reflective surface (like a white thermoplastic polyolefin roof or on the ground with light-colored stones), gains can be up to a 30% increase in power production just from the extra power generated from the rear [31]. Fig. 7 shows a PV system under installation at City Center Bishop Ranch, California, that I recently visited to learn the installation processes.

Fig. 7 Bifacial horizontal roof top solar arrays using 12 multi-busbar mono-Si modules (left); inclined half-cut 6 busbar Q-cells panels (right) under installation at Bishop Ranch, California. Single panel and cell schematics are shown. (Photo courtesy Distributed Solar Development)

PV Systems

A stand-alone PV system is made up of a number of individual PV modules (or panels) usually of 12 V with power outputs of between 50 and 300+ W each. These PV modules are then combined into a single array to give the desired power output. A small-scale PV system employs rechargeable batteries to store the electrical energy supplied by the PV array. Stand-alone PV systems are ideal for remote rural areas and applications where other power sources are either impractical or are unavailable to provide power for lighting, appliances, and other uses. In these cases, it is more cost-effective to install a single stand-alone PV system than pay for the costs of extending power lines and cables. Residential PV systems (4–10 kW) are grid connected and some may have additional battery storage (Fig. 8). Production and consumption is metered using net metering systems. Utility-scale PV systems are community shared via the grid. For hundreds of megawatt-scale solar farms, how will electricity get from large solar farms to cities? At present, the majority of high-voltage transmission lines are alternating current, but recent innovations suggest they are increasingly likely to be high-voltage direct current (HVDC) lines. HVDC are cheaper at longer distances over land and at very short distances underwater and underground [32]. This means that HVDC will enable electricity to travel long distances from renewable locations, connecting islands to the mainland and even continents to one another potentially.

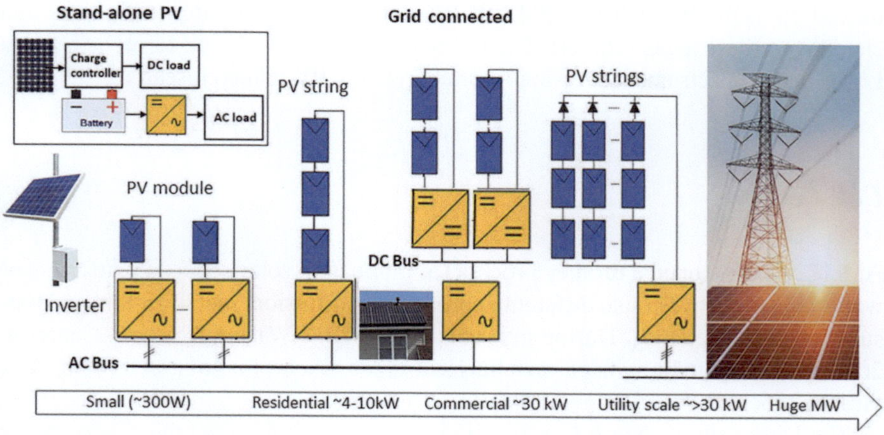

Fig. 8 PV systems ranging from stand-alone to grid connected residential, commercial, and utility scale systems

My Research at the Photovoltaics Microelectronics Intersection

While I was working on polysilicon for PV, the semiconductor chip industry was looking to use polysilicon as the gate material in complementary metal oxide semiconductor (CMOS) field-effect transistors for integrated circuits. In addition, interests in photodetectors and light-emitting diodes emerged for optical interconnects when the Strategic Defense Initiative "Star Wars" was announced (1983) [33]. Many PV researchers moved to these fields, including myself. Since then, my research and teaching has been in both semiconductor devices and photovoltaics.

The growth of innovative techniques that enabled the integrated circuit technology to become efficient in the high volume manufacturing of extremely small and complex systems (with nanometer-scaled billions of stacked devices) on large substrate, set up a sound base for the PV industry. Between the two, they share a common substrate—silicon and common thin film deposition techniques. PV contrasts itself from CMOS in being relatively simpler in device structure (essentially one large area ~ 250 cm^2 diode), relaxed in lithography and particle contamination controls. However, it differentiates in applications that require large area end-products, much larger than the flat panel displays, and is available at lower costs. Even though the PV industry inherits an experienced workforce trained in defining and following the roadmap driven by the Moore's law, engineering education needs to address developing the next generation of PV engineers.

I developed a cross-disciplinary course—"Photovoltaics Science and Engineering" at the graduate/senior undergraduate level in the college of engineering at Rochester Institute of Technology (RIT) in 2007. Students from various programs—electrical, microelectronic, chemical, materials science, industrial engineering, and physics—have enrolled in this course and many students have moved on to the PV industry, PhD in PV-related areas, and published papers and chapters in these areas [26, 34]. The following subsections ("Cell Level", "System Level" and "Spectrum Conversion") summarize my PV-related research.

Cell Level

At RIT, we developed a turnkey process for fabricating solar cells—a platform that was intended to investigate different approaches in diffusion, metallization, and top-surface reflectance [35]. During my sabbatical at IBM Watson Research Center in 2009, I investigated upgraded metallurgical-grade silicon for solar cells using this turnkey process.

The University of North Carolina and Georgia Tech founded the Silicon Solar Consortium, an Industry–University Cooperative Research Center in 2007. My group joined this consortium and worked on developing copper metallization to replace silver in silicon solar cells. Silver represents >48% of the metallization cost of a solar cell, or about 11% of the total raw material cost of a solar module. Copper presents a viable alternative to silver with the potential to reduce metallization costs by approximately 50% while maintaining device performance in solar cells. The primary concern with copper is its rapid diffusion in silicon, lowering the minority carrier lifetime, and therefore greatly diminishing the efficiency of the device.

IBM rocked the microelectronic industry by implementing copper as the interconnect metal in integrated circuits in 1997, as copper was undesirable due to its fast diffusion in silicon and difficulties in etch patterning. Special liners, diffusion barriers, electroplating, and patterning by chemical mechanical planarization processes were developed. Copper interconnects have since become the industry standard, enabling future generations of smaller and faster microprocessors.

Some metal silicides ($TiSi_2$, $CoSi_2$, NiSi) exhibit low resistivity and higher temperature process capability compared to aluminum which was used as the gate electrode in early days in integrated circuits. Self-aligned silicide process was developed in the mid-1980s that reduced gate and contact resistance by using metal silicides. Silicides are formed by metal deposition on silicon followed by annealing to form the desired low resistivity phase. Nickel monosilicide (NiSi) with a resistivity of 10–18 μΩ.cm consumes less silicon during its formation, and is widely used in ICs.

NiSi is a strong contender as a contact, as it shows promise as a copper diffusion barrier. My group studied the NiSi/Cu contacts for front metallization of silicon solar cells [36]. NiSi/Cu/TiN contacts with contact resistivities as low as 4×10^{-5} Ω-cm^2 were formed to the emitter of solar cells fabricated at RIT and 2×10^{-3} Ω-cm^2 on NREL fabricated tunneling contacts [37]. Contact resistance is related to the metal semiconductor specific contact resistivity, which is the fundamental property of a metal–semiconductor junction. We investigated the effects of transmission line measurements (TLM) geometries on the extracted value of specific contact resistivity [38]. This work, in collaboration with Professor Zhang, Michigan State University, and Professor Kris Davis, University of Central Florida, led to the development of standardized TLM structures for specific contact resistivity measurements.

System Level

System Monitoring

The Golisano Institute for Sustainability (GIS) building at RIT is equipped with mc-Si PV modules with an annual capacity of 45,241 kW-h. In our study (with my graduate student Felipe Freire), yearly power output of the GIS PV system was investigated using a mathematical model developed and comparing with real data collected from the monitoring system. The objective was to predict PV modules' performance with respect to changes in environmental parameters such as temperature, irradiance, and cloud coverage. The results were compared with the actual PV output data for the year 2014 and showed a very good correlation [39]. This mathematical model has been applied to several PV systems with known parameters—location, time of the day, panel orientation, and weather conditions.

Interestingly, it was used to explain the power dip in a residential PV system during the partial solar eclipse of August 21, 2017 [40]. The solar coverage was calculated using astro-imaging the Sun during the eclipse and calculating insolation over the eclipse time (Fig. 9). Strange coincidence, during the February 16, 1980, eclipse in India, as a student, I monitored a solar cell short-circuit current to provide insolation data to researchers investigating cosmic flux of gamma rays during the eclipse.

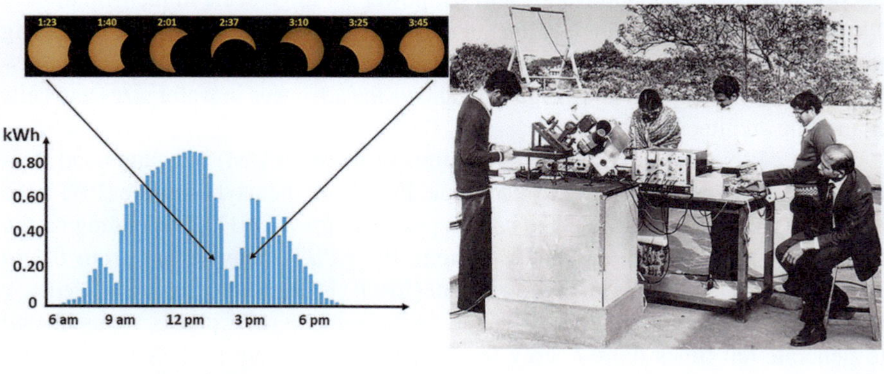

Fig. 9 Power output of a residential PV system during the partial eclipse of August 21, 2017 (left); monitoring solar insolation using a solar cell to calibrate cosmic gamma ray flux during the eclipse of February 16, 1980

PV Degradation

PV modules are considered the most reliable component of a photovoltaic system, and according to the manufacturers, have a lifetime from 25 to 30 years. However, some modules degrade or fail along their service time under outdoor exposure. To further understand the mechanisms behind the degradation and failure of PV modules, our study reviewed the governing degradation modes and investigated a 10-year-old operating crystalline silicon PV module installed at an active farm and animal sanctuary [41]. Field current–voltage testing and infrared imaging for hotspot detection were employed in examining this PV module showing localized damaged regions.

Spectrum Conversion

Currently, I am working on photon management techniques to enhance power conversion efficiencies. Development of special coatings that can shift the spectrum toward the favorable wavelengths would result in enhanced power output. For Si PV, down conversion of shorter wavelengths to longer wavelengths, which give higher quantum efficiency, is desirable. This is fostering innovations in photonic coatings. In collaboration with a startup company—SunDensity [42]—we have demonstrated increase in power output of a high efficiency PERC Si solar cell (Fig. 10) using a designed special coating.

Silicon Photovoltaics from Present to Future

The primary ambition of PV is to offer a cheaper, reliable, and resilient renewable energy alternative. The motivation and willpower for large-scale implementation of PV has been steering the innovations in materials, devices, and systems. Following the unprecedented growth and success of the semiconductor industry, the PV industry has its own "Moore's law." Figure 11 shows the advances in wafer sizes and cells in module architectures.

Table 2 lists and defines some key figures of merits (FOM) indicators and their current estimated values. The key economic FOMs are the payback time (PBT), the energy payback time (EPBT), and carbon payback time (CPBT). Depending on the technology and location of the PV system, the EPBT today ranges from 0.4 to 1.5 years [43]. Energy return on investment (EROI) is defined as the lifetime energy output to the energy invested. In general, on the average solar panels are estimated to generate ten times more energy than used to make them [44]. A typical solar panel saves over 900 kg of CO_2 per year resulting in a carbon payback period of ~1.6 years [45].

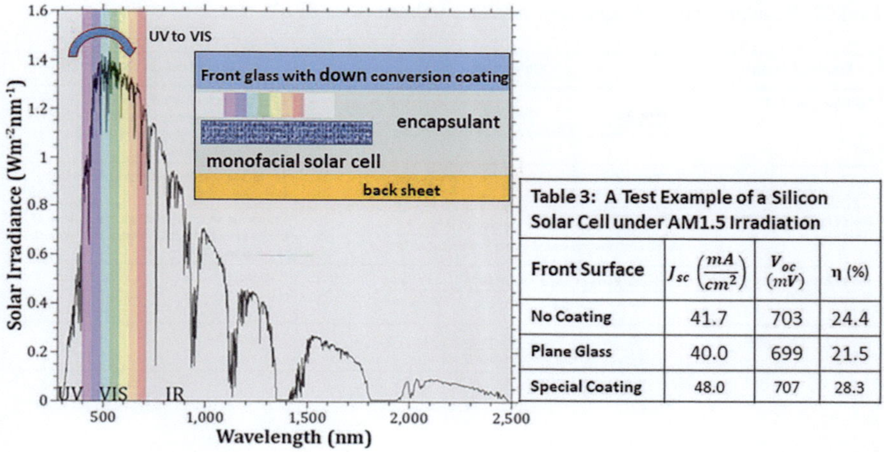

Front Surface	$J_{sc}\left(\frac{mA}{cm^2}\right)$	V_{oc} (mV)	η (%)
No Coating	41.7	703	24.4
Plane Glass	40.0	699	21.5
Special Coating	48.0	707	28.3

Fig. 10 Spectrum down conversion using a special coating (developed by SunDensity) tested for a high efficiency Si solar cell

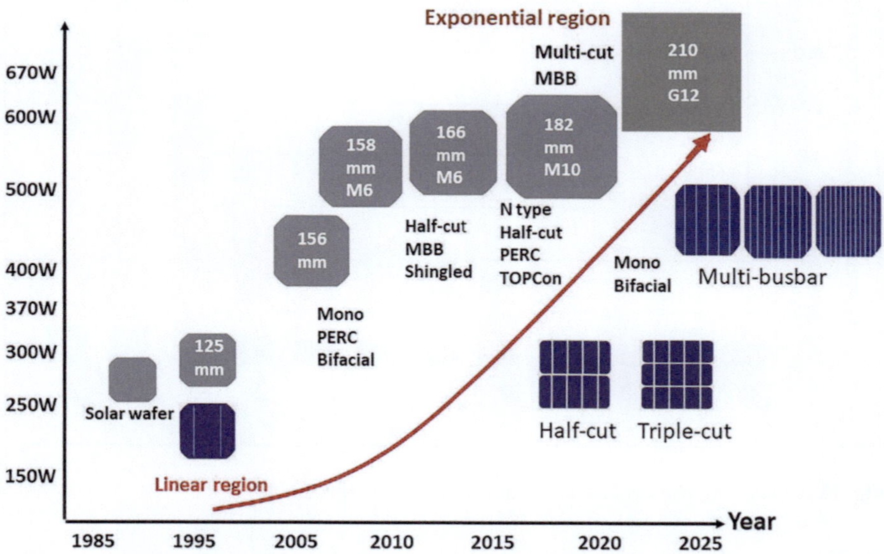

Fig. 11 Trends in silicon PV wafer sizes, emerging cell architectures toward bifacial, multi-busbar, and multi-cut modules

The cumulative PV installations from 2010 to 2020 are reaching 800 gigawatt-peak (GWp) (Fig. 12) [46]. A 2019 review in *Science* led by researchers from NREL describes an ambitious plan, in which 30–70 terawatt (TW) of PV capacity making it a central contributor to all segments of the global energy system by 2050 [47].

Most energy forecasters suggest that the installed cost of a complete PV system, including modules and balance of system (BOS) components will need to fall below

Table 2 Economic/environmental performance indicators for Si PV

Figure of merit	Definition	Current values (as of 2020)
Pay Back Time: PBT	$\frac{\text{Cost } \$/m^2}{\eta \times P_a \times \text{hours of peak sunshine hours/day} \times 365\text{days/year} \times \text{electricity cost } \$/W_e}$	3-7 years for residential in the USA Assuming (standard conditions) and 1,700 kWh/m² per year of available sunlight energy (the U.S. average is 1,800)
Levelized Cost of Energy: LCOE	Total Life Cycle Cost/ Total Lifetime Energy Production	\$32 to \$44 USD per MWh for utility scale PV [43]
Energy Pay Back Time: (EPBT)	$\frac{\text{Energy used to make the system } (MJm^{-2})}{\text{Annual energy generated by the system } (MJm^{-2}yr^{-1})}$	Depending on the technology and location of the PV system, the EPBT today ranges from 0.4 to 1.5 years [43]
Energy Return on Energy: Invested (EROI)	$\frac{\text{Lifetime Energy Output } W_i\ (MJm^{-2})}{\text{Embedded Energy } (MJm^{-2})} = \frac{LT\ (year)}{EPBT\ (year)}$	~ 10 [44]
Carbon Pay Back Time: (CPBT)	$\frac{CO_2 \text{ Emission Throughout Life Cycle [Kg]}}{\text{Annual } CO_2 \text{ Emission Reduction [Kg/year]}}$	1.6 years [45]

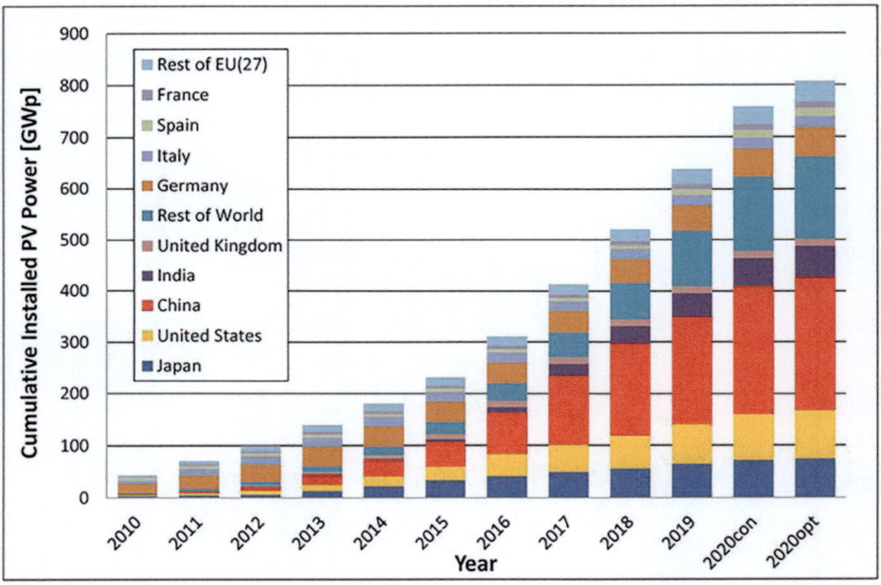

Fig. 12 Cumulative photovoltaic installations from 2010 to 2020 [46]

\$0.25 per watt for this goal. NREL's 2019 roadmap for continued innovations anticipates that the cost of crystalline silicon modules will decline to 0.24 per watt by 2030.

The International Energy Agency (IEA) Renewable 2020 reports that power generation from solar PV is estimated to have increased by 22% in 2019 to 720 TWh. Solar PV is well on track to reach the Sustainable Development Scenario level by 2030, which will require electricity generation from solar PV to increase 15% annually, from 720 TWh in 2019 to almost 3300 TWh in 2030 [48]. In 2019, PV generation overtook bioenergy and is now the third-largest renewable electricity technology after hydropower and onshore wind.

A major renewable energy project in Australia billed as the world's largest solar farm is in development. The farm will have a peak capacity of around 10-GW and will spread across an area of roughly 20,000 football fields! That is so large that it should be visible from space, once built. The project aims to transport its solar energy to Singapore using a 4500 km high-voltage direct current network, 3700 km of which will be undersea. It is projected that the project will be able to provide up to 20% of Singapore's power needs. Construction is expected to commence in 2023, with the farm expecting to begin exporting energy by 2027.

The intermittency of solar PV requires smart energy storage and dynamic power distribution systems. Hardware and software innovations will play a key role in managing storage resources, shifting capacity during peak periods, providing secondary services in the off-peak hours and standby power for emergencies.

Conclusions

Growing up during the Cold War, my interest in energy began with learning how nuclear reaction can be used for energy production. Nuclear energy technology with its high capacity factor and zero carbon emission, however, has the perception of safety risks. As we moved to the energy crisis era of the 1970s, interest in PV emerged. Subsequently, the growth of semiconductor technology dominated, leading to the Internet revolution. At present, energy consumption in computing is increasing super exponentially. My current work on semiconductors is also dedicated to energy-efficient computing devices. I foresee a future—using nuclear power plants for silicon production and solar cell manufacturing, and solar PV farms powering the energy-hungry data centers.

I have experienced the Si PV journey over the last four decades. Attention on Si PV got accelerated, slowed down, and recovered primarily due to economic circumstances. While major research innovations were achieved in the US, Europe, Japan, and Australia, China took the lead in manufacturing and currently shares 95% of the market share.

Today, Si-based solar PV cells are becoming more affordable and are being installed in large numbers with gigawatt (GW)-scale solar farms. Larger area mono-Si cells with advanced cell designs, multi-busbars, are being incorporated on multi-cut modules. Trends toward heterojunction with intrinsic thin layer (HIT), bifacial, and interdigitated back contact cells are promising. Just recently, scientists at The Australian National University (ANU) have produced a type of bifacial Si solar cell, using laser processing, setting a new world record effective power output of approximately 29%, well exceeding the performance of the best single-sided silicon solar cell [49]. On the other hand, major market players are likely to invest more in research and development to increase the efficiency of photovoltaic cells and find more effective material than silicon to build solar cells. However, these improvements will take time and are possible only if solar cells undergo bulk manufacturing and installations. In addition, although c-Si remains the dominant PV technology, it would need to continue to compete against evolving alternative PV technologies such as cadmium telluride (CdTe), copper indium gallium diselenide (CIGS), and perovskite modules. This competition will likely drive innovations for cost reductions across all technologies while presenting additional opportunities for system optimizations. Si PV may be augmented with multi-junction approaches with other semiconductors if the processes can be large-area centric and environmentally stable over the lifetime. Moving forward, new chemical approaches may be needed to make sand-to-Si reduction process more energy efficient [50]. Silicon is resilient. A 42 year-old, 42 W, 3.5 A, 12 V Kyocera solar panel exposed to environment is still reported to be working in Concord, NH [51]. *Silicon PV has become "too big to fail."* The future of solar cells is as bright as the Sun. Photovoltaics is set to become a dominant complementary energy technology of the world's energy portfolio.

Acknowledgments I gratefully acknowledge my PhD advisors Dr. G. C. Jain and Dr. B. K. Das; my NPL colleagues at that time when I worked on polysilicon solar cells, Dr. S. N. Singh, Dr. D. K. Sadana, Dr. R. Kishor, Dr. R. Kotnala, Dr. N. Goel, and Mrs. Balakrishnan, for their valuable insights and support. In my subsequent PV research, I acknowledge the collaborations from Dr. Michael Jackson, RIT; Todd Barnum, Linton Technology; Dr. Ajeet Rohatgi, Georgia Institute of Technology, Founder and CTO, Suniva, Inc.; Dr. David Young, NREL; Dr. Rajendra Singh, Clemson University; and Dr. Nish Sonwalkar, Founder, SunDensity. I highly value the contributions of our students, Felipe Freire, Sidhant Grover, Alex Marshall, Karine Florent, Ronak Kamat, and Archana Devasia in the PV research work conducted in my group. I overwhelmingly thank my collaborators for their wealth of knowledge and support. This work could not have been done without the funding support from agencies—National Science Foundation (via the Silicon Solar Consortium), Department of Energy, and from the industry. I express my sincere gratitude to my family for their steadfast support all along my career.

References

1. National Renewable Energy Laboratory, "Best research-cell efficiency chart," 2021 [Online]. Available: https://www.nrel.gov/pv/cell-efficiency.html
2. D. M. Chapin, C. S. Fuller, and G. L. Pearson, "A new silicon p-n junction photocell for converting solar radiation into electrical power," *J. Appl. Phys.*, vol. 25, no. 5, p. 676, 1954.
3. M. Wolf, "Historical development of solar cells," in *Proceedings of the 25th Power Sources Symp.*, pp. 120–124, May 1972.
4. R. L. Easton and M. J. Votaw, "Vanguard I IGY satellite (1958 beta)", *Rev. Sci. Instrum.*, vol. 30, pp. 70–75, Feb. 1959.
5. M. Green, "Silicon photovoltaic modules: A brief history of the first 50 years," *Prog. Photovolt: Res. Appl.*, vol. 13, no. 5, pp. 447–455, 2005.
6. H. J. Queisser, "Slow solar ascent," in *Advances in Solid State Physics*, 1st ed, vol. 44, Berlin, Germany: Springer, pp. 3–12, 2004.
7. M. Green, "Forty years of photovoltaic research at UNSW," in *Journal and Proceedings of the Royal Society of New South Wales*, vol. 148, no. 455/456, pp. 2–14, 2015.
8. Solar Energy Technologies Office, "The sunshot initiative." Energy.gov. https://www.energy.gov/eere/solar/sunshot-initiative (accessed Apr. 20, 2021).
9. G. C. Jain, B. K. Das, and S. Kumari, "On the origin of core losses in an Mn-Zn ferrite with appreciable silica content," *J. Appl. Phys.*, vol. 49, no. 5, p. 2894–2897, May 1978.

10. S. N. Singh, S. Kumari, and G. C. Jain, "Development of low cost polycrystalline silicon solar cells for terrestrial applications," in *Photovolt. Solar Energy Conference*, Cannes, France, pp. 615–619, Oct. 1980.
11. W. Shockley and H. J. Queisser, "Detailed balance limit of efficiency of *p-n* junction solar cells," *J. Appl. Phys.*, vol. 32, no. 3, pp. 510–519, 1961.
12. https://www.lmpv.nl/db/
13. A. Polman, M. Knight, E. G. Garnett, B. Ehrler, and W. C. Sinke, "Photovoltaic materials: Present efficiencies and future challenges," *Science,* vol. 352, no. 6283, p. 307–317, 2016. DOI: https://doi.org/10.1126/science.aad4424.
14. J. Liu, Y. Yao, S. Xiao, and X. Gu, "Review of status developments of high-efficiency crystalline silicon solar cells," *J. Phys. D: Appl. Phys.*, vol. 51, no. 12, 2018, Art. no. 123001.
15. X. Sun, M. R. Khan, C. Deline, and M. A. Alam," Optimization and performance of bifacial solar modules: A global perspective", *Applied Energy*, vol. 212, pp. 1601–1610, Feb. 2018
16. B. Kloter, D. Buß, A. Hussack, P. Kowalzik, S. Peters, M. Schutze, E. Stegemann, J. Muller, "Current status of high-efficiency Q.ANTUM technology at Hanwha Q CELLS", *2013 IEEE 39th Photovoltaic Specialists Conference (PVSC)*, 2013, pp. 1246–1248, doi: https://doi.org/10.1109/PVSC.2013.6744366.
17. M. Woodhouse, B. Smith, A. Ramdas, and R. Margolis., "Crystalline silicon photovoltaic module manufacturing costs and sustainable pricing: 1H 2018 benchmark and cost reduction roadmap," National Renewable Energy Laboratory, Golden, CO, USA, 2019. [Online]. Available: https://www.nrel.gov/docs/fy19osti/72134.pdf
18. Independent Commodity Intelligence Services, "Polysilicon (solar grade) methodology – September 2013," icis.com. [Online]. Available: https://www.icis.com/compliance/documents/polysilicon-solar-grade-methodology-september-2013/
19. G. Fisher, M. R. Seacrist, and R. W. Standley, "Silicon crystal growth and wafer technologies," *Proceedings of the IEEE*, vol. 100, Special Centennial Issue, pp. 1454–1474, May 2012, doi: https://doi.org/10.1109/JPROC.2012.2189786.
20. S. Schwinde, M. Berg, and M. Kunert, "New potential for reduction of kerf loss and wire consumption in multi-wire sawing," *Solar Energy Materials and Solar Cells*, vol. 136, pp. 44–47, May 2015.
21. TaiyangNews Webinar, "500W+ Solar Modules 2020", taiyangnews.info. http://taiyangnews.info/reports/500w-solar-modules-2020/ (accessed Apr. 25, 2021).
22. Vincent Shaw, "Big news as Zhonghuan Semiconductor unveils 12-inch super wafer", PV Magazine, Aug. 16, 2019. [online]. Available: https://www.pv-magazine.com/2019/08/16/big-news-as-zhonghuan-semiconductor-unveils-12-inch-super-wafer/ (accessed Jul 17, 2021).
23. Linton Crystal Technologies, https://www.lintoncrystal.com/ (accessed Jul 17, 2021).
24. J. Schmidt and K. Bothe, "Structure and transformation of the metastable boron- and oxygen-related defect center in crystalline silicon," *Physical Rev. B*, vol. 69, no. 2, Jan. 2004, Art. no. 024107.
25. N. E. Grant *et al.*, "Gallium-doped silicon for high-efficiency commercial passivated emitter and rear solar cells," *Sol. RRL*, vol. 5, Art. no. 2000754, 2021.
26. S. K Kurinec, C. Bopp, and H. Xu, "Emergence of continuous Czochralski (CCZ) growth for monocrystalline silicon photovoltaics," in *Emerging Photovoltaic Materials: Silicon & Beyond*, S. K Kurinec, Ed., New York, NY: Wiley, 2018, ch. 1, sec. 1, pp. 4–19.
27. "Module Materials," PVEducation.com. [Online] Available: https://www.pveducation.org/pvcdrom/modules-and-arrays/module-materials
28. H. Hanifi, J. Schneider, and J. Bagdahn, "Reduced shading effect of half-cell modules –Measurements and simulations" in *31st Eur. Photovoltaic Solar Energy Conference and Exhibition*, Sep. 2015. [Online]. Available: https://www.researchgate.net/publication/283488492
29. S. Braun, G. Hahn, R. Nissler, C. Pönisch, and D. Habermann, "The multi-busbar design: An overview" *Energy Procedia*, vol. 43, pp. 86–92, 2013, doi: https://doi.org/10.1016/j.egypro.2013.11.092.

30. B. L. Allsopp *et al.*, "Towards improved cover glasses for photovoltaic devices," *Prog. Photovolt. Res. Appl.*, vol. 28, no. 11, pp. 1187–1206, 2020.
31. L. Rodríguez, "Bifacial modules: the future of solar PV? A comprehensive guide on financial and technical performance of the next hot thing in solar: RatedPower, Feb 21, https://ratedpower.com/blog/bifacial-modules/# (accessed Jul. 17, 2021).
32. R. Rapier, "Transmitting solar power underneath an ocean," *Advanced Energy*, Nov. 2020. [Online]. Available: https://4thgeneration.energy/transmitting-solar-power-underneath-an-ocean/
33. U.S. Dept. of State, "Strategic Defense Initiative (SDI)," 1983. [Online]. Available: https://2001-2009.state.gov/r/pa/ho/time/rd/104253.htm
34. A. Devasia and S. Kurinec, "Teaching solar cell I-V characteristics through SPICE modeling," *Am. J. Phys.*, vol 79, no. 12, pp. 1232–1239, Dec. 2011, doi: https://doi.org/10.1119/1.3636525.
35. R. A. Kamat, "Engineered emitters for improved silicon photovoltaics," M.S. Thesis, Dept. Elect. Eng., Rochester Inst. Tech., Rochester, NY, USA, 2016. [Online]. Available: https://scholarworks.rit.edu/theses/9337
36. A. Kale E. Beese, T. Saenz, E. Warren, W. Nemeth, D. Young, A. Marshall, K. Florent, S. Kurinec, S. Agarwal, and P. Stradins, "Study of nickel silicide as a copper diffusion barrier in monocrystalline silicon solar cells," in *2016 IEEE 43rd Photovoltaic Specialists Conference (PVSC)*, pp. 2913–2916, 2016, doi: https://doi.org/10.1109/PVSC.2016.7750190.
37. A. Marshall, K. Florent, A. Tapriya, B. G. Lee, S. K. Kurinec, and D. L. Young, "Nickel silicide metallization for passivated tunneling contacts for silicon solar cells," in *2016 IEEE 43rd Photovoltaic Specialists Conference (PVSC)*, pp. 2479–2482, 2016, doi: https://doi.org/10.1109/PVSC.2016.7750089.
38. S. Grover, S. Sahu, P. Zhang, K. O. Davis, and S. K. Kurinec, "Standardization of specific contact resistivity measurements using transmission line model (TLM)," in *2020 IEEE 33rd International Conference on Microelectronic Test Structures (ICMTS)*, Edinburgh, Scotland, UK, 2020, pp. 1–6, Apr. 2020. doi: https://doi.org/10.1109/ICMTS48187.2020.9107911.
39. F. Freire, R. Dias, T. Trabold, and S. K. Kurinec, "Performance analysis of a "green" building photovoltaic system," in *2016 IEEE 43rd Photovoltaic Specialists Conference (PVSC)*, 2016, pp. 2657–2660, doi: https://doi.org/10.1109/PVSC.2016.7750131.
40. S. K. Kurinec, M. Kucer, and B. Schlein, "Monitoring a photovoltaic system during the partial solar eclipse of August 2017" *Eur. J. Photovolt.*, vol. 9, 2018, Art. no. 7.
41. F. Freire, S. Melcher, C. G. Hochgraf, and S. K. Kurinec, "Degradation analysis of an operating PV module on a farm sanctuary," *J. Renew. Sust. Energy,* vol. 10, no. 1, 2018, Art. no. 013505, doi: https://doi.org/10.1063/1.4994565.
42. SunDensity, https://www.sundensity.net/
43. Fraunhofer Institute for Solar Energy Systems, "Photovoltaics Report," ISE, Feb. 2020. [Online]. Available: www.ise.fraunhofer.de/content/dam/ise/de/documents/publications/studies/Photovoltaics-Report.pdf
44. Z. Zhou and M. Carbajales-Dale, "Assessing the photovoltaic technology landscape: efficiency and Energy Return on Investment (EROI)," *Energy Environ. Sci.*, vol. 11, no. 3, pp. 603–608, 2018.
45. F. Liu and J.C.J.M. van den Bergh, "Differences in CO2 Emissions of Solar PV Production among Technologies and Regions: Application to China, EU and USA," *Energy Policy*, vol. 138, 2020, Art. no. 111234.
46. A. Jäger-Waldau, "Snapshot of photovoltaics—February 2020," *Energies*, vol. 13, no. 4, pp. 930–937, 2020.
47. N. Haegel, H. A. Atwater, T. Barnes, and C. Breyer, "Terawatt-scale photovoltaics: Transform global energy," *Science*, vol. 364, no. 6443, pp. 836–838, May 2019, doi: https://doi.org/10.1126/science.aaw1845.
48. IEA, "Solar PV," IEA.com, Paris, 2020. [Online]. Available: https://www.iea.org/reports/solar-pv
49. https://www.anu.edu.au/news/all-news/anu-scientists-set-new-record-with-bifacial-solar-cells

50. S. Maldonado, "The Importance of new "sand-to-silicon" processes for the rapid future increase of photovoltaics", *ACS Energy Letters* 2020 *5* (11), 3628–3632, DOI: https://doi.org/10.1021/acsenergylett.0c02100
51. D. Brooks, "A solar panel in the New Hampshire woods is old enough to run for president", Concordmonitor.com, Jun. 26, 2018. [Online]. Available: https://www.concordmonitor.com/oldest-solar-panel-NH-power-18318376

Santosh K. Kurinec is a Professor in the Department of Electrical and Microelectronic Engineering at Rochester Institute of Technology (RIT). She is a Fellow of IEEE and a member of the New York Academy of Sciences. She received her Ph.D. in Physics from University of Delhi, India, and worked as a Scientist at National Physical Laboratory, New Delhi. She joined the Department of Materials Science and Engineering at the University of Florida as a Postdoctoral Research Associate. Prior to joining RIT, she was Assistant Professor of Electrical Engineering at Florida State University/Florida A & M University College of Engineering in Tallahassee, FL. Her current research activities include advanced integrated circuit materials and photovoltaics. She received the 2012 IEEE Technical Field Award for integrating research in teaching to prepare microelectronic engineers for future challenges. She was inducted in the Women in Technology International Hall of Fame in 2018. She has over 120 publications in research journals and conference proceedings. Her recently edited books are *Energy Efficient Computing & Electronics: Devices to Systems*, and *Nanoscale Semiconductor Memories: Technology and Applications*, and *Emerging Photovoltaic Materials: Silicon & Beyond.*

From Nuclear Engineering to Roller Coasters: The Ride of a Lifetime

Cynthia Sypher

How does one know that they are on the path to the right career? As a senior high school student, I considered taking theoretical mathematics in university but was unsure of what my career options would be upon graduation. My queasy stomach (later verified by a very embarrassing, short-lived attempt as an intern at a veterinary clinic) put a medical career off the table. My industrial arts (a.k.a. "Shop") teacher suggested that I apply to engineering. At that time, I had no idea what engineers did; I was the first person in my family to even apply to university. Upon researching the profession, I was captivated by the idea that you could have a full-fledged career after finishing your degree. The co-operative (co-op) program at the University of Waterloo, in Waterloo, Ontario, Canada would also provide the financial means for me to attend university without accumulating student debt. The co-op program also provides the opportunity to obtain valuable work experience during the five-year program.

What I did not know is that an engineering degree provides you with a framework to solve problems, even for those graduates who choose not to stay in engineering.

My last co-op work term in college was for an engineering consulting company that specialized in laboratory testing and engineering-procurement-construction for the nuclear industry. I was pleased when they provided me with the opportunity to start my engineering career there as a new graduate.

During my work term, I was hired to instrument a model steam generator with several thermocouples and instrumentation so that we could test the sample boiler for accident scenarios required in the design process of a nuclear power plant. In Canada, we had our own nuclear reactor technology called the CANDU reactor, illustrated in Fig. 1. Like all nuclear power plants, the splitting of atoms to produce a chain reaction, known as nuclear fission, is used to create heat. This heat is fed to

C. Sypher (✉)
Sypher and Associates Field Engineering Ltd., Toronto, ON, Canada
e-mail: cindy.sypher@safeconsulting.ca

M. Bailey, L. Shackelford (eds.), *Women in Mechanical Engineering*, Women in Engineering and Science, https://doi.org/10.1007/978-3-030-91546-9_16

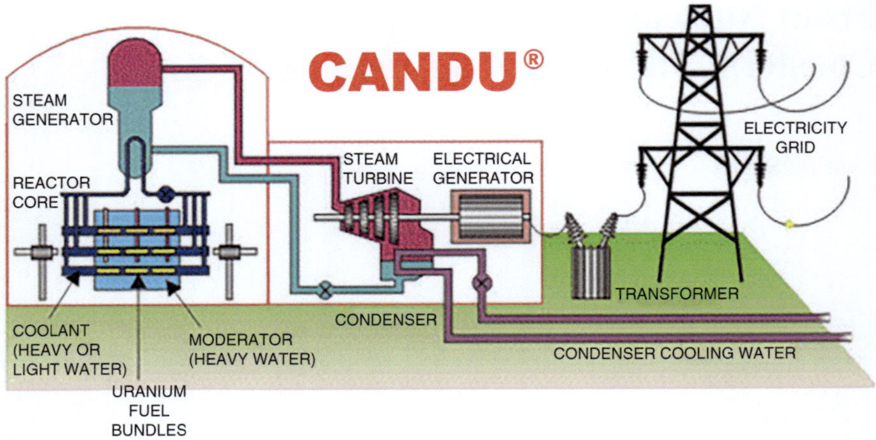

Fig. 1 Basic CANDU nuclear reactor design. (Image © Canadian Nuclear Association)

a steam generator and is used to turn water into steam. The steam rotates a turbine connected to an electrical generator, which produces electricity.

CANDU stands for Canada Deuterium Uranium. The CANDU reactor is a heavy water reactor design developed in Canada. Deuterium is the primary element in the heavy water moderator, and uranium is the fuel used in this reactor class [1].

While I initially found nuclear engineering interesting, as a young graduate it was very lonely to work for a small engineering firm of only seven people. An unusual part of the consulting services that we offered involved providing on-site safety engineering services to a major Canadian fair and the other events at the exhibition grounds where they were held. These services included providing amusement ride inspections, structural reviews, and inspections for all the events, life and fire safety reviews, and reviewing performances for general public safety. Our company had been providing these services since 1987. It was my dream job!! I begged my boss for an entire year to let me work there. Being old-school and traditional, he was not initially keen on this idea. He thought that I might be too sensitive for that type of fieldwork. (I am admittedly a very sensitive person, and even after all of these years, I still have "thin-skin.") My tenacity got the better of him; and when a position opened on-site as a full-time safety engineer, I was transferred there.

Working for this small consulting company allowed me to learn about the nuclear industry and to apply engineering principles to the design and regulation of amusement rides, ski lifts, special events, temporary structures, festivals, and concerts.

While seemingly very different, there are several commonalities between the design and testing of nuclear power plant safety components and the design review of amusement rides. These similarities include but are not limited to the safety performance level of the components, quality assurance programs, regulatory oversight, and public policy.

Safety Performance Level of Components

The Convention on Nuclear Safety, in addition to local regulatory requirements, requires that the design and construction of nuclear installations provide for several reliable levels and methods of protection (defense in depth) against the release of radioactive materials, and that the technologies incorporated in the design and construction of a nuclear installation be proven by experience or qualified by testing or analysis [2].

Qualification of equipment important to safety in nuclear power plants (NPPs) ensures its capability to perform designated safety functions on demand under postulated service conditions including harsh accident environment (e.g., loss of coolant, or LOCA), high energy line break (HELB), and seismic or other vibration conditions. The designer must ensure that safety-related components meet the design requirements for what is known as a "qualified life." The qualification process ensures that each component has the capability to perform designated safety functions on demand for the length of its "qualified life." The goal for this is typically a 30- to 60-year period. By the end of the qualified life, the components must be replaced. The design process helps to determine what the qualified life of a component will be. Environmental exposures such as elevated temperature, pressure, cyclical loading, and humidity all play a part. Potential occurrences/accidents such as an earthquake, chemical exposure, a "loss of coolant," or a "high energy or main steam line break" need to be planned for and incorporated into the design of safety-related components. Some components are required to function beyond the design-basis accident conditions.

The methods of qualification include:

1. Performance of a type test on representative equipment
2. Performance of a test on the actual equipment
3. Application of pertinent past experiences in similar applications

Engineering Analysis

In my early career, my position focused on the testing aspect of environmental and seismic qualification (EQ and SQ) of nuclear safety components. The EQ program ensures that all required equipment in a nuclear facility is qualified to perform its safety functions if exposed to harsh environmental conditions resulting from design-basis accidents. The SQ program ensures that all seismically credited structures, systems, and components are designed, installed, and maintained to perform their safety function during and/or after an earthquake event [2]. Special attention must be given to non-metallic components. Exposure to heat, radiation, and humidity will accelerate aging effects of polymers. Aging of polymers affects hardness, elongation-at-break, modulus of elasticity, compression resistance, insulation resistance, voltage sensitivity, sensitivity to chemicals, sensitivity to aggressive gases, sensitivity to vibration, color, dielectric constant, phase equilibrium, etc.[2].

In the testing environment, non-metallic components are exposed to humidity, radiation, and cycle aging that they would reasonably be exposed to during their qualified life. The effects are achieved by accelerated thermal aging. The acceleration is achieved through exposure to elevated temperature for a testing time. It is assumed that the relationship between temperature and the rate of degradation, r, follows the Arrhenius relationship:

$$r = Ae^{-\frac{E}{kT}}$$

where

A is a constant for the material tested
E is the activation energy for the process (in eV)
K is Boltzmann's constant
T is the temperature (in K)

The acceleration factor *F* is the ratio between the rate of degradation at the elevated temperature (the temperature at which thermal aging is completed) and the rate of degradation at the field temperature at normal operation, as follows:

$$F = \frac{r_2}{r_1} = e^{\frac{E}{k}\frac{1}{T_1}-\frac{1}{T_2}}$$

where

r_1 = rate of degradation in exposure to the temperature at normal operation
r_2 = rate of degradation at the test (thermal aging) temperature
T_1 = temperature (in K) at normal operation
T_2 = test (thermal aging) temperature (in K)

The qualified life t_{qual} is:

$$t_{\text{qual}} = F * t_{\text{test}}$$

where

$$t_{\text{qual}} = \text{time of exposure at the test}\left(\text{thermal aging}\right)\text{temperature}$$

A safety margin is also added.

Once a test component has been aged to reach its "qualified life," it is exposed to the environmental and/or seismic event(s) that it is expected to withstand. Our test lab had several thermal aging chambers, one boiler, a superheater, an environmental test chamber, and two seismic tables where we conducted the testing of nuclear safety components. Part of my early role was to instrument not only our test equipment but also special test set-up for various cycle-loading scenarios.

Amusement Ride Safety

As with nuclear safety components, for amusement rides, design and the end environment play a key role. There are three globally recognized standards for amusement rides: ASTM International F24 group of standards, EN 13814 "Euro Norm" Standards, and ISO 17842 standards. In most Canadian provinces, amusement rides are designed to the ASTM International F24 group of standards. The group of standards, shown in Fig. 2, is broken down into core standards, supporting standards, and ride-specific standards.

Rides must be designed for their environment. Wind, seismic, and rain/snow loads should all be considered. A ride designed to operate near the ocean will experience greater corrosion than one installed further inland. The significant and predictable acceleration that is generated by the ride is another key consideration in the design calculations. Dynamic factors are applied to impact loads. A pre-determined number of ride cycles is also considered in the design process. In accordance with ASTM F2291-20 "Standard Practice for Amusement Rides and Devices," all primary structures of an amusement ride or device (e.g., track, columns, hubs, and arms) shall be designed using calculations and analyses that are based on the minimum 35,000 operational hour criteria. The designer/engineer shall verify that the

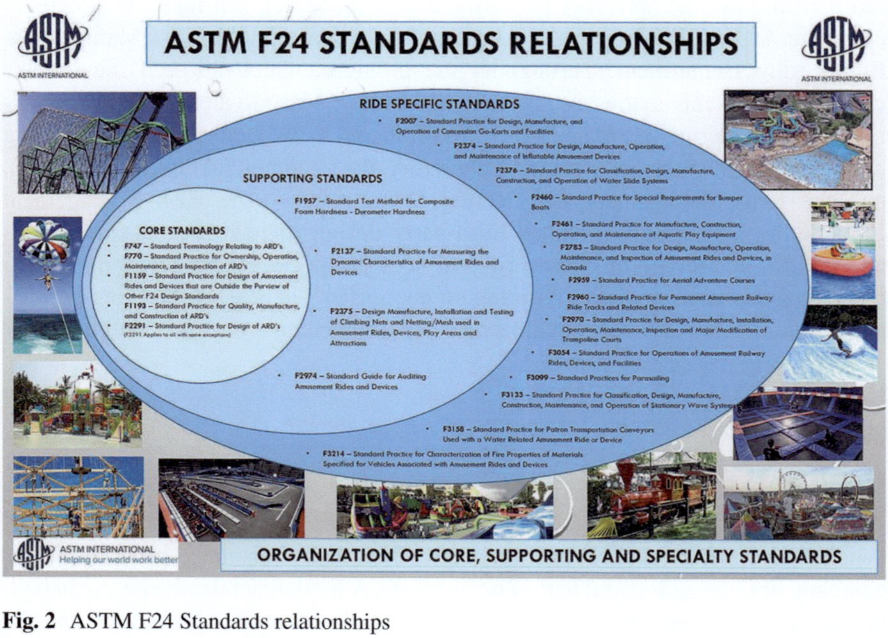

Fig. 2 ASTM F24 Standards relationships

calculations and analyses meet or exceed this minimum operational hour requirement. This requirement is intended to ensure that all primary structures within an amusement ride or device are designed for at least a minimum fatigue life [3]. A simplified definition of fatigue is how long a component will last before failing due to cyclic loading. There are a number of factors which will affect a component's fatigue life including: the type of material used, the temperature, and the shape of the component.

When designing an amusement ride, the designer is also required to consider a failure analysis for safety-related systems. The failure analysis shall include a fault tree analysis, a failure mode and effects analysis (FMEA), or other accepted engineering practices [4]. The purpose of these practices is to apply a systematic approach to consider all of the factors that may lead to failure of individual components of the ride or the ride as a whole. The Hazard Mitigation process outlined in ASTM F2291-20 is structured to consider the anticipated lifetime of components, based on their expected use, as well as the means and methods for detecting component failures. Detection methods include but are not limited to automatic detection by a control system, periodic operational testing, and periodic inspection of components including non-destructive testing.

The complex nature of both nuclear power plants and amusement rides led me to enroll in a master of engineering program. Under this program, I tailored my graduate courses with everything applicable to the challenges encountered in my career. This included courses on the Canadian Steel Code CSA S-16 "Design of steel structures," dynamics, vibrations, impact biomechanics, fluid power systems, nuclear reactor safety design, engineering fire mechanics, and finite element methods. A part-time, coursework's master's degree is not for everyone (one of my professors referred to it as "the hard way [to earn a master's degree]"). It took me 7 years to complete. At the start of my master's program, I was managing the Risk Management Engineering Department. Throughout the program, which allowed me to take courses at three universities, as needed, I had a baby, my son Valek. Shortly after my return from maternity leave in 2005, I was promoted to division manager overseeing three departments: the Engineering Laboratory, the Commercial Grade Dedication, and the Risk Management Engineering Departments. (The Risk Management Engineering Department included third-party design review of amusement rides and ski lifts; and providing engineering services for special events, temporary structures, festivals, and concerts.) Working on my master's degree while trying to manage an engineering division, having a new baby, and being a young mother made for an interesting journey.

Quality Assurance

When I was promoted to division manager, the quality assurance program of our company became an integral part of my purview. A fundamental cornerstone in both nuclear and amusement ride design is the quality assurance programs for the

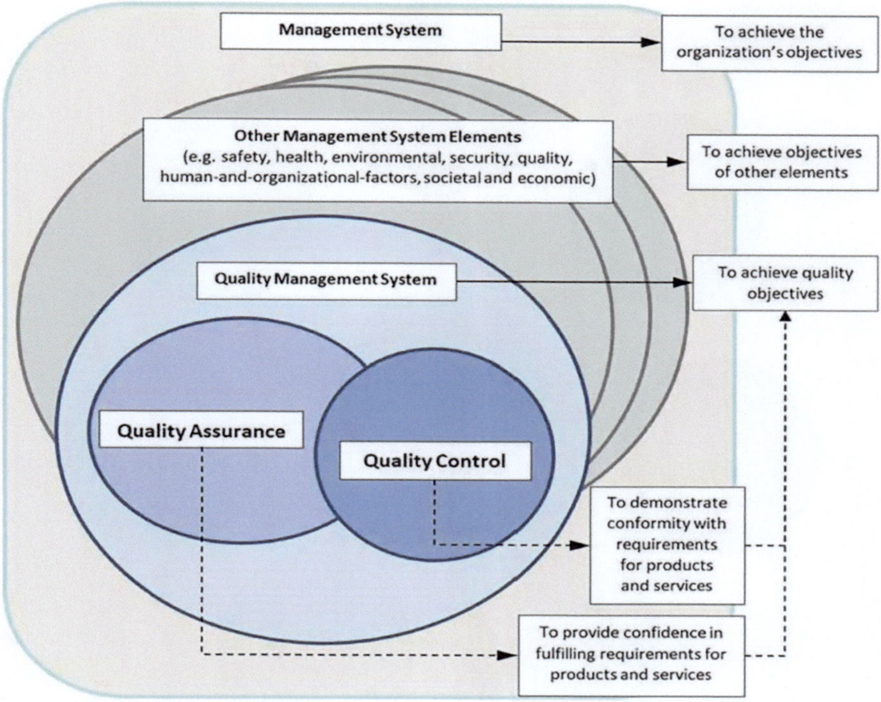

Fig. 3 Quality assurance and quality control as a part of the overall management system [4]

manufacture and testing/commissioning of the components. Quality assurance and quality control are a part of the overall management system in an organization. The relationship is shown in Fig. 3.

The successful implementation of quality assurance (QA) and quality control (QC) is essential to providing confidence in the nuclear industry. A high degree of reliability and integrity is required of products and services, and the requirements are particularly stringent for assuring nuclear safety. Failure of structures, systems, or components to perform their intended function, or their poor performance, could adversely affect the health and safety of workers and the public [4].

The QA and QC frameworks are shown in Fig. 4 and 5.

The parallel processes of QA and QC are shown in Fig. 6. Those shown in the red box were applicable to my role as the division manager of a supplier to the nuclear industry in Canada.

In the nuclear industry in Canada, there are quality assurance programs that fall under several nuclear standards. The four main standards are:

- ISO 9001 "Quality Management Systems"[1]

[1] From the mid 1990s to 2016, ISO 9001 became the commercial quality standard that was generally adopted by the nuclear industry. It was augmented by other standards to create industry-spe-

Fig. 4 Responsibilities for quality assurance activities [4]

- CSA N299 series "Quality Assurance Program Requirements for the Supply of Items and Services for Nuclear Power Plants"[2]
- CSA N285.0–2012/N285.6 Series "General Requirements for Pressure-Retaining Systems and Components in CANDU Nuclear Power Plants/Material Standards for Reactor Components for CANDU Nuclear Power Plants"[3]

cific quality management system. The previously withdrawn CSA N299 Series of standards was re-developed and released in 2016. This new group of standards acts to augment ISO 9001 to provide a comprehensive industry-specific standard for CANDU facilities.

[2] This standard is a group of standards that provide quality assurance for the procurement of items and services for CANDU nuclear facilities. N299 focuses on the entire manufacturing and planning process. N299 documentation outlines the expectations of the supplier and the customer during design, procurement, manufacturing, testing, and inspecting.

[3] This standard series specifies the technical requirements for the design, procurement, fabrication, installation, modification, repair, replacement, testing, examination, and inspection of, and other work related to, pressure-retaining systems, components, and supports over the service life of a CANDU nuclear power plant.

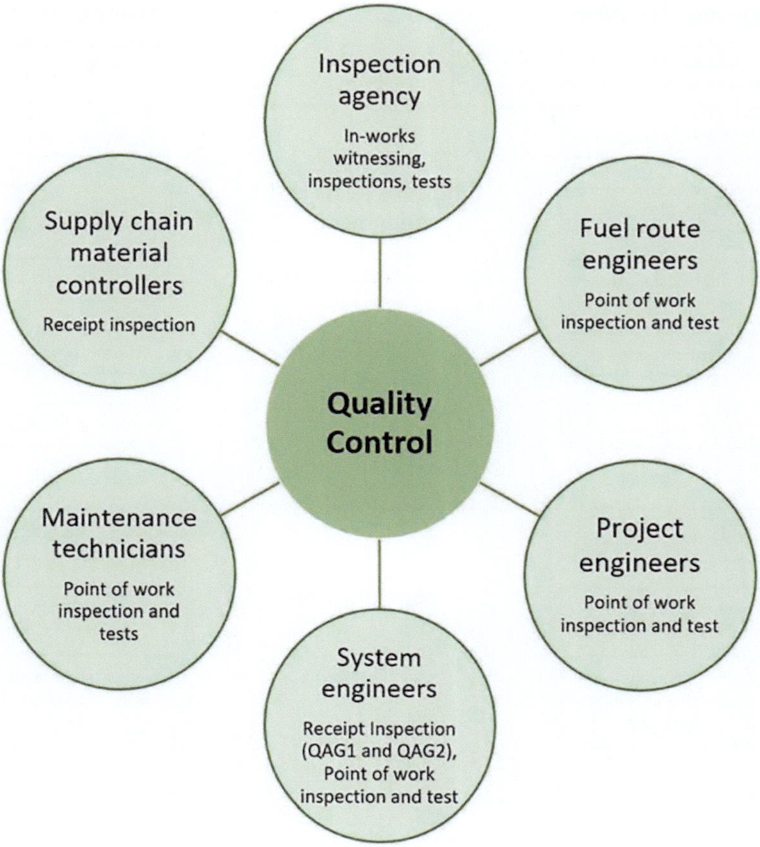

Fig. 5 Responsibilities for quality control activities [4]

- CSA B51 "Boiler, Pressure Vessel, and Pressure Piping Code"[4]

As my career progressed, that small consulting company of seven grew larger and was bought out by a US publicly traded company in 1998. With over a hundred employees and a growing focus on the nuclear sector, over time the differences between the nuclear part of the business and the small amusement ride safety-engineering division grew more apparent. On the nuclear side, the contracts were very large and could handle all the stringent overhead requirements imposed upon them by a nuclear quality program. The numerous but seemingly minute safety contracts became almost burdensome to the company, and with the associated liability, that business sector was seen in an unfavorable light. After a corporate

[4] Part 1 of this standard contains requirements for boilers, pressure vessels, pressure piping, and fittings. It is intended mainly to fulfill two objectives: first, to promote safe design, construction, installation, operation, inspection, testing, and repair practices; and second, to facilitate adoption of uniform requirements by Canadian jurisdictions.

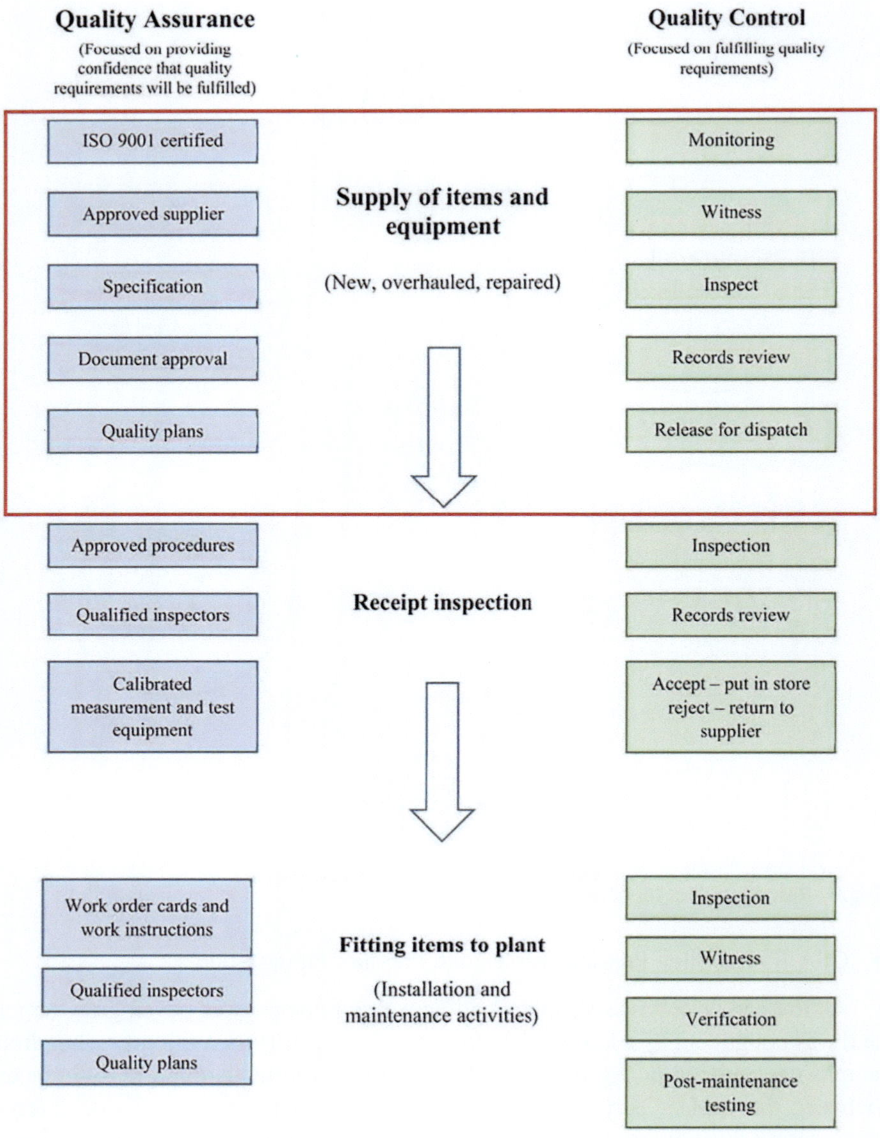

Fig. 6 The roles of QA and QC in a typical CANDU NPP [4]

restructuring in 2009, I was in a position where I would have to focus my future career on nuclear work only. This broke my heart, as my true passion was the amusement ride sector. I was left with a choice to make: the safe path of continuing to work in the nuclear sector or to start my own business and follow my passion. I chose the latter, and in 2009 I left my 17-year job and ventured out to form a boutique engineering company focusing exclusively on providing engineering services to the amusement ride and special event sectors. I followed the advice from my lawyer "not to burn any bridges," and I was upfront and honest with my employer about my intentions. They were initially offended but understood my position. In the end, they decided that amusement ride business did not fit within their corporate strategy, and they referred all those clients to my new company.

The knowledge gained while preparing for nuclear quality audits enabled me to prepare quality assurance plans for amusement ride/device manufacturers once I had my own consulting company.

Amusement rides are also subject to strict quality program requirements. Under ASTM F1193-18a, the manufacturer of an amusement ride shall have a written quality assurance program for use in conjunction with the design, manufacture, construction, modification, or reconditioning of the amusement ride [5]. The manufacturer is required to keep quality assurance documents including material certifications, test reports, and inspection reports. A drawing control procedure is required to track manufacturing drawings, engineering revisions, and related documents. The drawing control procedure ensures that proper distribution, movement, and keeping of all drawings is done and that those personnel requiring access to project documents will have the most up-to-date revisions and are aware of the document control process.

To ensure that materials, processes, and components (including raw materials) are in compliance with engineering specifications, the manufacturer shall have a material and component control procedure. Hold-point inspections are to be made on manufactured parts and subassemblies to verify conformance with the designer–/engineer-specified criteria. Changes to the design-/engineer-specified criteria are to be documented and approved. Non-conforming components are required to be dispositioned.

All tools and equipment used for the inspection and testing of materials and parts for the manufacturing of ride components/parts/subassemblies should be calibrated. All calibration shall be traceable to a national standard. Calibration should be completed on an annual basis or at a more frequent interval, as required by the equipment manufacturer.

In Canada, welding and welding procedures shall be completed by a certified welding company in accordance with CSA W47.1 and CSA W59 and shall be performed by certified welders. Supervision shall be provided by a certified welding supervisor. Canadian Welding Bureau (CWB)–approved Welding Procedure Data Sheets (WPDS) shall be used. The information contained in a WPDS shall include the type of joint, joint geometry, welding position, welding symbol, welding consumables, gas mixture, and the welding parameters that are to be used in making a weldment. CSA W117.2 shall be followed [6], as applicable for all welding procedures and processes. All CWB documentation for the certified welding company, qualified welder(s), welding supervisor, and WPS/WPDS shall be maintained.

Pressurized systems are also subject to the requirements of CSA B51 "Boiler, Pressure Vessel, And Pressure Piping Code."

Prior to shipping an amusement ride or component, the manufacturer shall generate a document certifying that the ride or component is in compliance with ASTM

Practice F1193. This certification shall be retained with other quality assurance documents for the ride or component. When requested by an authority having jurisdiction, purchaser, or owner, the manufacturer shall provide a copy of this certification document.

Regulatory Oversight

Regulatory oversight is a key component in the safe operation of both nuclear power plants and amusement rides. Nuclear programs have federal oversight in the United States and Canada, whereas amusement rides are regulated by the states/provinces.

Regulatory oversight is typically structured as shown in Fig. 7. Governments issue legislation, viz. laws, decrees, and orders. Regulatory bodies issue both legally binding regulations and non-legally binding guides. Standards are usually written by standards development organizations, and they may be mandatory or non-mandatory, depending on the local authority having jurisdiction. Other mechanisms that may make standards mandatory are contractual stipulations [4].

Nuclear power plants in Canada are subject to the Nuclear Safety and Control Act (NCSA) and its regulations. The Canadian Nuclear Safety Commission (CNSC) administers and enforces the NSCA and regulates all nuclear activities at nuclear-generating facilities, including their licensing and radiation protection programs.

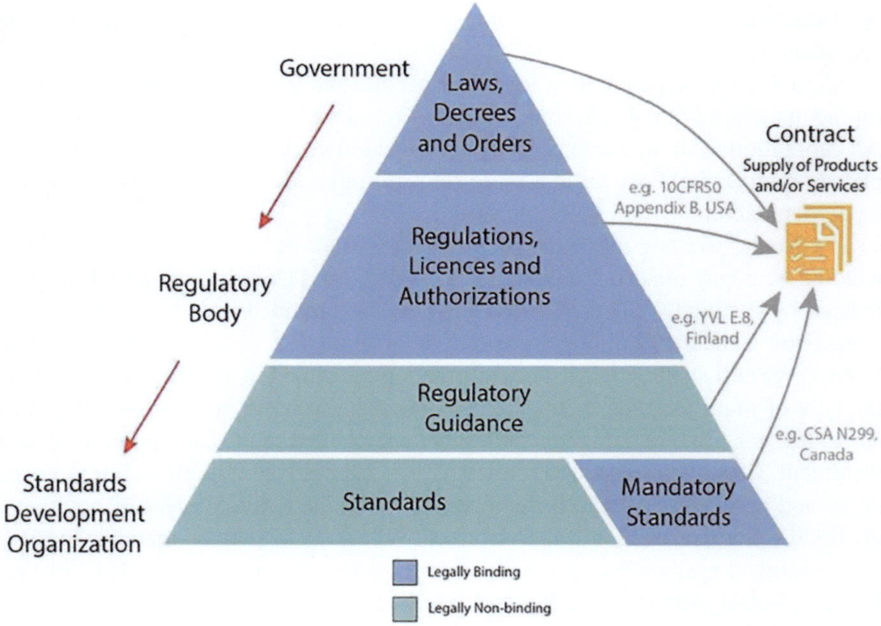

Fig. 7 Regulatory framework [4]

The CNSC is an independent administrative tribunal set up at arm's length from government. The Commission is supported by more than 800 scientific, technical, and professional staff. (In the United States, the Nuclear Regulatory Commission (NRC) is responsible for licensing facilities, commercial use of nuclear materials, facility inspection, and the creation of standards and regulations for nuclear power plant safety.)

In Ontario, Canada, where I reside and work, additional regulation for nuclear boilers and pressure vessels is done on the provincial level by the Technical Standards and Safety Authority (TSSA).

Amusement rides in Canada, the United States, and Australia are regulated on the provincial or state level. In Ontario, Canada, the regulator is also the TSSA. The adoption through TSSA's Code Adoption Document of ASTM F2783–20, "Standard Practice for Design, Manufacture, Design, Operation, Maintenance and Inspection of Amusement Rides and Devices, in Canada," provides a pointer document to draw in the core standards, supporting standards and ride-specific standards of ASTM F24 group of standards for amusement rides and devices (Fig. 2).

As a regulator, the TSSA encourages stakeholder engagement via its advisory councils and technical committees.

Stakeholder Engagement

Throughout my career, I have had numerous opportunities to participate in public policy development for both the nuclear and amusement ride industries. These opportunities include participation in standards development, acting on advisory/technical councils, in addition to stakeholders' boards such as the Organization for Canadian Nuclear Industries.

One area where I was able to participate in the development of nuclear standards was through the Canadian Standards Association on the technical committee N289.1 for the standard "General requirements for seismic design and qualification of CANDU nuclear power plants." This experience showed me the value of participating in standards development as a means to give back, with the benefit of networking with industry leaders.

During my tenure as the division manager of technical services, I was fortunate to hold a volunteer position on the board of directors of the Organization of Canadian Nuclear Industries (OCNI—formerly the Organization of CANDU Industries). The mission of OCNI is to promote companies in the Canadian private sector engaged in the supply of goods and services for CANDU and Light Water Reactor (LWR) nuclear power plants in domestic and export markets [7]. This volunteer experience allowed me to help to shape policy for the nuclear supply chain but also provided several personal benefits. These included: acquiring corporate governance skills, improving my strategic business thinking, enhancing confidence, providing first-hand experience of board-level operations, and offering invaluable networking opportunities.

In parallel to working on the CSA N289.1 technical committee on the nuclear side, I also worked on the (former) CSA Z267 technical committee for "Safety Code for Amusement Rides and Devices." After the 2000 edition of this standard was released, our technical committee was informed by industry stakeholders that the industry wanted Canada to adopt the ASTM group of standards for amusement rides and devices. Unlike the CANDU technology of the Canadian nuclear industry, there were not a significant number of amusement ride manufacturers in Canada (apart from two prominent water slide manufacturers). In 2006, the CSA Z267 task group on harmonization embarked on a partnership with the ASTM F24 Committee. CSA and ASTM struck a subcommittee to develop a new ASTM F24 standard, a comprehensive, one-volume document, that provides a mechanism to:

- Adapt F24 standards to Canadian needs
- Provide greater two-way communication between CSA and ASTM
- Facilitate adoption by Canadian provinces and territories

Over the 8 years we took to develop the first release of ASTM F2783, "Standard Practice for Design, Manufacture, Operation, Maintenance, and Inspection of Amusement Rides and Devices, in Canada", I have made many friends and colleagues through my involvement with both CSA and ASTM on this initiative. I am now the co-chair of the F2783 task group and am so proud of the contributions we have made to harmonize amusement ride standards globally.

The TSSA works with industry experts through advisory councils to gather information. This information assists the TSSA with making informed decisions related to their safety mandate. I was fortunate to have the opportunity to attend these meetings early on in my career in the amusement devices sector at times when my boss and engineering mentor was not able to attend. Upon his retirement in 1998, I was given the opportunity to join the industry advisory council for the amusement ride sector. Today I am the chair of this council. I value our council's ability to provide the TSSA advice and input in relation to safety strategies from stakeholders of our industry. Our primary focus is to:

- Identify strategic safety issues, current or anticipated, and to provide guidance for their resolution
- Provide input and advice regarding general enhancements in the service delivery by TSSA

I value immensely the volunteer opportunities I have been afforded for stakeholder engagement. As a means to give back, these opportunities have also greatly benefitted me by providing leadership opportunities, enhancing my confidence, building my business networks, and expanding my technical and business acumen.

Conclusion

My journey as a mechanical engineer began as a co-op student for an engineering consulting company in the nuclear sector. I am appreciative of the opportunity to practice in seemingly very different fields of the nuclear and amusement ride sectors. Seventeen years later, in 2009, I embarked on my next adventure when I opened my own boutique engineering consulting company providing risk management engineering to the amusement, special events, and manufacturing industries. In 2020, after building a successful consulting company in the face of the COVID-19 pandemic, I lost 90% of my business due to the ensuing lockdowns. I was able to pivot to work in the nuclear industry again, now with a focus on small modular reactors (SMR). I have come full circle from nuclear to amusement rides and back to nuclear. I am grateful that being a mechanical engineer has made that possible. Having this degree has enabled me to be resilient throughout my career and certainly through the past 16 months which have been most challenging in my life.

If, as a young student who loves math and science, you are not sure what career path to take, I recommend mechanical engineering. Even if you pursue other career options after graduating, this degree will provide you with the problem-solving and technical skillset to lay the foundation for success in all your future endeavors.

My journey continues, and I look forward to the lessons to be learned from this next phase.

References

1. W. L. Sunshine. “How heavy water moderates CANDU nuclear reactors.” ThoughtCo. thoughtco.com/candu-nuclear-reactor-is-moderated-with-heavy-water-1182652 (retrieved, Oct. 29, 2020).
2. Equipment Qualification in Operational Nuclear Power Plants*: Upgrading, Preserving and Reviewing*, International Atomic Energy Agency, Vienna, Austria, 1998. Available: Https://Www-Pub.iaea.org/MTCD/Publications/PDF/P052_scr.Pdf.
3. *Standard Practice for Design of Amusement Rides and Devices,* F2291-20, ASTM International., West Conshohocken, PA, USA, 2020.
4. *Quality Assurance and Quality Control in Nuclear Facilities and Activities*, International Atomic Energy Agency, IAEA-TECDOC-1910, Vienna, Austria, 2020.
5. *Standard Practice for Quality, Manufacture, and Construction of Amusement Rides and Devices1,* F1193-18a, ASTM International, West Conshohocken, PA, USA, 2018.
6. *Safety in Welding, Cutting and Allied Processes,* CSA W117.2-12(R2017), Canadian Standards Association, Toronto, Canada, 2017.
7. “About OCNI.” Organization of Canadian Nuclear Industries. www.ocni.ca/history/ (accessed Jan. 29, 2021).

Cynthia Sypher is a Designated Consulting Engineer and the President of Sypher & Associates Field Engineering Ltd. She is licensed to practice engineering in the provinces of Ontario, British Columbia, Newfoundland and Labrador, Alberta, and Saskatchewan. She has Bachelor's and Master's degrees in Mechanical Engineering from the University of Waterloo.

Her consulting interests include: temporary structures, mechanical and structural design, fire and life safety, emergency preparedness, rigging, general public safety initiatives, amusement rides, risk management, machine guarding, and playground design and inspection. She has consulted to major fairs, parks, and event promoters.

She is currently the Chair of the Technical Standards and Safety Authority's (TSSA) Amusement Devices Industry Advisory Council and is the Co-Chair of ASTM F24.80 Harmonization Task Group for Canadian Harmonization developing amusement ride standards for Canada. Ms. Sypher is a member of the International Association of Amusement Parks and Attractions (IAAPA) North American Safety Committee. She is the former vice chair of the CSA Z267 Amusement Rides Harmonization Technical Sub-Committee. She has developed and presented curriculum for the programs at the TSSA's NAARSO Safety Seminar, AIMS, IAAPA, and other industry associations.

Ms. Sypher also has 18 years of experience in the nuclear industry. She has participated in standards development on the CSA N289 series of standards for seismic design and qualification of CANDU nuclear power plants. She is a former member of the board of directors for the Organization of CANDU Industries.

Ms. Sypher has been consulted to render opinions for amusement devices litigation as well as on general mechanical and structural engineering issues. Ms. Sypher has been a member of the International Association of Amusement Parks and Attractions, ASTM International, the National Association of Amusement Ride Safety Officials, Amusement Industry Manufacturers and Suppliers International, and the Canadian and International Associations of Fairs and Exhibitions.

Section IV
Career Journeys

Unveiling My Engineering Identity

Yen-Lin Han

"Please introduce yourself."

"I am an associate professor of mechanical engineering at Seattle University. I teach energy systems and data acquisition courses this quarter. And I do research on how we can build a culture in academia that fosters engineering identity." This is how I often introduce myself nowadays. Interestingly, I rarely say, "I am an engineer," despite having an advanced degree in engineering. As someone studying others' engineering identities, I cannot help but ask myself what my engineering identity is and what it means to present myself as an engineer.

So, what is identity? Identity is both within the individual (personal) and without (social) [1]; it is situated in the self and in the groups to which we think we belong [2]. Identity is a consistency of the self across time [3] and a reflection of the individual's environment [4]. Our identities shape the experiences we embrace, and those experiences shape our identities [5–7]. We behave consistently with our identities [8, 9], choosing behaviors with meanings that match our self-conceptions [10, 11]. Identity may be an abstract concept, yet it is inseparable from the actions we take.

To answer the question of my personal engineering identity, I shall reflect on who I was, what I did, and whom I associated myself with. Through a retrospective examination of my identities and their development over time, I hope to better understand who I am today and whom I could become in the future.

Y.-L. Han (✉)
Mechanical Engineering, Seattle University, Seattle, WA, USA
e-mail: hanye@seattleu.edu

M. Bailey, L. Shackelford (eds.), *Women in Mechanical Engineering*, Women in Engineering and Science, https://doi.org/10.1007/978-3-030-91546-9_17

I Am Chinese

I was born in Taiwan in the early 1970s. My grandparents moved to Taiwan with Chiang Kai-Shek's troops during the "Great Retreat" when the communist party took over mainland China. When I was little, my grandfather would tell me our hometown was across the Taiwan Strait, and we were only in Taiwan temporarily. Even though he had not had any contact with his family in mainland for decades, to him, China was still home. We lived in a military dependents' village where almost all families had someone serving in the military. We were a special group of people eating different kinds of food and speaking other dialects. My grandparents considered themselves as Chinese to the day they passed. The strong identity they brought to the little island in the Pacific certainly caused tensions. To some Taiwanese, at the time, the existence of families like mine in Taiwan was a threat; we were "foreigners" to their land.

Growing up in a military village in the 1970s, we did not have very much, so we learned to live with what we had. Rather than hindering, the lack of resources promoted my imagination and competitiveness. As a little girl, I played with rocks to build anything I could think of. I ran faster and climbed higher than boys in my village. Raised by my paternal grandmother, who was widowed in her 20s, I witnessed her resilience and resourcefulness firsthand, and they certainly became a part of my traits. However, deep down, my grandmother was a traditional Chinese woman. She was happy to see me outrun boys, but she also cautioned me to behave like a girl. She stopped me from climbing trees so I would not get scars on my face, because a scarred face would not land me a good husband. In her mind, a woman's role was to be a good wife and mother. There was no self, only the family.

Despite my grandmother's traditional Chinese values and beliefs, she took from her own experience that a woman should also be capable of taking care of her family on her own. She never treated me unequally from my brothers when it came to education. She knew a good education would give me a better future. She regretted that she could not be formally educated so she wanted me to go to the best school I could. She borrowed money and skipped meals just to pay for my tuition.

With her devotion, I was able to attend a private junior high school. Going to a private school allowed me to focus on my academics so I could score high on my entrance exams to get into a good high school and then college. I would not get a quality education without succeeding in entrance exams. As a teenager, all I did was to prepare for exams with the only goal of scoring high in them. There was never a doubt in my mind that my life would change if I could get into the best college, but I had never thought of what that life would be, and I had no clue who I wanted to be. Those were never the questions I asked.

I studied hard to ace the entrance exam and got into the best girls' high school, where only the top 1% of girls of my age went. When I moved out of my military village home and to the capital city of Taipei to start high school, I noticed the disparities among people. While my village friends and I could only afford the "beef noodle soup without beef," some of my classmates could have pork chops every meal. The different levels of socioeconomic status became more obvious to me. During that time, Taiwan was also going through drastic changes economically and politically. There was more access to the outside world, and things that used to be luxuries (like apples) were suddenly available to many. With the significant growth in GDP, people in Taiwan certainly felt wealthier and freer than before. I remember my excitement when I finally had a different haircut (all secondary school children were required to have the same hairstyle before 1987 when the martial law was lifted in Taiwan), and the time I waited hours for my bus to make its way through roadblocks caused by protests (my high school was right next to the Presidential Office Building of Taiwan). It was an interesting era to live in, nonetheless, and these experiences allow me to appreciate the freedom and the resources I have today so much more.

I was proud to put on my green uniform every morning when I was in high school, but it meant even more to my family. My grandparents and parents did not get much education. They fought in the wars and became "refugees." They escaped their homes and settled on an island they had never been before. They lived through the hardships with nothing to spare. They had little to be proud of. Finally, the prestige of my high school gave my family the bragging rights they had been longing for, but it also put a tremendous amount of pressure on me. I had the pride of my entire family weighing on my shoulders. It was in our Chinese culture to "save face" for my family, so I could not disappoint them.

I Am a Girl

In the second year of high school (equivalent to 11th grade), I declared that I would take the college entrance exam for science and engineering majors. I could not recall why I chose this category, but if all my good friends chose it, why shouldn't I? Isn't it still how teenagers work nowadays? I honestly had no idea what engineering was about when I made this decision. I knew no engineers and, of course, no female engineers. I just went with my friends. I told myself, "They definitely know what they are doing. They are the smart ones."

I certainly enjoyed my days in high school. This group of smart girls challenged me to always do my best. Despite our differences, we became friends for a lifetime. Many of my strengths today are attributed to them: they were caring, funny, and dedicated to their work. They showed me that there was no limit to our potential, and we could make a difference if we put our mind to it. We were supposed to be the girls who changed the world, and yes, thirty years later, my high school friends are all playing important roles in their respective fields, and we continue to support each other no matter where we are.

My days in high school went by fast with all the challenges and fun. The time for the college entrance exam came and went. I scored high enough to get in one of the best colleges in Taiwan. Because I did not really have any specific interest in engineering at the time, I just filled in the bubbles on the survey for the majors I would be in based on my score. And there I was, majoring in materials science and engineering at National Tsing-Hua University.

Don't get me wrong: I had no complaints. It was a good major at the best engineering college, and I made my family proud again. It was the time that the semiconductor industry started to boom, and Taiwan Semiconductor Manufacturing Company (TSMC) was recruiting heavily from my department. It was almost guaranteed that I would have a great engineering job with my degree; many of my college classmates have been working at TSMC or other semiconductor companies ever since graduation. It was almost a no-brainer to take that route. However, despite all the objections from my family, I passed on that opportunity and went on an excursion to search for myself and my true interests. I simply could not imagine myself wearing white overalls working in a cleanroom the whole day. I literally needed more colors in my life.

College was very different from the all-girls' high school. I finally lived on my own and felt some relief from my family's expectation. As a young girl surrounded by boys, especially in an engineering college with a 4 to 1 boy-to-girl ratio, I was more interested in extracurricular activities than engineering subjects. My priorities had shifted. I no longer had to spend all my time preparing for exams. I thought "I had made it" (to college), so I cut myself some slack. Ironically, I lost my purpose to study hard and I felt like I was coasting without aims. It finally struck me that I did not really know who I wanted to be. I had no identity of my own.

Unfortunately, I did not find much inspiration in my engineering courses. All the faculty in my department looked the same, and most of my classmates looked just like them. I met no one like me who remained in engineering; I found no connection to engineering. One of my best friends entertained the idea of getting a minor in economics. I went along with her so I could possibly get out of being an engineer. I spent three years finishing the requirements for my engineering degree and the last year in college getting an economics minor. I enjoyed learning economics as it still required me to use math and logical thinking skills, but more importantly, I met more people who looked like me in the Economics Department. It was not odd for me to show up to my economics classes wearing heels and miniskirts as many others did. I felt the same support I had in high school again and the sense of belonging that had been lacking in my engineering classes. Looking back, however, it was not like my engineering classmates were alienating me; in fact, many of them were very supportive. But my stronger association with my gender identity and the discomfort I experienced due to the gender gap in engineering at the time contributed more to my decision of leaving engineering.

Besides, "engineers don't LOOK good and I want to look good when I go to work" – the younger me was definitely not above stereotyping engineers.

I Am a Fashionista

I always liked fashion ever since I had access to it. Flipping through fashion magazines in bookstores was my favorite afterschool activity in high school. When I got into college, I began to tutor grade school kids so I could afford new clothes and accessories. To me, putting on a look was a way to express myself and acknowledge my unique identity. With little surprise, I earned my popularity in college as a fashionista. I definitely showed more interest in fashion than anything else, and it gave me the recognition I desperately needed for myself at the time.

It seemed reasonable for me to find a job in fashion so I could dress up every morning to go to work. However, it was not a popular idea with my family who had high hopes of me being an engineer so my degree would not "go to waste." In the end, my desire of exploring other identities won over my role as a traditional Chinese daughter who would live by their family's wishes. I took actions that I would not if I had remained obedient.

I was offered a position as a manager trainee at a fashion retailer, which just started a program to hire college graduates and train them as store supervisors as it went through a rapid expansion. It was a move that aimed to shake the fashion retail industry, because traditionally, it was not a career path for college graduates. Looking back, I really appreciated this opportunity. I matured quickly and learned many essential interpersonal skills. However, the job really put a strain on my relationship with my family. In their minds, suddenly, I was no longer the sweet little girl who shouldered her family's honor. Instead, I turned into an independent woman who called her shots and took charge of her life.

I moved quickly up the ladder at work, and the adrenaline rush of success pushed me further into adulthood at light speed. I was soon married and became a mother within a few years of graduating from college. My college roommate once said, "*Yen just wants to cross all the milestones in life as fast as she can.*" Perhaps she was right. Being a mother to me was certainly less ambiguous than unveiling any other identity, and it was possibly the most natural way for me to feel a sense of accomplishment in life.

While raising a baby, I continued to thrive in the fashion industry. I got a chance to work for an international brand and began traveling to Europe for business. Being immersed in different cultures was eye-opening, and it made me realize there were more possibilities in life.

I Am Taiwanese

After almost five years in fashion, I reached a level that I needed to reposition myself. The repetitiveness of season after season in retail was tiring, and my personal growth had slowed down. Although I had been managing people and merchandise for years, I likely only did it by mimicking what others did. I was not equipped enough to shake things up in the fashion industry, so I wanted to educate myself more about management. The time I spent in Europe made me learn different perspectives, which prompted me to search for new learning opportunities outside of Taiwan. I packed my bags and went to Los Angeles to pursue my MBA degree.

"Why LA?" someone at Business school asked me a long time ago. "The Lakers," I said, and it was not a joke. I knew very little about LA and any knowledge I had was from pop culture. LA seemed like a vibrant city teeming with new experiences, so there I was. I took six classes per semester in business school with a toddler in tow. Every day was a busy day. But more importantly, being immersed in a multicultural environment like LA made me reflect on my limited knowledge of diversity. Interacting with people from different countries with different cultural backgrounds pushed me to deeply examine my own identities.

"Where are you from?" Since coming to America, I have been asked this question countless times. I did not think much about it at first, but nowadays, I would answer, "I am Taiwanese, and I live in Seattle," trying not to embarrass people who ask such questions. I was born and raised in Taiwan, and it was my home for half of my lifetime. Most of my family lives there. However, I am a naturalized US citizen, and getting asked this question now makes me wonder if I have been accepted into the "club." It is almost the same feeling I had when I was growing up in Taiwan – we, the people who lived in the military villages, were separated (and also self-selected) into a different category in society. The divide was caused by identity. The us vs. them mentality did not help in solving problems we were facing. There is no easy answer on how a society could be more integrated, but we know a society will function better with contributions from all different parties. Keeping our own identities should not mean excluding others who may or may not identify themselves the same way as we do.

To me, the "American Dream" really is to encourage everyone to explore, express, and develop themselves and to work towards the common good. I could not help combining my collectivistic cultural background with my individualistic views as an Asian American.

I Am an Engineering Student (Again)

I was fortunate to get the education I needed when I was growing up, which turned into a great gift for me so I could extend my knowledge and abilities quickly. I finally acknowledged that I was capable of using my abilities not just for myself, my family, but for more people in need. Going back to fashion after finishing my MBA degree simply did not sound appealing anymore. I wanted to do more than chase sales targets, but even with an MBA degree, it would still be difficult for me to land in any critical decision-making position to impact the fashion retail industry (or fast fashion in today's terminology). I did not agree with the business approach of fast fashion and could no longer align myself with it. I had a broader responsibility to the world we live in. Issues I witnessed before, such as poverty and access to education, turned into issues I wanted to address in the future.

"But how?" I asked myself. Still at the stage of exploring possibilities, "more education may help," I thought. With limited savings, it was more economical for me to take engineering courses with my BS engineering degree (less prerequisites) while figuring out what I could do to drive changes in our society. I enrolled myself in fundamental engineering classes again. This time, it was me who wanted to do it. Nobody else planned this for me and I made the decision based on my own curiosity. I was exposed to more projects that helped me connect everyday life with engineering, and I saw the potential of using my engineering skills to make a positive impact. When I experienced learning engineering as a way to train myself how to think and solve problems logically, I really fell in love with it. I finally had fun learning engineering, and I believed, with the right motivation and guidance, many others could too.

I Am a Researcher

While taking more engineering courses, I learned about MEMS (microelectromechanical systems), which reminded me of my materials science labs in college. Although it was tiring to spend days in the lab sanding and polishing the surface of metal specimens, I remembered my excitement when I finally saw the metal grain structure under the microscope. I guess I was drawn to things we could not see with the naked eyes.

Wanting to learn more about MEMS, I came across research from Dr. Phil Muntz at the University of Southern California (USC). I began my study at USC and took Dr. Muntz's Molecular Gas Dynamics class during my first semester. As he might tell you himself, Dr. Muntz was not the most organized lecturer, but he was the best advisor I could ever have. He was a brilliant researcher and an elected member of the National Academy of Engineers (NAE). He was also the "Mad Man Muntz," who played as a fullback in the Canadian football league in his 20s. He was full of life when he talked about research, and he was always generous in sharing his wisdom. It was quite common that he would get distracted by a research problem in his head and begin working on the board while students sitting in his class scratched their heads. But his random talks about research really got me hooked on molecular gas dynamics. The thought of working on a problem that nobody knew the answer to inspired me. I visited his office often and wouldn't stop asking questions until he finally accepted me as a volunteer in his lab. I was relentless in solving problems thrown my way. Dr. Muntz recognized my strengths and recommended I get my PhD in this area. Without a doubt, I was ecstatic to be taken under the wing of such a brilliant scholar.

My research focus was on the Knudsen Compressor, a micro-scale compressor with no moving parts [12–15]. Utilizing the rarefied gas dynamic phenomena of *thermal transpiration* (also known as *thermal creep*), the Knudsen Compressor can pump gas through a porous membrane by applying a temperature gradient across it. To operate the Knudsen Compressor, the membrane pore size must be close to the gas molecular mean free path and the membrane should have a low thermal conductivity to maintain the required temperature gradient. Using carbon doped aerogel membranes which absorbed radiant energy to heat up its surface, I was able to demonstrate the performance of a Knudsen Compressor operating at low pressures experimentally. In addition to the experimental results, I conducted numerical studies using Direct Simulation Monte Carlo (DSMC) technique to further investigate the thermally induced vortexes that impacted the performance of the Knudsen Compressor [16–23].

Getting a PhD was never an easy ride, but it was easier with a great mentor. With the support from Dr. Muntz, I learned many valuable lessons. He let me try things that I wanted to do but gave me pointers when I needed them. He respected my opinions and allowed me to develop my own thoughts. He taught me how to think critically and be curious every day. He taught me to be patient and enjoy the process, showing me that a researcher should not be afraid of the unknowns but should treat them as an opportunity for new discovery. He demonstrated how one should be passionate for their work and should generously share what they knew. I am eternally grateful to have had him guide me through the challenge and help me grow into an independent researcher. Although he is no longer with us, his influence lives on.

Having a PhD could open doors to many new opportunities. I certainly thought about what those opportunities could mean to me and my greater responsibilities, which pushed me forward to complete my degree. Having Dr. Muntz as my advisor, I naturally gravitated towards a faculty position where I could possibly replicate what he did in advancing science. However, when I finally finished my PhD degree, I could not wait to find a "dream" job. I needed the income to support my children after ending my previous marriage. Being a single mother with two young boys living in LA on a graduate student stipend was difficult. My kids probably spent more time in my lab than on the playground, and they could not have most toys others had because we simply couldn't afford much. I just could not see them sacrifice any more of their experiences for mine when I could finally provide them more.

In need of an income, I picked up a couple teaching assignments in my department at USC as soon as I was able to. However, my mind was still focused on research. Phil[1] remained a great supporter for me, but I gradually realized that—at least on paper—I was not as "prepared" as others who were in the faculty job market. Phil did not believe the number of publications was the main factor a young faculty candidate to be ranked on. He valued quality over quantity and believed good research sometimes could take years to see results. He thought that creativity and enthusiasm should be more important for a faculty's success, and that was why he trained his students to think, not just do. He also admitted that academia was different from when he started his career, and he did not prepare me enough for the "bean-counting" aspect of academia. Perhaps Phil was right in assessing the

[1] Dr. Muntz insisted I called him Phil because he saw us being equal once I got my PhD degree.

"modern" academic job market, and I did experience serious setbacks in my faculty job searching. There were times I was told that my publications just were not "enough" despite my solid performance in interviews. However, I was not discouraged, but more determined to pursue a career in academia. It took me some time to really understand the differences among various types of faculty positions (e.g., tenure track vs. teaching). But the time I took gave me an opportunity to really polish my independent research abilities that were built on the essence instead of the content of my training. Today, I can conduct research in autonomous vehicles and medical devices – fields with little direct relation to molecular gas dynamics in content – because I am using the same problem-solving skills I possessed through my training.

I Am an Educator

Teaching is an important part of the job as a faculty, but that is rarely a part of our training as an engineering PhD. There was no exception for me. I was the teaching assistant for several classes when working on my PhD, but that was the extent of my teaching training. Phil did not advise me much on how to be an effective teacher either, as this was not a priority of his. When I got my first teaching assignment, I had little idea how to really teach. All I had was my passion, my knowledge of the subject, and the dos and don'ts from my own learning experiences. I stumbled many times, but I learned from my mistakes and wanted to improve my teaching. I really enjoyed spending time with students. Witnessing the smile on their faces when they "got it" was such a special treat. Many of my students felt my motherly care and I could also sense their appreciation. The positivity I received from teaching allowed me to reexamine my career goal in academia. I wanted teaching to be an essential part of my job.

As a younger female faculty teaching mechanical engineering, I certainly had experienced difficulties. I was sheltered when I was teaching at USC—faculty treated me like their family because they were the ones who taught me and watched me grow in my career. I hardly noticed any bias towards me from students either. However, the situation was a lot more complicated when I took a teaching position at the University of Connecticut. My presence was a big change to the department; I was one of the few non-tenure track teaching faculty who was not retired from industry, and I was the only woman. As a teaching faculty, I taught three courses and advised several senior design projects every semester, which allowed me to hone my teaching and mentoring skills even more. Although it was not required, whenever time permitted, I still conducted research, switching my research focus to computational studies due to the lack of physical lab access. These experiences taught me how to utilize the resources I had to create new possibilities. All the positive outcomes aside, however, I had a hard time adjusting.

The bustle of LA was nowhere to be found in rural New England. I did not have to worry about traffic; the scenery was beautiful, and the atmosphere was serene.

People I encountered were kind and generous. However, I just did not feel I was a part of the community as one of few Asians in a small town. Driving more than an hour in the weekend to get ingredients I needed to cook was normal. I finally realized that I had only experienced a very special part of the United States and I had much more to learn about America. Better late than never though, I began to really pay attention to race and gender issues, especially in engineering. I could not stop thinking, "what is my job as a faculty?"

"There is no reason why I cannot do research if I also enjoy teaching," I wondered. The fact was that I didn't have to separate my passions in research and teaching. I started to dive deeper into engineering education research after I joined Seattle University (SU) as a teaching faculty. The mission of Seattle University, educating the whole person, drew me to this Jesuit institution. The small class size allows faculty to build close mentorship with each student. I was surprised when I first met with a couple of faculty members to discuss capstone senior design projects; they not only knew each student's name but were also familiar with each student's interests and characters and were able to recommend the best suited senior design project for each student. That was how I wanted to be as an educator—as Confucius said, *"Teaching students in accordance with their aptitude."*

My first engineering education research project was to use authentic engineering problems in my heat transfer classes. Utilizing the inverted (flipped) classroom format, I invited engineers from industry to my classroom to give students problems related to their areas of practice. Students worked on those problems during class time and presented their solutions to practicing engineers every other week. Named *Authentic Engineering Problem Centered Learning* (AEPCL), this new approach of teaching engineering subjects has been shown to improve students' problem-solving and design abilities [24–29]. This research experience broadened my views on pedagogy and gave me an opportunity to focus on engineering education research when I transitioned to tenure-track at SU in 2015.

My exposure to the American Society of Engineering Education (ASEE) and the Society of Women Engineers (SWE) also brought my attention to issues around underrepresented minorities (URMs) in engineering. Even in the mid-2010s, the percentage of female engineers was still below 20% [30]. I pondered what needs to be changed in engineering education so URMs are better represented.

Examining why I did not persist in my engineering path after college, my colleagues and I began to research on how identity could influence one's decision on whether to enter and remain in engineering. We began working on a five-year project, funded by the National Science Foundation (NSF) RED (REvolutionizing Engineering and Computer Science Departments) grant, to create a culture of inclusion at the SU Mechanical Engineering department, where students and faculty are immersed in doing engineering and, in turn, fostering their engineering identities [31].

Since the beginning of the RED project in July 2017, many changes have been incorporated in the department's shared vision, curriculum, faculty priority, and supportive policies [32–37] to enhance the connection between industry and academia. Practicing engineers regularly interact with students and faculty in various courses throughout the curriculum. Faculty are encouraged to participate in industry practice through summer immersion. Implementing these changes over the last four years, I included practicing engineers in many course activities and spent one month for my summer immersion at PACCAR/Kenworth Truck Company. By rotating through different divisions at PACCAR/Kenworth, I learned new things in practice such as caring for customers and investing in employees. This immersion experience allowed me to better understand the responsibilities of engineers and the operations of an industry, and because of it, I am better equipped to advise my students in their transition to becoming professional engineers.

Research has suggested that culture is especially important for women to persist in a field [38], and my own experiences can attest to that. Building a culture that fosters engineering identities could potentially promote URMs' participation and persistence in engineering. I am dedicated to this long-term effort of cultivating a more diverse and inclusive environment, something we all appreciate as important in today's climate. I wish my work in this project could provide a clearer vision of changes needed to build engineering identities and how such identities affect students' sense of belonging in a program and their persistence in the major.

I Am an Engineer

In an educational setting, a focus on identity encourages reflection and discussion about how students and faculty see themselves, their education, and their profession. Therefore, seeing myself as an engineer would allow me to be the bridge between engineering students and future engineers who are technically and professionally prepared to fulfill their duties.

I proudly identify myself as an engineer. I am curious and love solving problems using science and math. I take responsibility of others' well-being, and I hold paramount the safety of the public with my work. I conduct myself honorably and ethically. I live by the standard any engineer would.

Identity is not just a buzzword we hear; it influences the decisions we make every day. Examining my own identities has not only made me see clearly how each one of them has shaped who I am and what I do today, but also gives me a direction of what I want to do moving forward. Being aware of our own identities can make us more cognizant of our own biases and guide us to work better together. The conversation about identity should lead to a better understanding of how best to create an inclusive environment for all. I hope that this reflection on my identities has prompted you to think about your own identities, how they impact your journey thus far, and what you can build upon in the future.

Acknowledgment I cannot appreciate my support system enough—my husband, Nick, my children, Shang and Ethan, my family, colleagues, students, and friends. Without them, I would not have been able to accomplish what I have done.

References

1. K. Deaux, "Reconstructing social identity," *Personality and Social Psychology Bulletin,* vol. 19, pp. 4-12, Feb. 1993.
2. S. Stryker and P. J. Burke, "The past, present, and future of an identity theory," *Social Psychological Quarterly*, vol. 63, no. 4, pp. 284-297, Dec. 2000.
3. E. H. Erikson, *Identity and the life cycle*. New York: International Universities Press, 1959.
4. M. B. Brewer, "The social self: On being the same and different at the same time," *Personality and Social Psychology Bulletin*, vol. 17, pp. 475-482, Oct. 1991.
5. J. E. Dutton, J. M. Dukerich, and C. V. Harquail, "Organizational images and member identification," *Administrative Science Quarterly*, vol. 39, no. 2, pp. 239-263, Jun. 1994.
6. G. H. Mead, *Mind, self, and society*. Chicago: University of Chicago Press, 1934.
7. S. Stryker, *Symbolic interactionism: A social structural version*. Menlo Park, CA: Benjamin/Cummings, 1980.
8. J. D. Lee, "More than ability: Gender and personal relationships influence science and technology involvement," *Sociology of Education*, vol. 75, no. 4, pp. 349-37, Oct. 2002.
9. B. R. Schlenker, "Identity and self-identification: Toward an integration of the private and public self," in *The Self in Social Life*, B. Schlenker, Ed. New York: McGraw-Hill, 1985.
10. J. D. Lee, "Which kids can "become" scientists? Effects of gender, self-concepts, and perceptions of scientists," *Social Psychology Quarterly*, vol. 61, no. 3, pp. 199-219, Sep. 1998.
11. O. Pierrakos, T. K. Beam, J. Constantz, A. Johri, and R. Anderson, "On the development of a professional identity: Engineering persisters vs. engineering switchers," *Proc. of 2009 Annu. Frontiers in Education (FIE) Conf.*, San Antonio, TX: IEEE, 2009.
12. Y.-L. Han, M. Young, E. P. Muntz, and S. Shiflett, "Knudsen compressor performance at low pressures," in *AIP Conference Proc. 762- 24th Int. Symp. on Rarefied Gas Dynamics,* M. Capitelli, Ed. Melville, New York: American Institute of Physics, pp. 162–167, 2005.
13. Y.-L. Han, A. A. Alexeenko, M. Young, and E. P. Muntz, "Experimental and computational studies of temperature gradient driven molecular transport in gas flows through nano/micro-scale channels," *Proc. of the 2nd Int. Conf. on Transport Phenomena in Micro and Nanodevices*, Barga, Italy, 2006.
14. Y.-L. Han, M. Young, and E. P. Muntz, "Performance of micro/meso-scale thermal transpiration pumps at low pressures," *Proc. of ASME Int. Mech. Engr. Congr. and Expo.*, Anaheim, CA (IMECE 2004-61807): ASME, 2004.
15. Y.-L. Han and E. P. Muntz, "An experimental investigation micro/meso-scale Knudsen compressors at low pressures," *Journal of Vacuum Science & Technology B: Microelectronics and Nanometer Structures,* vol. 25, no. 3, pp.703–714, Mar. 2007
16. Y.-L. Han, A. A. Alexeenko, M. Young, and E. P. Muntz, "Experimental and computational studies of temperature gradient driven molecular transport in gas flows through nano/micro-scale channels," *Nanoscale and Microscale Thermophysical Engineering,* vol. 11. no. 1&2, pp. 151–175, Jun. 2007.
17. Y. L. Han, "Implications of imposing working gas temperature change limits on thermal creep driven flows," *Proceedings of 26th Int. Sympo. on Rarefied Gas Dynamics*, T. Abe, Ed. Melville, New York: American Institute of Physics, 2008.
18. Y. L. Han, "Investigation of temperature driven gas flows in 4 nm channels for applications of micro-scale compressors at above atmospheric pressure," *Proc. of ASME Int. Mech. Engr. Congr. and Expo.,* Boston (IMECE 2008-68170): ASME, 2008.
19. Y. L. Han, "Thermal-creep-driven flows in Knudsen compressors and related nano/microscale gas transport channels," *Journal of Microelectromechanical Systems,* vol. 13, no. 4, pp. 984–997, Aug. 2008.
20. Y. L. Han, "Simulation studies of micro-scale gas pumps driven by isolated heating elements induced thermal creep flows," *Proc. of ASME Int. Mech. Engr. Congr. and Expo.*, Lake Buena Vista, Florida (IMECE 2009-10856): ASME, 2009.
21. Y. L. Han, "Working gas temperature and pressure changes for microscale thermal creep driven flow caused by discontinuous wall temperatures," *Fluid Dynamics Research,* vol. 42, 045505, Apr. 2010.

22. Y. L. Han, "Performance model for optically driven micropumps with carbon opacified aerogel membranes," *Proc. of ASME Int. Mech. Engr. Congr. and Expo.*, San Diego, CA (IMECE2013-62197): ASME, 2013.
23. Y. L. Han, "Computational studies on the effects of non-linear temperature functions in thermal creep membranes of radiantly driven Knudsen compressors," *Proc. of ASME Int. Mech. Engr. Congr. and Expo.*, Houston, Texas: ASME, 2015.
24. Y.-L. Han, K. E. Cook, G. Mason and T. R. Shuman, "Enhance engineering design education in the middle years with authentic engineering problems," *ASME Journal of Mechanical Design,* vol. 140, pp. 122001-1–122001-9, Dec. 2018
25. K. E. Cook, Y.-L. Han, T. R. Shuman, and G. Mason, "Effects of integrating authentic engineering problem centered learning on student problem," *International Journal of Engineering Education,* vol. 33, no. 1(A), pp. 272–282, Sep. 2017.
26. T. R. Shuman, G. Mason, Y.-L. Han, and K. E. Cook, "A novel approach to educating engineers: learning in an inverted classroom through problems designed by engineering professionals," *Journal of Applied Engineering Science,* vol. 14, no. 3, pp. 329–334, Jul. 2016
27. Y.-L. Han, K. E. Cook, T. R. Shuman, and G. Mason," Development of authentic engineering problems for problem-centered learning," *Proc. of 2016 ASEE Annu. Conf. and Expo.*, New Orleans, Louisiana (10.18260/p.26821): ASEE, 2016.
28. T. R. Shuman, G. Mason, Y.-L. Han, and K. E. Cook, "Facilitating problem-based learning with an inverted classroom," *Proc. of the 6th Int. Symp. on Industrial Engineering – SIE 2015*, Belgrade, Serbia: SIE, 2015.
29. G. Mason, T. R. Shuman, Y.-L. Han, and K. E. Cook, "Facilitating problem-based learning with an inverted classroom," *Proc. of 2015 ASEE Annu. Conf. and Expo.*, Seattle, Washington (10.18260/p.24089): ASEE, 2015.
30. J. Bossart and N. Bharti, "Women in engineering: Insight into why some engineering departments have more success in recruiting and graduating women," *American Journal of Engineering Education*, vol. 8, no. 2, pp. 127–140, Dec. 2017.
31. National Science Foundation, Award Abstract #1629875, "ADVANCE Institutional Transformation at Seattle University". Washington DC: National Science Foundation, 2016. Accessed: Feb. 15, 2021, [Online]. Available: https://www.nsf.gov/awardsearch/showAward?AWD_ID=1629875&HistoricalAwards=false
32. Y.-L. Han, K. E. Cook, T. R. Shuman, G. Mason, and J. Turns, "Engineering with engineers: Revolutionizing engineering education through industry immersion and a focus on identity," *Proc. of 2018 ASEE Annu. Conf. and Expo.*, Salt Lake City, UT: ASEE, 2018.
33. K. E. Cook, Y.-L. Han, G. Mason, T. R. Shuman, and J. Turns, "Work-in-Progress: Engineering identity across the mechanical engineering major," *Proc. of 2018 ASEE Annu. Conf. and Expo.,* Salt Lake City, UT: ASEE, 2018.
34. Y.-L. Han, K. E. Cook, G. Mason, T. R. Shuman, and J. Turns, "Engineering with engineers: Revolutionizing a mechanical engineering department through industry immersion and a focus on identity," *Proc. of 2019 ASEE Annu. Conf. and Expo.,* Tampa, FL: ASEE, 2019.
35. K. E. Cook, Y.-L. Han, G. Mason, T. R. Shuman, and J. Turns, "Implicit engineering identity in the mechanical engineering major". *Proc. of 2019 ASEE Annu. Conf. and Expo.*, Tampa, FL: ASEE, 2019.
36. Y.-L. Han, K. Cook, G. Mason, T. R. Shuman, and J. Turns, "Engineering with engineers: Fostering engineering identity through industry immersion," *Proc. of 2020 ASEE Annu. Conf. and Expo.*, Montreal, Québec, Canada: ASEE, 2020.
37. Y.-L. Han, K. E. Cook, T. R. Shuman, G. Mason, and J. Turns, "Integrating electrical engineering fundamentals with instrumentation and data acquisition in an undergraduate mechanical engineering curriculum," *Proc. of 2020 Annu. of Frontiers in Education (FIE) Conf.*, Uppsala, Sweden: IEEE, 2020.
38. K. Tonso, "Enacting practices: Engineer identities in engineering education," in *Engineering Professionalism: Engineering Practices in Work and Education*, U. Jørgensen and S. Brodersen Eds. Rotterdam, The Netherlands: Sense Publishers, pp. 85–104, 2016.

Yen-Lin Han is an Associate Professor in the Department of Mechanical Engineering at Seattle University. Dr. Han received her B.S. degree in Material Science and Engineering from National Tsing-Hua University in Hsinchu, Taiwan, and M.S. degree in Electrical Engineering and Ph.D. degree in Aerospace and Mechanical Engineering from the University of Southern California. Her research interests include micro-scale molecular gas dynamics, micro fluidics, and heat transfer applications in Microelectromechanical Systems (MEMS) and medical devices as well as autonomous vehicles and robotics. She also holds the patent for the continuous trace gas separator and several pending patents in autonomous vehicles and robotic testing apparatus. Additionally, she has several provisional patents including a dynamic tumor ablation probe and an innovative design to replace the gladhand connector in semi-trucks. She is passionate about Engineering Education and experienced in developing inverted classroom lectures and facilitating students' learning through authentic engineering problems. She is currently the Co- PI for the National Science Foundation Revolutionizing Engineering and Computer Science Departments grant awarded to the Mechanical Engineering Department at Seattle University to study how the department's cultural changes can foster students' engineering identity with the long-term goal of increasing the representation of women and minority in the field of engineering.

Connected by the Environment: The Unique Yet Intertwined Journeys of Two Energy and Water Researchers

Arkasama Bandyopadhyay and Yael R. Glazer

Foreword: To Solve Global Energy & Environmental Challenges, We Need Women on the Team, by Dr. Michael E. Webber[1]

In the developing world, women are often victims of outdated energy systems and beneficiaries of modern energy systems. Primitive cookstoves and fuels such as wood or cow dung produce poor indoor air quality that leads to the premature deaths of millions of women annually. Access to better forms of energy such as electricity for water pumping and propane for cooking would alleviate the physical burdens of fetching and carrying water while reducing mortality from the avoided emissions. In the developed world, women are often the primary decision-makers for energy-consuming appliances.

Though women are the victims, beneficiaries, and decisions-makers of energy consumption, they are too seldom the engineers that design the solutions. Even less often are they the

[1] Michael E. Webber is the Josey Centennial Professor of Energy Resources at the University of Texas at Austin and CTO of Energy Impact Partners, a $2 billion cleantech venture fund. Previously, he also served as chief science and technology officer at ENGIE, a global energy company headquartered in Paris, France. His book *From Athletics to Engineering: 8 Ways to Support Diversity, Equity, and Inclusion for All* was published in 2021 with Johnnie Johnson as a co-author. Parts of this foreword were published as separate printed and online essays in ASME's *Mechanical Engineering* magazine in September 2019 and 2020.

A. Bandyopadhyay (✉)
Department of Mechanical Engineering, Texas A&M University, College Station, TX, USA
e-mail: arkasama@tamu.edu

Y. R. Glazer
Webber Energy Group, The University of Texas at Austin, Austin, TX, USA
e-mail: yael@utexas.edu

M. Bailey, L. Shackelford (eds.), *Women in Mechanical Engineering*, Women in Engineering and Science, https://doi.org/10.1007/978-3-030-91546-9_18

executives that deploy the capital and set the business direction about the future of energy. The energy industry is famous for the low participation of women in its workforce, with females comprising less than 25% of total employees [1]. Female executives seem especially rare: females make up 1% of the CEOs of the oil and gas sector [1]. Thankfully, trends are looking better—and there are some notable examples such as Vicki Hollub (CEO of Occidental Petroleum) and Lynn Good (CEO of Duke Energy).

We need more women engineers to help design solutions. Not just because women are often the victims of poorly-engineered systems or lack of access, but because we need the insights and capabilities of women to move the world forward. Good engineering teams bring together a range of skill sets and backgrounds to solve difficult problems, and gender diversity needs to be a central part of that approach.

To do this, we need inclusive institutions, starting with K-12 education, but also at the university levels and in industry. And members of the white, male power structure—people like me—have a responsibility to be allies and champions in this cause.

My perspective as an American expatriate working in France as ENGIE's (a global energy and infrastructure services firm headquartered in Paris) Chief Science and Technology Officer has given me some new confidence that it is possible to create inclusive educational and work environments. In the US, women make up less than 46% of the economy-wide workforce [2]. In France, women comprise more than half the workforce [3]. How did France get so far with gender balance among its workers? Because of other seemingly unrelated but important policies that are family-friendly: 12 weeks of paid maternity leave, three years of parental leave with social support (for both mothers and fathers), affordable and accessible healthcare, affordable and accessible daycare, and four weeks or more of generous vacation time. These protections, benefits, and support systems are much more robust than in the United States.

All in all, the French experience demonstrates that just saying you want women in the workforce isn't enough—those goals have to be backed up with a whole suite of policies. And it turns out that if you implement policies that are family-friendly in general, they will also be woman-friendly and mother-friendly.

Going forward, efforts to recruit and retain women in engineering in the United States (and globally) will require a consistent and system wide effort. A webinar on gender diversity once per year won't be enough: the entire system will need to be rethought top-to-bottom and implemented in a slow-and-steady way.

This book is an important step in that long march towards diversity, equity and inclusion. And when we get to our destination, we will all be better off for it.

Introduction

This book chapter chronicles the diverse yet intertwined journeys of two immigrant minority women and early-career academicians—Yael and Arkasama—who received their Ph.D.'s in engineering disciplines from the Webber Energy Group (WEG) at the University of Texas at Austin (UT Austin) in 2018 and 2020, respectively. Arkasama is currently a research assistant professor in the Department of Mechanical Engineering at Texas A&M University working on distributed energy resources, residential demand response, and building energy systems. Yael is a research associate in WEG at UT Austin. Her research interests span the realms of energy, water, sustainability, food waste, and the environment.

Women currently comprise only about 28% of the STEM (science, technology, math, and engineering) workforce in the United States and an abysmal 13% of the engineering personnel [4, 5]. These statistics are surprising since surveys have shown that half of current US college graduates are female [4]. Women, even those who are "STEM-ready" in high school, generally tend to gravitate towards majoring in business, psychology, education, or nursing [6]. Some of the factors contributing to these choices include personal interest (rather than ability), aspirations to work in fields with social impact, conditioning by family members, and lack of female role models [6–8]. Further, nearly 40% of women who major in engineering fields choose to leave (or never enter) their chosen profession because of inequitable compensation, demanding work schedules, uncomfortable work climate, sexual harassment, gender-stereotypical experiences during internships, insufficient professional recognition, narrow promotion routes, and lack of female mentors in the engineering community [9–11]. The dearth of women in the engineering career force poses a risk to creativity, innovation, and productivity and exacerbates the shortage problem of qualified engineers [12, 13]. Experts say that increased support and guidance from female mentors and peer network groups as well as upheaval of gender biases in families, schools, and workplace policies can potentially make engineering a more gender-diverse discipline [8, 14, 15].

While raised in different continents, our personal and professional growth stories share uplifting parallels, shaped in part by our local and global environment. We are daughters of highly educated parents who did not let gender play a discriminatory role in our upbringing or the opportunities presented to us. We were both taught to be independent and encouraged to pursue STEM fields, unlike many families, particularly in developing countries, where marriage and child-rearing are still considered to be the most important goals in a woman's life. We believe that our forward-thinking upbringing significantly impacted our ethos and played a major role in our academic careers as water and energy researchers. In this book chapter, we share experiences from our childhood, college, work (aka "the real world"), graduate school, and current career stage as well as the key motivators in each phase that led us to the next one. We hope that future engineers interested in a similar career path can glean some wisdom from our life journeys. We would also like our career stories to reach high school students and inform them about a career in engineering. Finally, by providing narratives of our successes and sharing lessons learned in moments of adversity, we hope to inspire young women to pursue careers in STEM fields.

"When I Was Just a Little Girl, I Asked My Mother, 'What Will I Be?'" [16]: The Formative Years

Arkasama was born and raised in Kolkata—a large metropolis in the eastern part of India. Her parents, who both have graduate degrees and work in STEM/medical fields, taught her from an early age that education is an important and effective way of climbing up the social and economic ladder. Although Arkasama was encouraged to cultivate hobbies when time permitted, her upbringing was centered primarily on family and education. Her family also travelled extensively across India as well as internationally, thus inculcating in her a passion for different cultures, cuisines, and people.

In a country like India where many families still lament the birth of a daughter and gender determination before birth is banned by the government to prevent sex-selective abortions of female fetuses, Arkasama's parents did not let gender play a discriminatory role in any aspect of her upbringing. Arkasama was always encouraged to pursue a STEM field—like many Indian kids, she jokes that her only options were either becoming an engineer or a doctor. While many of her peers in high school wanted to pursue a career in engineering, women were usually nudged towards computer engineering or chemical engineering instead of typically "hands-on" and "fieldwork intensive" majors like mechanical engineering. However, Arkasama found a mentor in her high school physics teacher who imbibed in her a love for mechanics, and fueled by her parents' support, she decided to move to the United States in August 2011 at the age of eighteen to pursue a bachelor's degree in mechanical engineering.

As a part of an upper middle class family, Arkasama led a relatively sheltered life growing up in India. However, she was perceptive to the struggles of households in her vicinity that did not have access to clean water or uninterrupted supply of electricity. As an immediate consequence of the Indian economy opening up in the early 1990s, there was a sea-change in the energy sector. The burgeoning middle class could now afford air-conditioners, and the huge spike in demand stressed the electricity grid, leading to frequent blackouts in residential communities, often lasting entire days. Little did Arkasama know that these childhood experiences would later motivate her to tackle the energy, water, and sustainability problems of the world.

Yael was born in Israel and immigrated to the United States with her parents as a preschooler. Although raised in a different part of the world, her upbringing was similar to Arkasama's in many ways. Yael's parents are both engineers who worked in the Silicon Valley and nudged her towards math and science from an early age. She was never treated differently from her older brother in terms of education, opportunities, and expectations. Yael recounts that she found it absurd when one of her friends (say 'M') casually mentioned to her that M's parents were saving for M's marriage instead of planning to help with college tuition—simply because M was a girl. We both agree that it is extremely important to have parents who advocate for their children and "walk the talk" when it comes to gender equality. It is easy to verbally support women's empowerment but much more difficult (as well as inspiring) to lead by example. Further, we think that a child's beliefs and opinions are significantly shaped by the people they grow up around. For example, if a girl is raised by parents who believe that marriage and child-rearing are the most important goals in a woman's life, she might be more likely to make choices that prioritize those aspects of her adult life. This is not to say that twenty-first-century women should focus only on their careers, but we believe that the decision must not be biased by age-old doctrines.

Just like Arkasama, several of Yael's childhood experiences motivated her to research water and sustainable energy systems in graduate school. Having Israeli parents and growing up in drought-ridden California (of the late 1980s and 1990s) meant that Yael learned the importance of water conservation both at home and at school. Furthermore, Yael experienced her share of brownouts during the California Electricity Crisis of 2000–2001 which revealed to her that when resources are not managed well, the living conditions in a developed country like the United States can quickly resemble those in the developing world. Thus, while both of our upbringings were shaped by the environments (and countries) where we grew up, having parents who worked in STEM fields and prioritized family and education were key components to our solid foundations.

"To Face This on My Own, Well I Guess This Is Growing Up" [17]: The College Experience and First 'Real-World' Job

At the age of eighteen, Arkasama moved to the United States to pursue a bachelor's degree in mechanical engineering at Oklahoma State University (OSU). Although she had grown up travelling extensively with her parents, studying abroad (thousands of miles from home and anything familiar) presented its own share of challenges. However, Arkasama quickly learned that people were generally willing (and eager) to help; one simply needed to venture out of their comfort-zone and ask for assistance. During her sophomore year, Arkasama was enrolled in an "Introduction to Engineering" class in which students had to work on projects involving the use of a sawmill and lathe machine. On the first day of class, the machine shop manager quickly showed the students how to use the machines and went back to his office after announcing that he would be available for clarifications and any emergencies. Arkasama looked around and strangely all her American classmates, regardless of gender, had a clear idea on how to complete the projects. Most of them had grown up in a DIY environment, watching their parents fix stuff around the house or helping their parents build small household items in their makeshift garage workshops on weekends. In contrast, Arkasama was used to her family calling a carpenter when a small piece of furniture needed to be built, a painter to freshly paint the house, a plumber to fix leaky faucets, or an electrician to change light bulbs. It was easy and inexpensive to find and hire professionals from these selected professions in India. Arkasama had no idea how to use the machines, and the 15-minute introductory explanation of the process was clearly inadequate. She questioned her choice to pursue mechanical engineering and might have changed her major if not for her dad who, after patiently listening to her cry for hours, encouraged her to reach out to other students in her class asking if she could quietly observe as they worked on their projects. Arkasama hesitantly emailed a few of her peers who were more than willing to help out. She spent hours upon hours standing next to them as they worked—keenly observing and taking notes. Thankfully, this story has a happy ending; Arkasama was not only able to complete the assignments but also got an A in the class. While she never wants to be near a sawmill or lathe in her life, Arkasama hopes to communicate to future generations of engineers that hard work, perseverance, and seeking out help can almost always make up for lack of experience.

Arkasama also considers herself blessed to have found several informal mentors during her undergraduate career who selflessly invested their time and effort in her success. In her first freshman Calculus class at OSU, her professor Dr. David Wright patiently interpreted the British mathematical terminology she had been taught in high school in India and "translated" those to the American system multiple times throughout the semester. Dr. Jennifer Glenn, who taught engineering economics during Arkasama's junior year, met with her several times even after the class ended to help her polish her resumé and refine her statement of purpose while applying to graduate school. In a particularly demanding major like engineering, Arkasama found a role model in Dr. Glenn who had a successful career in academia while raising three children with her husband. Arkasama wanted to be just like Dr. Glenn "when she grew up."

Arkasama enjoyed her undergraduate classes on thermodynamics, fluid mechanics, heat transfer, and renewable energy and was motivated to pursue doctoral studies in the thermal fluid sciences. She knew that she wanted to be a university professor and a doctoral degree was a necessary stepping-stone in achieving this goal. She also loved tutoring physics and calculus at OSU as well as working as an undergraduate research assistant; these experiences further strengthened her desire to follow the academic career path. At this point in her life, Arkasama still didn't know that she would later become an energy and sustainability researcher.

Yael received her bachelor's degree in bioengineering from UC Berkeley. Looking back at her undergrad days, she has three pieces of advice for future engineers: (1) Surround yourself with peers who are "better than you"; (2) reach out to professors and professionals to seek knowledge (as her mom says, the worst that people can say is no); and (3) make efforts to demonstrate your commitment. During her time at UC Berkeley, Yael's closest friends were the brightest students who always scored better than her in exams. While it was frustrating (and humbling) to be outperformed by her friends every single time, those experiences motivated her to work on herself, continually strive to improve, and learn with the most brilliant. She also quickly learned that while she might rarely be the best student in any class, she was an incredibly hard worker and that quality would help her succeed in any path she took. Yael strongly believes that the college years are not only meant to receive formal education but also a great time to seek out experiences outside the classroom. If she found a class particularly interesting, she would approach her professor to learn about their research. Those conversations and interactions helped her figure out the topics that she might be interested in pursuing professionally in the future. Additionally, Yael pursued several undergraduate research opportunities to learn more about life in academia as well as internships in biotechnology companies during the summers to acquaint herself with a career in industry. She also went out of her way to make professional connections with engineers who worked at companies she was interested in. These efforts helped her build a strong professional and peer network even before she had stepped out of college. To this day, Yael remains in regular contact with many of her friends, acquaintances, and professors from her college days and internships.

After graduating from UC Berkeley in 2004, Yael worked at Genentech, one of the world's leading biotech companies, for seven years, developing and optimizing manufacturing processes to bring medicines to patients with unmet medical needs. The dependence of the industry on clean water and the usage of specialized water filtration systems—especially in contrast to the lack of clean water and sanitation facilities in many communities and countries around the world—motivated her to go to graduate school at UT Austin in the fall of 2012 and research water and sustainable energy systems. During her time at Genentech, Yael worked on a variety of projects whose ultimate goals were to allow for, or maximize, the product yield of the therapeutics the company was developing. In one project, she was tasked to build a system to test cell shear stress under different conditions. With little direction—and alongside a peer—they built a system to conduct the testing and spent months working with the system they designed together. This project—and many like it—began as an overwhelming and daunting task. Where would she begin? How long would it take? What if she failed? But by keeping a positive mindset, patiently working through the challenges, tweaking the design, and collaborating with her peers and superiors, Yael successfully completed the project. And it was these types of projects that she is most proud of because the ambitious goals and steep learning curve made their successful completion that much sweeter. During this time, Yael also sought out mentors whose careers were inspiring to her and who were willing to regularly meet with her to discuss their career trajectory, help her objectively think through her own career aspirations, and overcome professional challenges. Similar to Arkasama, she found several people eager to invest in her success; she simply had to be bold and ask for help.

"At First I Was Afraid, I Was Petrified…I Will Survive" [18]: Navigating the Ups and Downs of Graduate School

Arkasama moved to Austin, TX in the summer of 2015 to start graduate school at UT Austin later that fall. In addition to settling in and exploring the bustling city during the three months before school started, she reached out to several Ph.D. students in the mechanical engineering program to learn about their research. Those conversations helped her realize that theoretical research in the core thermal fluid sciences did not appeal to her. Instead, she was interested in solving real-world problems and making a direct impact on energy and society. Arkasama wanted to work towards mitigating the water scarcity and electricity blackout issues she had grown up seeing in India around her. The research goals and mission of WEG, led by Dr. Michael Webber, aligned perfectly with Arkasama's values and aspirations.

After joining the research group, Arkasama quickly realized that despite her interest in renewable energy and electricity markets, she didn't have enough knowledge to partake in meaningful discussions on those topics with her lab mates. Rather than letting herself feel unworthy of belonging to a top tier graduate program, she came up with a plan of action to improve her grasp on those concepts. Arkasama set aside 30 minutes every day to read general interest articles about sustainability and renewable energy. Slowly but steadily, her energy-related vocabulary and conceptual knowledge improved; she could more confidently discuss sustainability issues with her peers along with being able to understand research topics that her lab mates were working on. As she made progress on her research, Arkasama sought out conferences, symposiums, and research competitions (like the Three Minute Thesis Competition, UT Austin Energy Week poster competition, etc.) where she could present her dissertation work. While it was often uncomfortable to receive criticism during her presentations or not have the knowledge to answer follow-up questions from the audience, those experiences taught her to be a confident speaker.

Arkasama also wants to communicate to future engineering students that taking action (often by going out of one's comfort zone) and making progress is the key to success in graduate school—a lesson she learned from her own experiences. During the summer after her first year of graduate school, Arkasama got the opportunity to work as a graduate engineer intern at Austin Energy—the local municipal electric utility in Austin, TX. One of the technical tasks assigned to Arkasama during this internship was a planning-related analysis investigating the effects of the interconnection of a large number of small, distributed generation sources to the Austin Energy Distribution System. As a mechanical engineering major, it was initially daunting to not understand electrical engineering terms and not have any background on the software needed to conduct the analysis. However, Arkasama overcame this challenge by undergoing formal training, asking clarifying questions to her co-workers and supervisors, and utilizing a variety of resources like user manuals and YouTube tutorials for self-training. At the end of the summer, her supervisors were extremely pleased with her performance and this led to Austin Energy sponsoring the remainder of her Ph.D. studies.

Arkasama's dissertation focused on developing novel techno-economic methods for analyzing the effects of solar, storage, and demand response on electricity systems. Her research required her to combine her background in thermal fluid engineering with operations research methods and electrical engineering concepts—areas that she has not been trained in previously in her graduate coursework. Intimidated at first, Arkasama reached out to several (formal and informal) mentors to gather technical knowledge. Her Ph.D. co-supervisor at UT Austin, Dr. Benjamin Leibowicz, and electrical engineers from Austin Energy, Lisa Martin and Ralph

Daniel, were some of the people who were integral to shaping her dissertation research. Interestingly, during the last couple of years of Arkasama's Ph.D., when Yael re-joined WEG as a research associate after her own graduation, Yael (along with postdoctoral research scholars Isabella Gee and Emily Beagle) was able to provide invaluable feedback on Arkasama's dissertation research and academic progress. Thus, their friendship, which had been fueled by compatible personalities and similar interests since the semester Arkasama joined the research group, evolved into the relationship between a mentor and a mentee.

Yael had a challenging transition from industry to graduate school after a gap of seven years. She had to relearn many of the mathematical concepts she had been taught in college but had not used since and had thus forgotten. However, as Yael started to check out relevant books from the library to relearn concepts and spend extra time during office hours, she found the process of refreshing her memory much easier than she had expected. Thus, one of the key takeaways from her first semester at UT Austin was that she could learn or relearn anything as long as she confidently took action and sought out resources. Yael knew that it was extremely important to pick a research group in graduate school that aligned with her goals and values: she wanted to solve water and energy problems, and have smart and kind colleagues and an empathetic advisor who would advocate for her success as well as her physical and mental well-being. WEG was a perfect fit for her. Yael hopes to communicate to future Ph.D. candidates that the most important aspect of graduate school is choosing the right advisor and research group. Graduate school inherently has many challenges—demanding classes, qualifying exams, dissertation proposals, and an often brutal defense to name a few—and spending five to seven years working for a difficult advisor can make life miserable. Yael's years at UT Austin were made more challenging by the fact she became a mother (twice) during this period. In spite of having an understanding advisor and a supportive husband, the act of balancing motherhood and graduate school was exhausting for her. The sleepless nights, feeding, diapers, crying babies, conference presentations, assignments, journal articles, and research meetings seemed to blur into one another. However, Yael figured out that she could get through this phase by (1) being open and honest with her advisor and (2) seeking help with childcare and outsourcing household chores whenever possible.

Our graduate school years were made enjoyable, in particular, by the culture of WEG. The vibrant, brilliant, and kind group of researchers regularly discussed energy and water issues, brainstormed possible solutions, collaborated on projects, and partook in happy hours. We also celebrated each other's achievements and acted as pillars of support during difficult times. Additionally, the team was gender diverse (almost 50% female)—a striking characteristic for a mechanical engineering research group. A man ahead of his times, Dr. Webber understood the many benefits of creating a diverse workforce and purposefully chose to build his team this way.

"Today Is Where Your Book Begins...The Rest Is Still Unwritten" [19]: Life as Early Career Academicians

After graduating with her Ph.D. from UT Austin in August 2020, Arkasama joined the Mechanical Engineering Department at Texas A&M University as a research assistant professor. Both her professional and personal aspirations led her to the quaint town of College Station, where Texas A&M is housed. Unclear as to whether to pursue the conventional academic tenure-track route, Arkasama wanted a transitory academic position that would give her a taste of both research and teaching and help her figure out where she would be happy (and provide the most value) long-term. Additionally, she was particularly looking for a position at Texas A&M where her husband is also a faculty member. This temporary position was the perfect mix of all that she was looking for. In the first few months of being in the "real world," Arkasama quickly learned the importance of being an independent researcher and was incredibly grateful for having a hands-off Ph.D. advisor who had encouraged her to build this skill in graduate school. While it was often frustrating to not have an advisor to turn to for help whenever she was stuck at the initial stages of her Ph.D., she was forced to learn how to seek out resources, approach her peers for help, "Google" or "YouTube" complex concepts, and set internal schedules to meet sponsor deadlines. Each of these skills became her forté and demonstrated her professionalism and competency to colleagues at Texas A&M. Arkasama currently works on research projects involving ground source heat pump systems, occupant-centric building energy management, IoT-based residential comfort control and energy efficient solutions, and disaster-resilient sustainable energy systems. She also mentors first-generation LatinX engineering students as well as other graduate students in the Mechanical Engineering Department and volunteers to judge graduate research competitions. One key lesson that Arkasama has learned from her short stint as a faculty member and one that she hopes to communicate to future generations of engineering students is that it is better to strive for completion instead of perfection. A 'B+' performance is usually acceptable for most things in life, and it is more important to meet deadlines with "good enough" reports instead of jeopardizing one's mental and physical health trying to perfect every single sentence.

Yael completely agrees with this philosophy; as she likes to say, "A good dissertation is a done dissertation." After graduating with her Ph.D. from UT Austin in August 2018, Yael re-joined WEG as a postdoctoral research associate. Yael's current research interests span the realms of energy, water, sustainability, food waste, and the environment. She works on analyzing techno-economic opportunities for mitigating environmental liabilities (e.g., flared gas and wastewater) associated with oil and gas production, assessing the biogas potential of consumer-level food waste, and overseeing the development of an open-source electric grid modeling tool for capacity expansion modeling. In addition to academic research, Yael enjoys occasional consulting projects where she helps clients assess the technical and economic feasibility of emerging innovations and energy projects. She also mentors undergraduate and graduate students within WEG—meeting with them weekly to discuss progress on research projects, helping them break down complex concepts, edit research articles, and prepare reports to meet sponsor deadlines. A key aspect of academia that she enjoys is the ability to maintain a flexible work schedule while leading and conducting research that is valuable to society.

During the Spring 2020 semester, Yael had the opportunity to teach an undergraduate "Energy and Society" class. She found the process of preparing course materials; assigning and grading homework, quizzes, and exams; and lecturing in a way that captures the attention of the students for the entire class duration to be stressful and tiring. Further, as if the normal problems faced by a first-time lecturer were not enough, all classes at the university were moved online in the middle of the semester because of the COVID-19 pandemic. Instead of getting flustered, Yael reached out to Dr. Webber for advice (remember: don't be afraid to seek out help) and utilized this opportunity to redesign the class in a way that was conducive to online instruction. She requested colleagues to give guest lectures and modified the assignments significantly to keep her students engaged and involved. These endeavors were time-consuming and took time away from her research, but the end-of-semester reviews from her students praising her efforts and the notes telling her how some of them were interested in pursuing energy research in the future made all her efforts worthwhile. An important lesson that Yael gleaned from this once-in-a-lifetime experience is that adversities can turn into opportunities with advice from mentors, creativity, a positive mindset, and the willingness to invest time and effort. She hopes to communicate this wisdom to young engineers and future first-time instructors.

Conclusion

During a lecture at the Energy and Nanotechnology Centre in Houston in 2003, the late Nobel laureate Richard Smalley identified energy, water, and food as the top three problems for humanity in the next fifty years [20]. As engineers, we strongly believe that our job is to attempt to play some small role in helping solve the world's problems using our analytical skills and technological know-how. Thus, we plan to continue working on Smalley's compelling list. Further, while we plan to focus our efforts on the energy and water sector in the near future, we are confident that we will continue making a positive impact on society even as our research directions and careers evolve. We also hope to utilize our knowledge to improve societal and environmental conditions in the developing world through collaborations with scientists and engineers from those countries.

In addition to doing research that moves the world forward, we aspire to train the next generation of scholar-leaders who will also do their part. We both acknowledge the significant roles that several (formal and informal) mentors have had on our personal and academic success. We are motivated to play a similar role in the lives of young engineers so that they too can help address humanity's biggest challenges in the future. Thus, we continually strive to give back to the profession by engaging and cultivating young minds through teaching, volunteering in outreach activities aimed at increasing gender diversity, judging undergraduate research competitions, and mentoring graduate students. We hope this book chapter will serve as another asset in this endeavor and motivate the next generation, particularly women, to pursue STEM fields.

References

1. "Women in Energy - Gas, Mining, and Oil: Quick Take." Catalyst.org. https://www.catalyst.org/research/women-in-energy-gas-mining-oil/ (accessed April 5, 2021).
2. "USA: Labor force: percent female." TheGlobalEconomy.com. https://www.theglobaleconomy.com/USA/Labor_force_percent_female/ (accessed April 5, 2021).
3. "France: Female labor force participation." TheGlobalEconomy.com. https://www.theglobaleconomy.com/France/Female_labor_force_participation/ (accessed April 5, 2021).
4. "NSF Science and Engineering Factors." NSF.gov. https://nsf.gov/statistics/2018/nsb20181/report (accessed April 5, 2021).
5. R. Rincon. "SWE Research Update: Women in Engineering By the Numbers." Alltogether. SWE.org. https://alltogether.swe.org/2018/09/swe-research-update-women-in-engineering-by-the-numbers/ (accessed April 12, 2021).
6. J. Speer. "Bye Bye Ms. American Sci: Women and the Leaky STEM Pipeline." Appam.confex.com. https://appam.confex.com/appam/2019/webprogram/Paper31669.html (accessed April 12, 2021).
7. P. Meiksins, P. Layne, K. Beddoes, M. Lewis, A.S. Masters, J. Deters. "Women in Engineering: A Review of the 2018 Literature." Alltogether.SWE.org. https://alltogether.swe.org/2019/04/women-in-engineering-a-review-of-the-2018-literature/ (accessed April 12, 2021).
8. S. Bhatia and J. P. Amati, ""If these women can do it, I can do it, too": Building women engineering leaders through graduate peer mentoring," *Leadership and Management in Engineering*, vol. 10, no. 4, Oct. 2010, pp. 174–184, doi: https://doi.org/10.1061/(ASCE)LM.1943-5630.0000081.
9. N.A. Fouad, W. Chang, M. Wan, and R. Singh. "Women's reasons for leaving the engineering field," Front. Psychol., vol. 8, June 2017,doi: https://doi.org/10.3389/fpsyg.2017.00875
10. K. Quirk. "Why women leave engineering." UWM.edu. https://uwm.edu/news/why-women-leave-engineering/ (accessed April 12, 2021).
11. S.S. Sibley. "Why Do So Many Women Who Study Engineering Leave the Field?" HBR.org. https://hbr.org/2016/08/why-do-so-many-women-who-study-engineering-leave-the-field (accessed April 12, 2021).
12. "Fostering Innovation Through a Diverse Workforce." Images.forbes.com https://images.forbes.com/forbesinsights/StudyPDFs/Innovation_Through_Diversity.pdf (accessed April 12, 2021).
13. T. Merritt. "Why we need more female engineers." Concreteconstruction.net. https://www.concreteconstruction.net/business/management/why-we-need-more-female-engineers_o (accessed April 12, 2021).
14. S. Sirimanne. "How can we STEM the tide of women graduates leaving science?" Weforum.com. https://www.weforum.org/agenda/2019/09/stem-women-gender-equality-science-technology-engineering-mathematics/ (accessed April 12, 2021).
15. T.C. Dennehy and N. Dasgupta. "Female peer mentors early in college increase women's positive academic experiences and retention in engineering," in *Proc. of the National Academy of Sciences*, vol. 114, no. 23, June 2017, pp. 5964–5969, doi: 10.1073/pnas.1613117114.
16. J. Livingston and R. Evans. *Que Sera, Sera (Whatever Will Be, Will Be)* (1956) [Recorded by D. Day for The Man Who Knew Too Much.]. Accessed April 21, 2021. [Online Video] Available: https://www.youtube.com/watch?v=xZbKHDPPrrc.

17. M. Hoppus, S. Raynor, and T. Delonge. *Dammit* (1997). [Recorded by Blink-182] . On Dude Ranch. Accessed April 21, 2021. [Online video]. Available: https://www.youtube.com/watch?v=sT0g16_LQaQ
18. F. Perren. and D. Fekaris. *I Will Survive.* (1978) [Recorded by G. Gaynor]. Accessed April 21, 2021. [Online video]. Available: https://www.youtube.com/watch?v=gYkACVDFmeg
19. N. Bedingfield. *Unwritten.* (2004) Accessed April 21, 2021. [Online video]. Available: https://www.youtube.com/watch?v=b7k0a5hYnSI
20. M.E. Webber. "The progression of progress," *Mechanical Engineering-CIME*, vol. 136, no. 7, July 2014, p. 16, Available: https://go.gale.com/ps/anonymous?id=GALE%7CA373477402&sid=googleScholar&v=2.1&it=r&linkaccess=abs&issn=00256501&p=AONE&sw=w

Dr. Arkasama Bandyopadhyay is a Research Assistant Professor in the Department of Mechanical Engineering at Texas A & M University (TAMU). She previously earned a B.S. in Mechanical Engineering with a minor in Mathematics from Oklahoma State University and a Ph.D. in Mechanical Engineering from the University of Texas at Austin. Her research interests span the areas of distributed energy resources, residential demand response, and building energy systems. Arkasama enjoys combining her background in thermal-fluid engineering and renewable energy with operations research methods to lead novel research that benefits society. During her Ph.D., she worked with Austin Energy, the local municipal electric utility in Austin, TX, to analyze the techno-economic impacts of adding distributed solar and battery storage to the electricity system. Some of her current projects at TAMU include forecasting of residential electricity demand, occupant-centric building design and control, long-term performance of ground source heat pump systems, and smart building services. Arkasama is also interested in engineering education and sustainable energy systems in developing countries. In addition to academic research and teaching, she enjoys mentoring graduate students and first-generation undergraduate students in engineering disciplines within and beyond TAMU.

Dr. Yael R. Glazer is a Research Associate in the Webber Energy Group at the University of Texas at Austin (UT Austin). She previously earned a B.S. in Bioengineering from the University of California at Berkeley and an M.S. in Environmental and Water Resources Engineering and a Ph.D. in Civil Engineering from UT Austin. Prior to graduate school, Yael spent seven years working at Genentech, one of the world's leading biotech companies, developing and optimizing manufacturing processes to bring medicines to patients with unmet medical needs. Yael's current research interests span the realms of energy, water, sustainability, food waste, and the environment. She works on analyzing techno-economic opportunities for mitigating environmental liabilities (e.g., flared gas and wastewater) associated with oil and gas production, assessing the biogas potential of consumer-level food waste and overseeing the development of an open-source electric grid modeling tool for capacity expansion modeling. Through her projects, Yael has collaborated with a variety of stakeholders across the private, public, and nonprofit sectors including Equinor, the US Department of Energy, and the Cynthia and George Mitchell Foundation. In addition to academic research, Yael also consults with investors to assess the technical and economic feasibility of emerging innovations and energy projects.

From Spacecraft to Biocomposites: The Story of a Shuttle Launch, a Recession, a Surprise Doctorate, and Motherhood

Jessica Vold

"I Can't!"

These are the words my five-year-old daughter screamed out in frustration as she tried so hard to accomplish a task. I looked at this perfect tiny replica of myself and thought of all the times I felt the same way. In that moment, I knelt down beside her and asked if I could help her. In true Jessica fashion she said, 'No.' she really is a tiny version of myself. I pulled her into a hug and told her I never wanted to hear the words *I can't* come out of her mouth again. There is nothing she cannot accomplish as long as she really wants to. I told her that it is normal to feel frustrated when things are hard, but we have to keep trying until we finish. She dried her tears, turned around, and finished the task at hand like she had been doing it forever. Sometimes, you just need to be told ***you can***. One day I will tell her my story.

As a child, I did not realize my own parents were instilling this same mentality in me. I watched as my mother went back to school as a non-traditional student to pursue her bachelor's in elementary education while raising a young family. There are many nights I remember "helping" her with school assignments by testing out lesson plans she had carefully crafted. In my eyes we were just having fun together; I had no idea she was actually teaching me how to balance motherhood and accomplishing the goals we set for ourselves. My father was an electrical engineer working to develop cutting-edge technology for his employer. He is still an example of what innovative really means, with eleven granted patents and a twelfth one filed as of this publication. I watched him encourage my mother to accomplish her dreams all while they raised my brother and me. Naturally, if you had asked me at an early age what I wanted to be when I grew up, I wanted to be just like my parents. Like a sine wave I would swing back and forth between wanting to be a teacher or an

J. Vold (✉)
Department of Mechanical Engineering, North Dakota State University, Fargo, ND, USA
e-mail: jessica.l.vold@ndsu.edu

M. Bailey, L. Shackelford (eds.), *Women in Mechanical Engineering*, Women in Engineering and Science, https://doi.org/10.1007/978-3-030-91546-9_19

engineer. Never in my wildest dreams did I see being both at the same time. *I can't do that!*

The older I got the more I fell in love with math and science, so naturally I gravitated towards engineering. I knew I wanted to be an engineer, but I had no idea what kind. It was my 13th birthday and John Glenn was returning to space on STS-95. Having spent most of my childhood in Ohio, John Glenn was a big deal. I had come to love all things NASA and a shuttle launch on my birthday turned my love into an obsession. I can still remember being angry that NASA had "decided" to launch in the middle of a school day. How was I going to watch when I was in school all day? It would be a few more years before I understood the importance of launch windows and finally forgave NASA for such an indiscretion. Shockingly, I was not allowed to skip school to watch a shuttle launch so I settled on forcing my parents to record it for me so I could watch it after school. The whine of the VHS player still plays in my head when I think about anything space related. This was the moment I knew I was destined to be an aerospace engineer. The STS-95 flight patch hangs in my office over my desk to this day as a reminder that *dreams do come true, just not always how you planned them.*

At eighteen I felt like I was on top of the world. I knew who I was, I knew what I wanted to do in life, and I had the confidence to take on the world. More importantly I had a five-year plan, I was more than prepared when asked in interviews where I saw myself in five years. I entered my undergraduate studies ready to tackle aerospace engineering and graduate at the top of my class to land a job with NASA. Academics came easy to me; I never had to try all that hard to do well and college was eye opening to say the least. It was not as easy as I thought it would be. There were many nights when I would let the phrase *I can't* flow through my thoughts.

Thankfully, when it came time to choose where I wanted to live my freshman year, someone gave me the advice of requesting to be in the Women in Science and Engineering (WISE) House. WISE was a dormitory floor of all female science and engineering majors. That would later turn out to be the best decision I made going into freshman year. I had a built-in study group and friends who all knew what it was like to be entering a male dominated field. I am positive we had far more fun than we should have that year. Our resident assistant was thoroughly annoyed with our impromptu study sessions in the hallway using mirrors as white boards. This group of women was the best thing that happened to me; they were always ready to say *yes you can* when I needed it the most.

I could have really used the WISE women in my senior year when I once again found myself saying *I can't do this*; except this time, it was not in my own thoughts it was out loud to my major advisor. In aerospace engineering, two extra courses that also counted as technical electives would get you a minor in astrophysics. I found the idea of working in an observatory looking at celestial bodies romantic and one of the reasons why I fell in love with the field of aerospace engineering. Naturally, I chose to take introduction to astrophysics and an upper-level mathematical modeling course to complete my astrophysics minor. They both counted as technical electives for my major, so it was a logical decision. I can still remember my sweaty palms and racing heart as I sat down in my advisor's office to discuss how I was going to graduate. I was in my final year and my mathematical modeling course was not going well; in fact I was failing. I had pushed very hard to finish my degree in four years, and as such there was no room in my final semester to fit another technical elective. At this point I really did not care if I finished my astrophysics minor; I just needed to finish my aerospace engineering degree. He looked at me and said, "Just take a breath. Let's think about what has to happen to ensure you graduate." In reality I just needed to get a C but *I did not think I was smart enough* to do that. *I was not good enough* to get through a complex math course. I could and did design a two stage to orbit hypersonic scram jet, but I could not tackle complicated math. Then my advisor said the one phrase that I needed to hear, *"You can do this."* I share this story with my students in my early general engineering courses to show them that STEM is hard for everyone, even their faculty members. We all have doubts about our own abilities and knowing that others share our struggles can be very empowering.

Ten years after STS-95 launched, I walked across a historic stage to receive my bachelor's in aerospace engineering and mechanics. In the depths of the Great Recession, I was a young, just-out-of-college aerospace engineer desperate for the job I had always dreamed of: designing anything that flies for NASA. At the time of my graduation, a bubble of retirements was about to burst in the aerospace industry. The recession meant that companies were promoting from within and not backfilling to save money. I took a permanent position for a small company I interned for during my senior year. I had an aerospace engineering degree and had accepted a position as a computer programmer. This was the first time I truly felt lost in my adult life. I was going to work each day, going through the motions and dreading every minute of it. *I can't do this.* Do not get me wrong, I learned so much during my time at this company; it just was not for me.

I started applying for jobs only to receive rejection letter after rejection letter. A ten-minute phone call with a recruiter would change my entire career path. I remember sitting at my desk the morning of the phone call; I was nervous about someone finding out I was looking for a new job. A few minutes before the scheduled call, I snuck out the side door nearest my desk. It was a gorgeous fall afternoon, and our building was nestled in the woods, so I planned to take a walk around the building while chatting with the recruiter. My phone rang and I could feel my stomach drop: could this be the moment I was waiting for? I answered the phone, and the first words out of her mouth were, "I want to be upfront; we are filling this position from within the company." My heart sank, *I can't do this.* She continued on to say, "I wanted to connect with you because I was really impressed with your resume. For the position you applied for we would prefer a mechanical engineering degree. Maybe if you go back to school and get a mechanical engineering degree, when you are done, we will have the money to hire you." Maybe *I can do this.* I hung up the phone and applied to graduate school to get a master's degree in mechanical engineering.

Now, this was a move that those closest to me questioned: why would I want to leave a good paying job to get another degree I "did not need"? The money meant nothing to me; having a job that did not feel like a job was much more important. In hindsight, this may have been because I applied to one school that I happened to have a boyfriend at. Do not worry. He is my husband and father of our two beautiful daughters now. What others around me did not know was I had a plan. I was going to get a master's degree from a school that was thought of highly in our area and reenter industry as a design engineer. Nothing can go wrong, right? *I can do this!*

I spent the first few months of graduate school meeting the faculty within our department and learning about the research they did. I wanted to focus on aerospace applications, and there were several faculty in that space. I met with each of them and heard about innovative research in micro-UAVs. I could see myself working on any of the projects I heard about. Something kept pushing me to keep looking though, to find my real passion. Everyone kept telling me to go meet with one more faculty member; he had a great research group doing unique materials development. I was hesitant; materials science was the one class in undergraduate that I loathed. I never wanted to think about materials science again. What on Earth did it have to do with aerospace engineering anyway?

I happened into the office of my soon-to-be graduate advisor not knowing what a composite material was, other than they existed and the aerospace industry was just figuring out how to widely utilize them. His passion for his research was infectious. The excitement he showed for the work he and his students were doing was what I was missing in my previous job. It took all of 30 seconds into our meeting to know this is what I wanted to do. Maybe I could learn to tolerate materials science. I left the meeting with his promise to let me know if he had a position open up. Not long after I got the email I had been waiting for, he had some funding, and it was mine if I wanted it. *I want to do this, but am I smart enough?*

I remember the cold metal stool I sat on and the exact location in the lab when my inner monologue of *I can't* turned into *maybe I can do this*. In our weekly research meetings, we would always spend time introducing ourselves and our research focus to new members of the team. It was my turn that day and I rattled off my standard introduction: "I'm Jessica. I'm a master's student, and I work on thermally stabilizing biomass for introduction into high temperature plastics." Then my advisor would normally crack a joke about me being a University of Minnesota Golden Gopher before moving on. Today though, he had something more profound to say, it may have been ordinary to him but for me it was life altering, "and you're sticking around for your PhD, right?"

I sat in silence thinking this was not my plan, *I can't, I'm not smart enough*. My plan was to graduate with a master's and go out into industry not into academia. What else is there to do with a doctorate besides teach anyway? This was the point where I started to think *maybe I can do this*. As I shook my head no, the word yes came out of my mouth. If someone I respected so much thought I could do it, *maybe I could*. It took me a few more weeks to retrain my inner monologue to think *I can do this*.

A few years before graduation, I was asked to help teach a course within our department. Up until this point, I never really experienced the discrimination that females often talk about within engineering. Sure, I had my own self-doubts but never an external voice adding to those doubts. The next conversation runs through my thoughts like it was yesterday. I was sitting in the same room where I had finally realized *I can do this* when the instructor teaching the other section of my course walked in. We had planned to meet to discuss goals for the course before the semester started. He opened with: "Well since you are a novice at teaching and a novice at [the course content], I think we should just plan to co-teach." I had to fight the urge to say something I would later regret so I took a breath to pause before saying anything. "While you are correct, I am a novice at teaching, I am what you would call an expert in [the course content]." What he did not know was I left a full-time job doing exactly what I was asked to teach to come back to school; how could he when he wrote me off as a nobody that was just filling an empty void. Now, I am not sure if he actually said anything after this; I was too furious to care or listen. How could he think so little of me before knowing anything about me? He did not know one more tiny fact about me; tell me *I can't,* and *I will* do everything in my power to prove you wrong.

Six years after walking into his office having no idea what a composite material was, my advisor hooded me as I graduated with a doctorate I never thought I was smart enough to complete. In the middle of those six years, I was awarded a master's degree and married my best friend. To this day, I have moments where I doubt my abilities and wonder how I managed to graduate. While I was excited to be graduating, I was even more excited to be starting a job with a company that had licensed the technology I helped develop in my graduate studies. I had found my passion and was excited to see where I could go. Two weeks after I walked across that stage, I started working. Was this it? Was this my dreams coming true? The short answer was no, and this is where my journey takes some rather unexpected turns. Remember earlier when I said, never in my wildest dreams did I see being both an engineer and a teacher at the same time? Now, I saw how I could do that and it was all I wanted to do.

As I was approaching graduation, my husband and I had many conversations about what I wanted to do with my career. After I had come to the realization that *I was indeed smart enough* to get a doctorate, I started to seriously look into what I could do with a doctorate outside of academia. I had fallen in love with research and started to lean towards a research and development position within a larger company, preferably one that got us closer to our families as we looked to start our own family. He asked me periodically if I wanted to teach at the university one day. Of course I did; I felt at home in my classroom, but I never thought that was a possibility unless someone within the materials curriculum left. I had become comfortable with the idea of working in industry once again until I might have the opportunity to come back to academia. Secretly though, *I can't* was still running though my head. *I am not smart enough* to be a faculty member. *I am not innovative enough* to start my own research program. *I am not good enough* to teach anywhere else. We knew my position was temporary, so I continued to apply only to receive rejection letters.

We were also starting our family. As I walked across the stage at graduation, I was fighting back waves of morning sickness. We were excited and scared all at the same time but for different reasons. My husband was scared to be a first-time dad; I was petrified of what a family would do to my brand-new career. I was working on a contract to do some specific grant work, and the fall after I graduated, I was asked to also do some teaching as an adjunct faculty member. Teaching as an adjunct was a far different experience than teaching as a graduate student, and I enjoyed it even more than before. While I enjoyed my work, I felt unsettled. Life was about to drastically change for me and my husband, and my career was more uncertain than it was established. I could not see a path forward but knew I had to keep going. I kept applying for jobs and still could not even get a phone call. *I'm not smart enough.*

My employer was more than supportive of me as I navigated a new career, and my husband and I welcomed our first daughter. When I left for maternity leave, I was working on some very interesting projects that I knew would be finished by the time I got back. It was disheartening to see someone else finish the work I had started. While I was on leave, the company picked up a new project that was a great combination of my aerospace and materials engineering skills. I was excited at the thought of a new project that was all my own, but I was terrified of leaving my daughter all day. *I can't do this.* My first day back from maternity leave felt like my first day on the job. I felt like I was starting over.

Life was starting to settle into place finally. I was feeling more and more comfortable in my new role as a working mother. Part of my job was to write small business grant proposals. I had written more proposals than I could count and had yet to secure any funding. I once again felt inadequate. Like all the other well-timed moments of encouragement in my life, I was going to get one more that would forever change my passion and career path. We had learned about the Innovation Corps program that lays out the process of validating academic research for commercialization. The entire program is seven weeks long with a kick-off and wrap-up conference. This would be the first time I was away from my daughter more than a working day; I was terrified of leaving her. Would she forget who I was? In between conferences, the team had to conduct interviews to vet out customers, strategic partners, and suppliers. Our goal with the program was to strengthen our commercialization pitch to secure grant funding, but in the process, I found a new passion for entrepreneurship. I loved talking with potential customers to understand their pain points and devising a plan to help alleviate them. It felt good to be helping others; I could see the difference I was making for our customers. *I can do this.*

I had gone from being an aerospace engineer to a materials engineer focused on entrepreneurship. I was enjoying my job and watching our daughter grow, but something was missing. I still felt unsettled in my career. I was once again asked to do some teaching and again felt at ease, dare I say settled, in my classroom. I never realized how much joy teaching brought me. Unlike in my engineering roles, I could easily see the good I was doing for my students. I knew the impact I was making. I had not allowed myself to think about it as I did not see a path to being a faculty member while allowing my husband to stay with his employer. About this time, we started thinking about adding to our family.

As we welcomed our second daughter, I once again left behind projects I had started at work and watched someone else finish them. I knew I would be coming back to all new projects and had come to terms with it. Little did I know I would be coming back to a whole new job. While I was on leave, another small business with roots in academic research needed a scientist to help them complete some government grant work. This company focuses on bio-based chemistries, something I knew nothing about. I knew the basics of conducting research and working in a lab, but the chemistry part was not my strong suit. *I'm not smart enough for this, but I will try.*

Once again, my first day back from maternity leave felt like my first day on the job because it was. Now I really could not see where I was going; my "new career" felt like it was restarting every year and I needed to find new traction on my path. I was splitting my time between two companies, and while I loved all the new opportunities to learn, I just did not feel like myself. I started actively applying for jobs once again only to receive rejection letters. Had I taken a wrong turn somewhere along this journey or was the right opportunity coming still? The fall after our youngest was born, my graduate advisor reached out to tell me about an opportunity he thought might be a good fit for me. The university had been gifted some funds to create a new position focused on encouraging entrepreneurship and innovation from the students, staff, and faculty within the college of engineering. This was my chance to become a faculty member.

The following spring, I got the call that I had made the initial cut for phone interviews, the first phone interview I would do since I was applying for jobs in undergraduate. The interview went well, but when I was asked about grantsmanship, a rather important aspect for a tenure track faculty, I did not have a good answer. I was worried this would disqualify me for the job, but to my surprise, I soon got the request to schedule a campus interview. *Maybe I am smart enough.* My campus interview went as well as it could have. Interviewing with faculty and staff I had worked with off and on for nine years felt more like catching up with old friends than a job interview. The more time passed without hearing from the university the more lost I felt. I knew, after about a month of no news, I had not gotten the job; I was just waiting for the rejection letter. It eventually came, not that I was surprised but it stung just the same. I even got a phone call from the department chair I had interviewed with saying how impressed he was with my ideas and he wished he had a position for me. I was on my way into the office after dropping my kids off when he called; I remember hanging up the phone and trying not to cry the entire way to the office. How was I going to pick up the pieces and move on? I had to do something, so I finally took the plunge and applied to my undergraduate alma mater. This would require moving our young family, but if we were going to move, this was the time to do it before our kids were in school. We could finally move closer to family, and I was a little more confident in my qualifications to be a faculty member. I continued to apply for jobs in town, so we could stay where we were but was overqualified for most of them. I just really wanted a place to call home where I felt I belonged.

Out of the blue, I was contacted by two companies in town to set up interviews. I was excited about the opportunity for growth and change. A few days before my first interview, I got an email that the university wanted to discuss an opportunity with me. I scheduled a meeting between my interviews and planned it during a trip to campus to do some testing; why not kill two birds with one stone. I remember walking into my chair's office to chat, my current co-worker waiting outside to go do our testing. Never in my wildest dreams could I have predicted what transpired next. After catching up on how work and my family were doing, he started describing the person they had hired for the entrepreneurship position I wanted so badly. I sat there in silence wondering why on Earth he is telling me this; it was like pouring salt on an open wound. I was half listening trying not to show the emotion that was boiling up inside; *I can't do this again.* The conversation then took an interesting turn; the new hire never showed up. It was about this point that I was given several pieces of paper. Had my chair not verbally described what was written on the paper, I never would have realized exactly what was happening. My head was spinning. I was being given an offer for the position only months ago I had been turned down for and they wanted me to start as soon as possible. *I really can do this.*

If you asked me back at my undergraduate commencement where I would be in eleven years, doctorate, faculty member, and entrepreneurship would not have been part of my answer. This was not part of my well-crafted five-year plan. After a lot of thought, I accepted the offer and started a few weeks later. *I did it; I really was smart enough to do it.* I just needed something to point out that I could do it to drown out my own self-doubt. So often as women we allow our non-data driven self-doubt to define our path in life. Yet, as engineers we are focused on the data and facts at hand to show where our research should go next. What does the data say? We need to let the data of life show us where to go. Allow the external nudges of encouragement and the data of work well done show us the path to take; let it drown out the non-factual inner self-doubt. *Keep on trying and collect enough data to define your own unique path.* A few weeks after I started, my husband looked at me and said, "You're a whole new person. You're happy again." I had not realized just how unhappy and lost I had become trying to find my path. An invisible weight had been lifted that I never realized was holding me back. *Do not let your own self-doubt cloud the data that says you are capable of accomplishing anything.*

I openly share my journey with my students these days to show them *STEM is hard for everyone and sometimes you end up with a job you never thought you would have*. I remember the classes I dreaded going to or worried about my final grades in. They all chuckle when I tell them I hated materials science as an undergraduate or that I nearly failed a math course that of all things was a technical elective. They sit in awe when I tell them; these are the classes I love teaching the most now. Looking back, the passion my instructors showed for their course material had a huge impact on how I felt about the courses. This is something my graduate advisor showed me when I walked into his office twelve years ago vowing to never think about materials science again. Here I am today with a doctorate I never thought I was smart enough to get focused on advanced materials for some of the most out of this world applications. I consciously work every day to share my passion with my students. This comes easy when I feel settled in my classroom and know I am exactly where I am meant to be finally.

"I Did It!"

Dr. Jessica Vold is the Engineering Entrepreneurship and Innovation faculty member for the College of Engineering at North Dakota State University. Dr. Vold teaches entrepreneurship, engineering mechanics, and materials engineering courses within the Mechanical Engineering Department. Dr. Vold earned her Bachelor of Aerospace Engineering and Mechanics from the University of Minnesota—Twin Cities in 2008. After graduation, Dr. Vold worked for Aerosim Technologies programming fixed base, flat panel, and virtual flight deck flight simulators for pilot training. She obtained her master's degree in 2013 and her doctorate in 2015 in Mechanical Engineering from North Dakota State University. Both her master's and doctoral research were focused on high temperature engineered thermoplastic biocomposites. After graduation, Dr. Vold went on to work for two small businesses spun out of academic research at North Dakota State University, c2renew Inc, and Renuvix LLC. In October 2019, Dr. Vold left private industry and joined the faculty of North Dakota State University. Her research focuses on advanced materials for additive manufacturing, biobased composite materials, and how engineering education can be transformed through the incorporation of the entrepreneurial mindset. Outside her classroom or lab, Dr. Vold can be found gardening with her family, crafting with her kids, or playing fetch with her dogs.

My Most Valuable Lessons in Energy Engineering

Jessica F. Bull

It was late August in 2009, and I'm sitting on my couch with a former roommate, turned close friend, drinking a glass of wine and lamenting about my dead-end job working in insurance. I lived paycheck to paycheck and my job left me wanting more for my life and from my career. Christie said, "You like math. Why don't you become an engineer?" I looked at her inquisitively and responded, "What does math have to do with engineering? Better yet, what does an engineer do?" Christie laughed and explained simply, "They build stuff!" That night changed my life forever.

I had a bachelor's degree in psychology from Binghamton University, and I was licensed to sell personal lines property and casualty insurance. Neither my degree nor my career sparked any passion in me. That night, my friend and I got to work and Googled all the different engineering curriculums from local colleges, and we found mechanical engineering and sustainability. I was hooked; I felt like I could have a career I was passionate about, and I could make a decent living doing it! Without a second thought, I put wheels in motion that would lead me to quit my job, go back to school full time, and dedicate the next 5 years of my life to getting three more degrees, all in engineering.

I remember in the beginning of this journey I heard doubts and concerns from people close to me. My boss at the time scoffed when I put my notice in and told me, "HA! They are a dime a dozen." My friends and family asked me, "Are you sure you can do this?" It's hard to describe the resolve I felt, the sheer will and determination to get where I wanted to go, but I held no doubt. Going back to school as an adult has its advantages and its disadvantages. On the one hand, I had the experience of working full time and the maturity that comes from that. On the other hand, I wasn't sure I knew how to interact with my peers who were mostly 10 years younger

J. F. Bull (✉)
CHA Consulting, Rochester, NY, USA

M. Bailey, L. Shackelford (eds.), *Women in Mechanical Engineering*, Women in Engineering and Science, https://doi.org/10.1007/978-3-030-91546-9_20

than me. In the end, I made it out alive with an associate's degree in engineering science from Broome Community College, a bachelor's in mechanical engineering, and a master's of engineering in mechanical engineering with a focus on sustainability from Rochester Institute of Technology (RIT). I'd like to tell you some of my memorable and pivotal stories as they relate to my experience in engineering. Along with these stories, I will share what I have learned both technically and about life as a woman in engineering. My chapter is written for the person who may be interested in changing careers, re-entering the workforce, and starting out fresh in college or for academic educators looking to understand the experiences of women in the industry compared to their own experiences. What I have taken away from my own experience is that engineering and technology is always changing, just as much as your life will change. Finding your passion and following your heart (a.k.a. your gut) makes a substantial difference in the success that you experience in both your life and your career.

College Time

It was my third year at RIT, and I was walking down the hallway when a professor I had worked for as a research TA stopped me and asked if I had the chance to apply to the Rochester Engineering Society (RES) scholarship; with the deadline being that week. Sigh. I didn't get a chance since all I had been doing was trying to keep up with all the assignments, all the extras, all the projects, and meetings. Majoring in engineering (regardless of discipline) is no walk in the park. It's demanding, but if you have the passion and determination, it's an achievable and worthwhile endeavor. On this particular day, Professor Smith happened to catch me stressed out and my first feeling was annoyance. The scholarship required an essay, two letters of recommendation, and all my transcripts. I said to him, "I'm not sure I have the time." He offered to write one of my letters and I realized this was an opportunity I should not ignore. This person took the time to support me in a way that professors don't always have the ability or desire to do. I promised to complete the requirements and sent it off for review. I was called into a local engineering firm, CHA Consulting, to interview for their scholarship. I thought an interview for $1000 was a little strange, but sure, why not! When I sat down, the two gentlemen explained that the scholarship was in memory of a manager in their group who was tragically killed in a drunk driving accident. Every year, they interview three candidates that have the same passions and spirit, and I was one of those selected. (How nice!) They continued to tell me more about the firm and that their group (Mechanical Energy) was dedicated to energy efficient solutions for large industrial and commercial customers in the area. I was stunned since that was precisely the type of work I was looking to do with my degree! I originally thought I would work in the residential sector, but as they say, "Go big or go home!" We had a great conversation, and by the end, I asked them for an internship. I had one remaining co-op, and if they had room for me, I was extremely interested in working for their firm.

It turns out that I won that scholarship and they gave me the internship. That internship turned into a full-time job, and that full-time job turned into several promotions in a rewarding career that I love. My career counselor at RIT said that it was not likely I'd find my dream job right out of college, but lucky for me, I found it while I was still in college. The thing that sticks out most in my mind is how I almost let this opportunity go because I had been consumed by the work and demands right in front of me. Sometimes, you need to stop and remove yourself from the current situation to be able to see the gifts right in front of you. Professor Smith probably saw more in me than I did at the time, and his persistence changed my life. There will be moments where you will want to keep your head down and ignore that gut feeling. Don't. Follow your heart and allow the people around you to see your potential and accept their help when it's offered to you. You never know, it can change your life.

While at RIT, I focused my studies on energy and the environment. As I mentioned, my passion was for energy efficiency and the future of renewable energy. While I was at Broome County Community College (BCCC), I took all the typical prerequisite engineering courses such as statics and dynamics, but never got into anything specific to my interests in energy. At RIT, I was able to take courses in heating, ventilation, air conditioning, and refrigeration (HVAC-R), advanced thermodynamics, renewable energy, and heat transfer. These courses were the basis for all the work I would end up doing in the field. I loved Thermodynamics so much that I became a teaching assistant (TA) for two semesters after taking it myself. Becoming a TA for a subject so crucial to my degree helped me master the concepts and foster good working relationships with my peers. I was a mentor and a friend to students that could come to me when they needed help in thermodynamics, or other areas. For the students out there reading this, I encourage you to take opportunities for leadership positions if they are available to you. These experiences enrich your college career and allow you to stand out to future employers.

While my experiences at both BCCC and RIT were exceptional and left me with a comprehensive set of skills and knowledge, I will admit that the majority of true learning is done on the job. If you are in school, please take any opportunity possible to obtain internships or cooperative work experiences. My co-op at LORD Corporation taught me that aviation manufacturing was not my true passion. I learned a lot from my experience there, but ultimately it was my co-op at CHA which really aligned with my passions and interest. You don't know until you try!

My Career at CHA Consulting

After completing my summer co-op, I was asked to stay on with CHA part-time through graduation. This work allowed me to pay for my living expenses and continue to grow within the organization. By Thanksgiving 2013, I was offered a full-time position as an Engineer II upon graduation. One of our major clients is New York State Energy Research and Development Authority (NYSERDA). We had a contract to perform technical review for their pre-qualified incentive program, and during my time as an intern, I functioned as the lead engineer and project manager for this contract. This experience taught me a great deal about the engineering consulting world. While the work was reasonably straightforward and procedural, the hours fluctuated significantly and required us to respond with flexibility. Through communication to upper management about the workload assigned to us on a weekly basis, I was able to project the required hours needed to complete the work in the timeframe expected by the client. Consulting work is centered around staff being utilized with billable hours. I would come to learn more about utilization later in my career, but even in this early stage, I was able to illustrate the need to hire one more staff for this effort.

An engineering consulting firm is comprised of project management groups and technical groups. The project managers are responsible for business development and client advocacy, while the technical groups are responsible for meeting the deliverables outlined in the proposals and contracts. It's important to note that this is a generalized statement of roles. Truthfully, a good engineer is going to do everything while focusing on their specific responsibilities to the project. Establishing a good rapport with your client will lead to repeat business. In consulting, no matter what role you are in, you will interact with your clients. In the beginning of my career, I did not realize or appreciate how important this relationship is. Repeat and referral business can easily be 90% of the billable work for you and your team. As a consultant, it's important to focus your energy on the value that you bring to the client. Just like you learn in school: identify the problem, devise a solution, and present your results. If you would like to simplify each project, think of it in terms of the following:

- A given set of variables (technology, reliability, quality, cost, etc.)
- Methods used to develop a solution
- Viable solutions to problems with pros/cons
- Present results to clients with supporting evidence

Clients seek out consultants to solve problems outside of their expertise. In the mechanical energy division, we are experts in energy efficient solutions for various clients and various energy-consuming systems. When I was first starting out, I needed all the hands-on training I could get, but at the time we had three new hires in Rochester, including myself.

While I appreciated the initial experience with the pre-qualified incentive program, I was excited to expand my technical knowledge to new areas. CHA was actively performing multiple steam system energy studies and completing steam trap surveys for our clients. My two new colleagues informed me that Spirax Sarco was hosting a steam system basics training at their facility in Pennsylvania and our local representative was supporting our attendance by sponsoring the registration for only $100 out of pocket. This was a two-day event and extremely informative and valuable for a new energy engineer. My colleagues informed me that our manager allowed them to sign up and would pay for the travel and lodging. I assumed that I would also be going, but to make sure, I dropped into my manager's office. I told him that I was excited for the opportunity to get the steam basics training since I didn't have nearly as much experience as my two new colleagues. He had a sad look on his face when he told me that there was only enough budget to send the two gentlemen in our group and he was sorry but promised to send me next time. I felt truly disappointed and left feeling deflated.

In the time, it took me to get from his office to my desk, several thoughts ran through my mind. When am I going to get this opportunity again? Why wouldn't I go since I have so much more to gain from the experience? Was this gender discrimination? Why isn't my manager willing to support my professional growth? My disappointment rapidly grew into anger. I have not worked this hard to find my dream job, only to allow someone else to control my path. I know this was only a training seminar, but if I am going to pave my way, then I'll just have to advocate for myself. Now that I had this great job, I realized a few things. We get vacation time, I have money for both the registration and hotel, and lastly, I can carpool with the other guys. I marched right back into my boss's office and I let him know I'll be taking those days off to attend the training. I told him how important it was for my professional development and then left feeling that I made the right decision. There were no other pressing projects that would prevent me from attending, so I had no reason to feel bad for taking that time off.

Ten minutes after that interaction, my manager sends me a note to come back to his office. This time, he's not happy. He tells me that my "stunt" makes the company look bad for sponsoring two junior engineers and not the third with similar level and responsibilities. His hand was forced, and the company would pay for my trip as well. I was genuinely surprised, as if I had no idea how funds could magically appear in just 10 min. It was not my intention to get the company to pay; I only intended to get a fair and equitable experience.

I tell this story to young women looking to head into a career in engineering because I feel that we may question our abilities and what we deserve at times. Maybe this is my own unconscious bias, but sometimes I feel we are reluctant to stand up and advocate for ourselves. It is a skill to know when to challenge authority and the status quo. I don't mean to advise you to reject all your manager's decisions, but if you feel there is an injustice towards you that will limit your professional growth, you will have to work to overcome these obstacles. That manager is no longer with the company; and my current management is open, receptive, understanding, and supportive of my growth. Having respect for your colleagues and management will help them see you for what value you bring, and they will open doors for you to succeed. Find these people and learn as much as possible from them. Develop professional relationships with experts. Always ask questions, even if you think it's dumb. If you were wondering, that steam training was actually an incredible experience that taught me a great deal that I still remember and appreciate to this day.

Life Happens (Sorry for the Interruption)

After my experience with my first manager, things quickly changed for the better. Many other people were recognizing my contributions and my drive to accelerate my career. My passion to change the world (even a little bit at a time) was driving my desire to learn more and help others coming up behind me. It was just days after I returned from a vacation with my husband when something happened that would change my course. I woke up early and wasn't feeling right, so I woke up my husband and he told me not to worry. I went back to sleep and went about my day. Later that morning at work, I still wasn't 100% right; I asked a female colleague if my symptoms seem normal to her and her face said everything she did not. She just said, "No and you should call your doctor." I called the doctor and tried to phrase my experience as a no-big-deal situation, but she also wasn't buying it. She told me to make my way to the ER. Three grueling days later, I received a diagnosis that would change my life. Even now, I am not inclined to talk about the specifics, but needless to say, it was serious enough. I was facing a life-long diagnosis, but fortunately, not a terminal illness. This was my first experience with grief. I just spent 5 years of my life reinventing my career, was just married, and was in the best shape of my life. Everything I had worked for felt like it was just taken away in one fell swoop. I was scared and didn't know how this was going to affect my career.

Luckily, I had a great support system. I had friends, family, and especially my husband there to tell me that nothing was actually lost, even if I felt that way at the time. I used my successes at work to help me feel confident again, and I started to pull out of the fog that follows a tough diagnosis. There were several things I learned. My career was always my priority, but now I needed to make room for others as well. Live every day like it matters, because they all do. If anyone should ask me about my five-year plan, I will try not to laugh out loud in their face. I like to plan out my plans; I'm an engineer. Ultimately, this had to teach me that life is going to happen, and you don't know whether it will be a good or a bad shift in your trajectory, but it's wise to be prepared for the unprepared. This lesson feeds directly into my earlier advice: "Follow your heart." All my life, I never wanted children, but I felt as though I had to stop and consider what experiences I wanted to have while I'm here and breathing. Kids are challenging, but I did want to experience the joy of having a child, now that joy was one of my new priorities. Luckily, my husband agreed to at least one, no more than two…

My son just turned three and our lives are, in fact, changed. It is most certainly harder while also rewarding and fun. I found that joy I thought I was missing, and I have no regrets. Starting a family did change the work-life balance, but I can talk about that more later.

My Career Continued…

When I got back from maternity leave, I dove right in. At the time, we were conducting a compressed air study for a manufacturing plan and I was assigned as the lead. I haven't had the chance to mention, but I love compressed air systems. I have completed some large-scale manufacturing compressed air studies and recommended significant energy savings opportunities to our clients. Since I came into this project after it had already begun, I was not clear on the timeline and client expectations. Another junior engineer and I were waiting on the client for some missing data to complete our analysis. Once that was delivered, I was busy with another assignment and my colleague was going through a family crisis. No one communicated to the client and we were late with our deliverable. Even though the work product was of high quality, the fact that the communication was poor, and the deliverable was missed was a crucial error on my part. In engineering consulting, the client relationship is one of the most important aspects of the business. It did not matter whose responsibility it was to communicate changes or a delay; it is always important to be accountable and reliable as I would come to learn in the most important project of my career to-date (Figs. 1 and 2).

Accountability comes in all shapes and sizes and it leads to integrity and trust. The more an individual exhibits these qualities, the more opportunity for experience and advancement there will be. At this point in my career, I was invited to join an interview with a potential high-profile client. Our company was short-listed for a

Fig. 1 Proud selfie with 1950s' vintage Ingersoll Rand reciprocating air compressor in operation at client site in Canandaigua, NY (September 2016)

Fig. 2 Client site visit at a large industrial complex in Rochester, NY (July, 2018)

major energy study in one of their highest energy-consuming buildings, and I was asked to sit at the table to represent our experience and abilities. We won the project, and were set to begin fieldwork in the building within a month. The audit was to be completed on a very prestigious laboratory building on an Ivy League campus, so this project had high visibility. If we were to do a great job, we would surely establish a great relationship with this client and win more business. I was assigned as the lead engineer on the project. After 2 weeks in the field, we came back with a ton of data, and I began to get to work. The major energy user was the HVAC system, since most of the building was served by 100% outside air handling units. This is relevant to energy consumption, as when you bring 100% fresh air into a space, it will need to be conditioned to the space setpoint which can be a huge change in temperature based on the season. Alternatively, recirculated air, or return air systems, use maybe 20% of outside air for ventilation, significantly reducing the load on the cooling and heating systems.

This project was my first real introduction to airside systems, and the scope included approximately 400 labs and over 300 offices in a building with 7 floors. Due to the tight controls needed for these rooms, there is an extensive building management system (BMS) that monitors and records airflows, heating hot water flow, space temperature, etc. After all the data was collected, it was my task to create the energy models that would be the baseline to any potential modifications that would save energy throughout the building. I pulled in a few junior engineers to help where I could, but the truth was that I was in over my head. I could crunch numbers better than most, but knowing how to model energy conservation measures (ECMs) for a building like this needed some expert assistance. Due to my past experience, I didn't feel as comfortable asking for help. I just plowed through feeling as though it may all just come together in the end. When my boss asked how things were going, I would always answer positively. I had forgotten how to ask for the help I needed for this project to succeed. Due to this oversight on my part, we needed to pull in a large group of people working long hours to fill in all the gaps at the 11th hour. The failure here was that I was not accountable in admitting my inexperience. This trickled down to affect numerous crucial pieces of this very high-profile puzzle. We were late in delivering the product, were missing some items in the contract, went over budget, and created a tense relationship with the client. This was one of my largest and hardest lessons to learn.

Fortunately, due to the high quality of the final energy audit product, we went on to win the design proposal despite the previous issues. Most of my experience has been modeling energy, and this was my first real taste of detailed engineering design. The difference here is that in energy studies, the energy savings solutions are presented in a high-level manner. As an energy engineer, the solutions we come up with are vetted for feasibility, but the details of installation are not completely mapped out. A design engineer would take those initial recommendations and create the drawings and specifications specific to the modifications or upgrades for the client.

In the aforementioned high-profile energy audit and design, I learned about variable air volume (VAV) boxes, venturi-style air valves, diffuser style and throw, fume hood operation, and space conditioning requirements for a vivarium. At the time of this writing, we have recently submitted the 100% construction drawing (CD) set which will go out for bid to the mechanical and electrical contractors who will construct the proposed changes. They are relying on our expertise in these drawings to ensure what we are proposing is constructible. My boss and I remained deeply involved with the project, even after the audit phase. We were able to bring our past lessons learned into this design phase. When we were behind schedule, we called the project manager for help. When I didn't know something, I immediately reached out to a colleague with more experience. Lastly, we communicated regularly with the client to ensure all the deliverables were known and being provided when they were expected. The relationship with the client has improved substantially, and we look forward to earning their business again in the future. I feel the largest lesson I learned here is to hold yourself and others accountable. Even if it is not your exact role on the project, own the outcome and be proud of your work. Some projects will need a team to meet the deliverable, but if you see something that is running amiss, always take ownership and speak up.

During the design process for this project, there were many pieces I wasn't an expert on, as mentioned earlier. In order to ensure the project is constructible, you will need to learn about new equipment and how it works within a given system. There are several excellent sources of technical information, and one of them is your product vendors. When pricing out new equipment, you will find that many product sales engineers have been in the business long enough to see more unique situations than you. Do not just take information from a cutsheet without verifying its feasibility in the solution you are proposing. For example, if you are looking to replace an air-cooled compressor with a water-cooled compressor, you will have to consider the condenser water requirements. Talk to the cooling tower representative to understand if the existing unit has enough capacity and where you will be able to tie into the system. What ancillary equipment would be needed? Being an energy engineer means you also need to be a design engineer and a proper cost estimator. Never propose a solution that is not feasible or does not consider extra cost to make it feasible. If you aren't sure, check with a manager, a design engineer, or the vendor.

My manager approached me about a year after I came back to work (during the previously mentioned audit) with an opportunity to become a section manager. At my organization, a section manager is responsible for interviewing, assigning projects, supporting professional development, assisting with staffing and utilization goals, and coordinating with project managers on proposals. At the time, I was struggling with postpartum anxiety, and trying to figure out this "work-life balance" business so many talk about. My time at work was already dedicated to technical work and project management work with NYSERDA. Adding this role is something I desperately wanted to begin my role as a leader, but my gut told me it wasn't the right time. Understanding that the opportunity would not be available for long, I decided to take it despite my reservations. My transition to management did not have a specific structure. I was given direct reports to manage and I would ask my manager what I needed to do. He would help me work on proposals and add me to other management level conversations. I was involved in my company's first cohort of their new "Principles of Management" course. This helped me understand how to interact with my team and how to be the best leader I can be. It covered a range of topics on personality types, motivation, and performance management.

There was a lot to cover in my new additional role and, as you may have expected, I was overwhelmed. I tried to be a great manager to my new staff, and it caused my dedication to my current projects to fall below what I felt was acceptable. This was likely a contributing factor to the issues I had in the major energy audit I was working on. How can I be a good leader if I can't also lead by example? I had a hard time finding the time to do everything I signed up for while still maintaining the family life I was enjoying. While my son was that young, I felt as though every minute counted. Everyone tells you how fast they grow up, and it's true. While I'm not interested in stepping away from my career, I do want to keep balance between work obligations and family time. The only advice I can give is to be realistic about expectations of what you will be able to accomplish. Saying "yes" makes you agreeable, not necessarily reliable. Decide what is most important and focus on completing this work most efficiently. Ultimately, I am grateful for the opportunity to become a manager and I'm thankful I stuck it out. The role and responsibilities do get easier over time and more rewarding as you are able to see your direct reports have their own successes with their careers.

Prior to starting a family, I spent much of my extra time involved in professional organizations and charities where I could. I've been a member of the Society of Women Engineers (SWE), American Society of Heating, Refrigerating and Air-Conditioning Engineers (ASHRAE), and Association of Energy Engineers (AEE). Through the local SWE professional chapter, I have held positions as the Outreach and Membership chair. Through the Outreach initiative and coordination with local engineering facilities, we were able to offer presentations for high school girls on the different engineering disciplines, showing them engineering in real time and sparking their interest to pursue a career in STEM. I have volunteered and fund-raised for the Women's Build through Habitat for Humanity and joined many other events for heart disease and Alzheimer's disease. It's a passion of mine to support the community, whether it is inside or outside of engineering. I've scaled back significantly to balance out my time during these precious early family years but plan to go back when my gut tells me it's the right time. It's important to understand what drives you, what inspires you.

What's Next?

For me, my career is still young and there is an incredible amount of opportunity for growth. I am looking to expand my networking and business development skills, as well as my technical expertise. The type of work is changing as technological advances are surging forward. With the effects of COVID on the economy and the environment, more people are focused on both energy (cost) savings and indoor air quality (IAQ). The new question is, "How can I improve the IAQ of my facility while keeping costs down?" I want to be ready to answer this for our clients. How are smart buildings changing the future? How will we get to a carbon-free future? What new renewable technologies are emerging as the next champion to get us there? How will I have time for all this while I'm waiting for my second child to arrive?

As of this writing, the world is starting to open back up as we navigate the post-COVID world and economy. Many things will change. The organization you are involved in will change direction and policies based on the market or needs of the people they serve. Your life may change, willingly or unwillingly. One thing will not change, and that is following your heart. I want to close out my chapter by saying that if you are accountable, determined, and true to yourself and others, you will find success in anything you do. If you are a woman or another minority within engineering reading this, you may have to try harder to be seen, to be heard, and to be valued. But I believe if you stay honest, ethical, and accountable, you will be able to overcome any obstacle. Keep trying, keep failing, keep celebrating the wins, and always keep learning. Make your career a meaningful journey instead of just a job, and you'll find success (Figs. 3 and 4).

Fig. 3 At 2018 World Energy Engineering Conference as a speaker on measurement and verification in Charlotte, NC (October, 2018)

Fig. 4 Family Selfie at Corbett's Glen in Brighton, NY (November, 2020)

Jessica F. Bull, P.E., CEM is a Mechanical Engineer with CHA, a full-service engineering and construction management firm servicing clients all throughout North America. She has over 8 years of experience working with manufacturing, industrial, and commercial customers to identify energy conservation measures, including calculating investment payback from energy saving capital upgrades, operational and maintenance modifications, and renewable solutions. She is a section manager for CHA's Mechanical Energy division, leading energy audits and assisting with subsequent mechanical design opportunities. As a contracted NYSERDA project manager, she has vast experience with technical review of many different energy-related projects completed by other engineering firms.

Jessica received her B.S. and M.Eng. Degrees in Mechanical Engineering from Rochester Institute of Technology in 2014. She is an active member of the Society of Women Engineers (SWE) and the American Society of Heating and Air-conditioning Engineers (ASHRAE). She participates in other organizational and charitable events such as Habitat for Humanity, Association of Facility Engineers (AFE) and Association of Energy Engineers (AEE). Within CHA, she is a lead member of the Inclusion and Diversity Committee and a new member of their Toastmasters chapter. Jessica, her husband, and her two boys (3 years, and 5 months old) live in Webster, New York.

I Never Claimed to be a Lady

Mary Bonk Isaac

The Priming Years

When I graduated from high school in 1973 as the second oldest of a brood of eight, I had no idea what I wanted to be. Playing with dolls was boring because I had several of the real thing to burp and feed and change from the age of three on. My early spare time was spent reading, solving puzzles, and building doll furniture from cereal and shoeboxes when I wasn't tending a sibling.

Life was mostly pretty idyllic in suburbia. Boys played Little League; girls were lucky if a couple of dads got together to coach an informal softball team, which happened the year I turned 12. Girls set the table; boys took out the garbage. Boys became engineers, policemen, and soldiers; girls became stay-at-home moms, teachers, and nurses, including most of the really smart girls who graduated from high school with me. Betty Friedan's book *The Feminine Mystique* [1] was published in 1963, and I read that, plus Susan Collins' *Valley of the Dolls* [2] by the time I hit adolescence between 11 and 12, so I was one confused girl.

My mom had my last sibling when I was almost 10, and I still remember thinking it would be a funny Halloween costume for my mom to dress up as Mary, Jesus' mom; she did not think it was funny. I switched to public school at a new junior high after the diocese raised tuition from a family rate to a "per student" rate. Most of the kids from our neighborhood went to the same school, so we sallied into our teens with a pretty cohesive group, albeit me being somewhat of an outsider due to spending eight years as a "Holy Roller."

M. B. Isaac (✉)
Principal Consultant, HEDGE, San Diego, CA, USA

M. Bailey, L. Shackelford (eds.), *Women in Mechanical Engineering*, Women in Engineering and Science, https://doi.org/10.1007/978-3-030-91546-9_21

The Transformational Years

Although I was good at math and science and my dad was an electrical engineer, no one ever seriously suggested that I consider studying engineering, or if they did, I blocked it out. Engineering seemed too linear and it was the Age of Aquarius! I started playing guitar and singing in high school for folk mass, and my first aspiration was to be a rock star.

I joined choir and enjoyed performing in front of others, so I figured rock star was perfect for me, especially since I didn't have to work that hard at singing and performing, right? By now, I started accepting my female-ness and knew when it played to my advantage and when someone was trying to take advantage of me.

I did have a state Regents scholarship, but I dropped out after a year of liberal arts at a community college to search for myself (and maybe play a little music). I finally settled down in 1976 when I got my first "real" job after being a singer, waitress, caretaker, nurse's aide, and/or maid, at the ripe-old age of 21. My dad mentioned they were looking for "girls" for a job at General Electric (GE) as an apprentice machinist, starting at $11.58 an hour (>$50 an hour today), plus pension and insurance benefits. I applied and got in.

I was one of less than a few handfuls of women who went through the apprentice machinist program in Schenectady, NY, learning everything there was to learn about manufacturing turbine generators before I left to go back to college in 1979. I was one of less than 100 apprentices enrolled in the program at the plant, which would dwindle significantly over the years as the manufacturing landscape shifted.

We were a motley crew, my female colleagues and I; two of us got married and/or pregnant during my time with the program. Most of us attended the vocational program, to save the college slots for the men. There was one woman who lobbied for the college program and she won; I was fine where I was, where expectations were lower.

One of the other machinist apprentices was a real girly-girl; me, not so much. My nails were short, my face naked; she had fake nails and wore full makeup, touching up her lipstick regularly. My hair was in a ponytail; she brought hot rollers to work. I wore work boots; she wore platform shoes with open toes until she was ordered not to. Many of the other women and I thought displaying such stereotyped femininity made it challenging for the men to accept us as equals and told her that over beers. She didn't care.

It became a little harder to squelch our femininity for those of us who got pregnant, especially with regard to physical changes such as morning sickness, thickening body, and hormonal swings. Our jobs involved a lot of physical activity, so when I had to go out on short-term disability in the fourth month of my pregnancy due to complications, it was a first in the program.

I went to full term (he turned 42 this year) and returned eight weeks after having a normal childbirth, which forced me to have to address childcare needs at a time when the choices were very limited. Initially, my mother and mother-in-law helped fill that role, as moms and mothers-in-law like me STILL do in too many cases today.

After I returned to the apprentice program, as part of coursework, I argued that the union include a childcare program in the upcoming contract negotiation, leveraging the community college that offered the apprentice associate degree, and solving the growing childcare crisis. Although they liked the idea, it would "be the first thing the union would give up, so what was the point?" Some things never change.

The few black men in the program and I gravitated together since we both experienced a similar shunning from our white male colleagues and could commiserate about "the Man." One was a superb artist who was in the drafting program; another a radical who was perceived as an activist with a chip on his shoulder. I suppose we were all seen as rebels of one sort or another by the 99% white males around us. We certainly took courage from each other.

In early spring 1979, the Three Mile Island nuclear accident occurred about the same time as a new gas crisis, effectively throwing my work world into a tailspin. Steam turbine and generator orders got canceled, and for the first time, I started seeing 30–40 foot long wind turbine blades in the shop instead of turbine and generator forgings.

I was not adapting well to marriage and a new child with both of us working, and had an existential crisis as the person responsible for so recently bringing a new life into the world at a time when the future was not looking so bright. I missed the creative outlet I had found in music and was feeling burdened with having to care for both a child and a spouse who expected certain traditional things of a wife that I would never be willing to fulfill.

Three Mile Island, plus a 3rd-shift sweeper job staring me in the face, was the trigger to return to school so that I could finish my degree and take care of my son on my own if I had to. Although I was accepted into a music program at Skidmore, I chose engineering at Union College. I chose it because I was concerned that: (a) I was not good enough to make a decent living at music, and (b) I thought I might learn to hate it if it was my lifelong career.

I was not worried about hating engineering because I had never particularly liked it; my dad turned me off to it when younger with his process focus, drilling down, and making me think. The machinists I worked with had respect for very few engineers, so my attitude had nowhere to go but up. I pragmatically considered my degree a stepping-stone in learning self-sufficiency and making a reasonable living while doing it.

Hubby #1 and I separated and I attended school fulltime year-round using scholarships and student loans. Welfare and food stamps covered the basics, along with a great roommate who stepped up to help with childcare, too. I bartered childcare for writing papers, fixing meals, and giving haircuts; when I could, I did music gigs with a partner in upstate New York.

Every quarter I went to the county Social Services office to turn in an affidavit that I was still enrolled and passing. Sitting in that waiting room hearing snippets of sad stories opened my eyes to who the primary recipients of welfare were, and why. It was a humbling experience but made me grateful that I was finding a way out; I figured I paid my benefits back in income tax after just a few years as an engineer.

The Performing and Quasi-Conforming Years

Having been away from any kind of serious coursework for nearly seven years, I averaged a C in my first three courses. I almost quit, but told myself that C just meant I was average. I had to learn to "settle" a lot in those years to keep from losing focus on the future—being less than perfect was a bitter pill to have to chew on from time to time over the years, but you learn how to succeed from your failures.

By the time I graduated, my average was up to a B (for balance), but not good enough for the elite entry-level programs at GE. Utility interest in nuclear power was picking back up and GE's nuclear boiling water reactor business was looking for field engineers for construction, then field service. I passed muster, was rehired, and went to San Jose for a short training program with a dozen others, before heading to one of the last nuclear power plants under construction in the Midwest, Perry Nuclear Power Plant in Ohio.

Since the training program was relatively short and the challenges of taking my four-year-old son with me to northern California felt monumental, he stayed in New York with my older sister who had two other young children. In later years, I would hear from him that his memory was that I was gone for a whole year, but really we were apart only from July until October, when we moved to Ohio.

The next three years in construction as a nuclear field engineer were a blur. I rotated through four different engineering assignments conducted from trailers at the power plant, which had both GE boiling water reactors AND turbine generators, quite the coup. There were initially two female engineers among the 40 or so GE employees on site, and a third joined us later on. My first assignment was under a very outspoken redheaded woman who processed extra work authorizations with the customer; we became best buddies when she learned I could cuss almost as well as she did.

My next assignment was in technical support, preparing drawings, specifications, and procedures to cover the extra work and rework, probably the least interesting, but very important in the nuclear business. Some might say the nuclear business suffers from an excess of such process, but unless it was signed in black ink in triplicate, it never happened.

During the three years I worked at the power plant, my shift morphed frequently, which meant, as a single mom, my childcare arrangements also had to shift at a time when there weren't many choices except church schools or a good network. My youngest sister came out to help me the first summer, which was a life-saver as I settled in to a routine. Over the years, making new childcare arrangements became a recurrent struggle every time I moved to a new assignment.

As a production engineer, I oversaw the installation of the "Inclined Fuel Transfer System," used to transfer nuclear fuel bundles back and forth between the spent fuel pool on the reactor vessel head level and the lower pool for treatment, handling, or storage. I climbed ladders and scaffolding from the top to the bottom all day, approximately seven flights of stairs from top to bottom, with no elevator.

Our work in the lower fuel pool also included mounting sensors to the pool wall. One day there was a big scuffle between trades on site that shut down work. The boilermakers considered any welding in the reactor to be under THEIR contract since it was just a big pressure vessel, but the pipefitters had the contract for all non-pressure vessel pipe welding, while the millwrights contract covered welding things to the pool wall, and the electricians had the contract for all wire-pulling and sensors.

I had been told by one of my boilermaker foremen that their general foreman didn't like me much because I was a girl and pushy to boot. The day of the work stoppage, I was in the bottom of the fuel pool talking to one of the millwrights while we looked at drawings hanging on the wall, which was about 40' deep. I took off my hardhat to get closer to the paper on the wall so I could read a number when there was a loud BANG! within six feet of where we were standing. We both looked up and saw the boilermaker general foreman 30' above, grinning down at us before his face disappeared. On the bottom of the pool was a two-pound stainless steel bolt and a small dent in the floor. The millwrights reported it, but fortunately I never had to work with him. He was just one of too many.

My last assignment was working with a millwright to complete as-built drawings of our work under contract, which took me to all sorts of places in the power plant and showed me how important it was to document changes as you go, especially related to systems that required sign offs. Once I completed the as-built drawings for the second, partially completed unit, it was time for me to move on to a different job or transition to a traveling field engineer.

As a single mom with a seven-year-old, I was not excited about arranging child-care while I was on the road, so I headed back to NY as a turbine service engineer for GE's service shop division, with repair shops worldwide where GE fixed broken turbines, generators, and motors. The service business was positioning itself to focus on upgrading and uprating existing gas turbine fleets using specialty parts with much healthier profitability than steam turbine components.

Over the course of two years, I worked on understanding the nature of (and who to go to for help in solving) typical service issues for gas, steam, and generator equipment, including several months at shops during outage season. My first job after the training program was turbine engineer in Cincinnati, where I applied what I had learned to plan and manage work to support gas and steam turbine generator outages.

Working in Cincinnati was the first time I truly felt unwelcomed as a female engineer. Most of the hourly workers accepted me as qualified once they learned about my machinist experience, but the lower level management team was not quite so ready for me. I learned some hard lessons about the expectations that go along with being a woman in a very gendered environment, not least of which was that I was expected to always behave like a lady, which I frequently failed at. It was in Cincinnati that I first uttered the phrase that is the title of this story.

Outage season is different depending on where you are and occurs during the time of the least energy demand, typically fall or spring. Utilities and industry schedule planned outages for these times versus when demand would be expected to be high because of ambient conditions requiring heating or cooling energy. By scheduling, they can arrange to buy the cheapest power elsewhere to support any base demand normally provided by the equipment on outage.

Forced outages can happen anytime, and every second that the equipment is down runs up cost that wasn't planned for. Just a few months before I left Cincinnati for my next job, we had some equipment in on a forced outage, plus some that was on a planned outage from a different customer. I left the night before thinking that my turbine rotor would be balanced by morning and could be prepped for shipment that day, but when I went out to the shop floor, it was sitting in the same spot as the day before.

I was told by the balance guy that the production planner had told him to do another job instead, so I stormed up to the planner, put my hands on my hips, and said "W_T_F???" He said, "I won't have a lady talk to me that way!" and I said, "Well, I never claimed to be a f*%@ing lady! That's all in your head and you better get used to it!" Apparently ladies didn't cuss in his world or maybe it was the talking back part. He was considered for my job before I came along, so that may have contributed to his pulling rank; but my tolerance for double standards based on legitimized myths, especially related to how women should behave in business, was waning.

My next few jobs at GE all had to do with getting involved with the commercial side of the business, while still keeping a foot in the hands-on side. A position as a marketing specialist taught me some of the basics of commercial selling: learning the product, how to schmooze, and how to ask for the order. I got to travel to Europe and all over the United States to negotiate equipment deals on behalf of my company.

Well, actually, for the first few years, I got to carry the bags of the MEN doing the negotiating, then began doing the negotiating myself, then I started teaching other people how to do it, and finally got the opportunity to be an executive hand-holder just prior to my early retirement. Along the way, my rough edges began to soften and, although many thought my language was still too colorful for a woman, I learned to better control my impulse to express it in public.

In the intervening years, there were a few bumps in the road, the first one work-related just before my 20-year work anniversary and the second just a few years after.

In 1996, five years before Jack Welch, longtime CEO of GE (aka "Neutron Jack" for taking people out of the businesses and leaving the buildings [3]), retired, the energy business was in a bit of a slump. It was immediately prior to the "gas bubble" which peaked early in the new century. The utility power market in the United States had been deregulated (1992), with many private utilities divesting and re-aligning their generation assets under investor-owned utility (IOU) companies.

As the price of gas increased, IOUs and independent power producers (IPP) began somewhat speculatively looking at gas-powered equipment as a cheaper source of power than old, less efficient, coal-fired steam turbine plants, driven by the concept of "spark spread," the real-time difference between the wholesale market price of electricity and its cost of production using natural gas.

Many US-based energy companies had re-focused their investment strategies outside the United States, and I led a commercial team supporting equipment and power plant sales in the Indian subcontinent. Money was tight and the field sales team submitted many proposals that went nowhere. The sales director in India moved on during this period and I threw my name in the hat for his role, but it went to someone I had trained who had less than five years of experience.

Initially, I just worked harder and stuffed my anger down so that I could continue working with the guy I had trained, who was now looking to me to tell him what to do. He did the best he could and we continued to function as well as could be expected as a team, but at 41 years old, the fire in my belly dimmed with that rejection. A few months later, the business issued a call for voluntary job eliminations (VJE) to reduce operating cost, and I applied and was accepted.

I had nearly half a year to find another job, but I was burned out and feeling undervalued, so I nursed my wounds for the first few months. I explored what I REALLY wanted to do with my skills and talents and also played a lot of walk-up golf.

I also stayed in touch with the guy who got the job and knew he needed (and wanted) me back, so when pressed for what it would take to make that happen, I told him WAY-Y-Y more salary and working from home part of the time. He went to bat for me and I returned to essentially the same job I had eliminated six months earlier, only being paid more and feeling more valued. I was reminded by Human Resources upon my return how unusual it was for someone to return to the same role they had eliminated prior, which made me feel both special and cautioned as a person who did not play by the rules.

My life would have been different if that opportunity to take a break hadn't come up or if I hadn't taken it, but I do recall feeling so undervalued and used that I didn't feel like I really had a choice once it presented itself. In retrospect, many of my best life moves have been made taking risks that most would not.

One bump I couldn't plan for, though, was discovering in 1999 that I had a brain tumor pressing on my right optic muscle and nerve.

Fortunately, it was a benign meningioma, which are apparently somewhat common in women in their early 40s, but still required surgery. Preparing for the possibility that I might be different afterward, or possibly not even make it off the table, was a reflective experience. Work was very supportive, and my main legal colleague joked in a key customer meeting just prior to my surgery that "At least we have PROOF (MRI) that she has a brain!" Also very fortunately, I was in good health other than the tumor, so I sailed through unscathed other than a reduced peripheral field in my right eye.

Three weeks after surgery, I celebrated the closing of a key $350 million deal in London, still not 100% recovered, but 'In It To Win It' and not one to miss a party. Whether it was the brilliant health recovery or major deal closing or perhaps the stars aligning, from that point to when I finally retired in 2007, my career at GE took off. Certainly, my lifelong attitude that life is too short to drink cheap beer prevailed, but now I was more driven to do things that made a difference and not spend my time spinning my wheels.

I headed back to the field a year later as the general manager of GE's largest US service sales region. I LOVED that job—I inherited a great team and in four years changed its face from one that was 95% white male to one that reflected all the faces of the consumers our energy customers served. By my next gig, my staff was 35% female or non-white. If someone came to me with a candidate slate, it had to be diverse or they were sent back to the recruiting board. Sales practices like using expense accounts to take clients to gentlemen's clubs for lunch or dinner were abandoned.

Unfortunately, one lesson you learn over a lifetime is not to count on the bosses you love to stick around forever. The first question the new manager (soon to become a VP) asked me at our introductory tête-à-tête was whether I had ever noticed that all of the women in our business who were in the executive band had husbands in lower roles. I hadn't. Husband #2 also worked at GE, in the service side of the nuclear business. For the 25+ years we were married, we cohabited the same spaces only about half the time due to work; eventually, when we started spending 100% of our time together after his early retirement, we realized a mutual incompatibility and divorced.

I was both impressed that my new manager bothered to find out something personal about me and concerned that he had picked something that was not particularly relevant to me, but apparently was relevant enough to him to bring it up. I tucked his comment away, but within a year, I knew it was time to find a new job. Whereas past staff meetings had always been very collaborative and productive, as the numbers got softer, they became ordeals laden with bullying and unprofessional behavior, all modeled by our leader.

One of the things GE tended to do well (or at least frequently) was re-engineer their businesses. When they decided to merge five US sales regions into four, I saw that as an opportunity to move on without major repercussions about why and asked Human Resources if I could start a search for my next role. They probed at Why much more diligently than I expected.

I was in a hard spot: if I called out the abuse being piled on another when it occurred, the bully might turn on me, but if I said nothing in defense of a colleague who was the target of the day and ignored the offensive behavior, where was my sense of right? I told them I just wanted to be away from the situation, to escape what I felt was a "hostile work environment" due to the bullying. I did not realize what those three words would trigger.

He was investigated and my comment determined to be valid; he would get additional "coaching and monitoring," while I went through one final humiliation before escape: I would only be allowed to search outside the department after I went through "The Beauty Contest," where my four male peers and I competed for one of the four reorganized GM roles.

Reading the writing on the wall, several key staff decided to find new roles, some outside GE; I was accused of poisoning the waters and not being a "team player." After not being chosen (as planned, though it still hurt to be rejected), I found a customer quality executive role working out of the same facility as my husband, who was nearing retirement, something I wanted to do before age 60.

I had no idea how many years before 60 it would actually be, proving once again that when opportunity knocks, it pays to listen.

The Re-forming Years

In the engineering design process, new product solutions often come from revamping an existing one. The next opportunity to re-form myself happened when my executive hand-holder job was eliminated after my customer decided to buy equipment from our competitor. Customer quality managers were only assigned to key customers and buying equipment from the "other guy" quickly removed that "key" honor.

At age 52, I jokingly asked the Human Resources person if I could retire, not knowing that once you reach the "magic number" (service plus age: 80 for GE), any benefits get carried through to age 60, including pension and health insurance. They told me they couldn't stop me, but I had to sign a separation agreement.

Taking the package meant that I did not have to worry about being able to afford retirement after age 60, and just needed to figure out how to afford paying bills for the next eight years until I started! Money would be a little tight for a few years between when my husband started his pension and I started mine, but still manageable.

Unfortunately, the bottom fell out of my 25+ year marriage, at the same time as the market started to tank. Hubby #2 and I separated in 2007 and I headed back to grad school remotely for secondary school teaching credentials. My son and daughter-in-law in California needed some childcare help, so I headed West to do both for a few years.

By the time I finished my master's in 2011, I knew I would not have made as good a stay-at-home mom as my son seems to think I should have been, nor did I have the stamina to be a middle school teacher. While the middle school students I taught were mostly good kids, teachers spend more time on social work than teaching, which didn't equate to a stress-less slide into retirement.

As chair of the Society of Women Engineers' (SWE) outreach committee, I was in the perfect playground for research exploring female engagement in STEM fields, so I ratcheted it up a notch and found a PhD program that wouldn't involve moving at Old Dominion University in Norfolk, Virginia with their STEM and professional studies doctorate [4].

While not a groundbreaking dissertation, evaluating the validity of data produced by a survey instrument that SWE used to measure middle school female interest in STEM was both a valuable exercise and opportunity for me to exercise some critical thinking muscles that had gotten flabby while overdosing on *Blue's Clues* and *Tele-Tubbies* as a grandnanny.

The most challenging part of completing my dissertation was synthesizing my experiences—having to prune and trim them (plus MORE reading) into something resembling a word bonsai, compact, and succinct. It took me three tries (and multiple committee chairs) over 14 months to get my prospectus approved [5]. There is definitely something to be said for being on campus versus remote when it comes to academic relationship building.

In terms of personal relationship building, although husband #2 and I had separated in 2007, we stayed loosely connected until our home in North Carolina sold in 2012 and we could finalize our divorce amicably. Having no kids together definitely made it a little easier. By now starved for adult companionship, I joined an online dating site; and just before I deleted my account because it seemed everyone was looking for either a Sugar Mama or a nurse, I saw husband #3's profile and asked if he would meet for a dog walk. He said yes and that was that. While at the county offices for a marriage certificate 14 months later, the clerk asked, "Why him?" I ticked off five reasons: (1) he was healthy, (2) had a good job and his own health insurance, (3) loved dogs, (4) had a great smile, and (5) adored me.

Somehow, husband #3 and I fumbled through my doctoral program, with him nursing, nurturing, and commiserating with me throughout. Since graduation in 2015, although I am now officially on Medicare as of 2021, my mission and passion are still intertwined with several organizations focused on engaging more girls in engineering and technology careers through role models, including SWE and, since 2019, the American Association of University Women, as camp director for their Tech Trek summer STEM camp program in San Diego and virtually across California [6].

In 2020, I planned and piloted a virtual robotics program in collaboration with Qualcomm's Thinkabit Lab [7] to 40 girls who would have attended the in-person camp in San Diego had Covid 19 not occurred. Ten such camps in California (and several elsewhere) went dark in 2020, but in 2021, the pilot camp was scaled up in California to make it possible for 640 girls to be served virtually, with three different weeks of camp in June and July, and over 250 volunteer coaches and volunteers that helped make it possible.

The most challenging aspect of the camp (besides buy-in from branch leaders with an average age of 60+ who were NOT digital natives or particularly techie) was moving a paper process involving minors to online in a very short period of time. Due to the complexity of the application and selection process that had been developed over the last 20+ years since the program was first delivered in California, available camp software programs would have required significant customization to meet our needs.

So I took on designing and creating a NEW system and database using REDCap [8] in serendipitous collaboration with Monica McGill, the founder of CSEd Research in Tennessee and host to our database [9]. At the time, she was looking for people to learn and test REDCap's potential usage in the non-profit world outside of the pure research realm so we were a great test case, especially given that I knew nothing about the tool when I started. The experience has been likened by the AAUW California leadership team to building the plane while flying it, which, while definitely a bumpy ride, got us to the destination with our passengers (minor campers) safe and secure.

In retrospect, my path may appear planned or prescribed, but I had no idea where I was going when it was happening, although I was confident, that I had it under control. What I hope I've left you with is the knowledge that it is OK to not know who or what you want to be or where you want to go in life, and sometimes you need to loosen the reins and sometimes hang tight. Engineering is all about solving problems—if you like and are pretty good at it, you undoubtedly will find it is who you are and that is not a bad thing.

Even if you are a girl.

References

1. B. Friedan, *The Feminine Mystique*, New York: Norton, 1963.
2. J. Susann, *Valley of the Dolls*, New York: Bernard Geis Associates, 1966.
3. Wikipedia, "Jack Welch," 2021. [Online]. Available: https://en.wikipedia.org/wiki/Jack_Welch.
4. Old Dominion University, "Department of STEM Education & Professional Studies," 2021. [Online]. Available: https://www.odu.edu/stemps.
5. M. Bonk Isaac, "Priming middle school females' engagement in engineering and technology," 2015.
6. American Association of University Women , "AAUW Tech Trek," 2021. [Online]. Available: https://www.aauw.org/resources/programs/tech-trek/.
7. Qualcomm Thinkabit Lab, "Qualcomm Thinkabit Lab," 2021. [Online]. Available: https://thinkabitlab.com/program.
8. Vanderbilt University, "About REDCap," 2021. [Online]. Available: https://projectredcap.org/software/.
9. CSEd Research, "About CSEdResearch.org," 2021. [Online]. Available: https://csedresearch.org/about-csedresearch/.

Dr. Mary Bonk Isaac is a passionate practitioner and researcher interested in improving female engagement and retention in engineering and technology through formal and informal communities of practice.

She practiced in the energy business at General Electric (GE) for 30 years with a B.S. in Mechanical Engineering (Union '82), retiring from an executive position in 2006 to start her own company focused on engaging more females in engineering (HEDGE Co.). She has an M.A. in Teaching Technology Education (NCAT '11) and Ph.D. in Education (ODU '15). In addition to SWE, she collaborates actively with several organizations committed to equity in STEM fields' education and assessment, including ASEE, FIRST, AEA, AAUW, and ITEEA.

Her research interests lie in examining the socio-cultural-economic aspects of engaging and retaining females in engineering and technology fields, specifically as it relates to the community of practice of female engineers/potential engineers, including SWE, FIRST, and AAUW Tech Trek. She has volunteered as a SWE sponsor, inspector, and referee annually for FTC and FRC since 2015 in the San Diego region.

She serves on National Science Foundation panels and regularly evaluates proposals for funding, awards, and conference sessions in her work with others.

Severing the Links of the "Gordian Knot": Envisioning Doctoral-Level Engineering Education and Workforce Sustainability as a Key to Environmental Sustainability

Catherine G. P. Berdanier

Trapped Within My Own Gordian Knot

Unfortunately, engineering has a "look" that often drives some people, particularly women and marginalized racial and ethnic groups in the United States, away from even considering it as a field of study. The face of engineering as predominantly white, male, and heteronormative persists. This imagery takes root in our psyches early. Researchers interested in K-12 engineering education use a "Draw an Engineer" prompt to capture these dominant narratives: One study [1] showed that when prompted to draw an engineer, only 18% of elementary schoolchildren (1st–5th graders) drew women: Of the 400 children studied, no male students in grades 1, 2, 4, or 5 drew female engineers (three male 3rd graders, however, did). This "default" image isn't confronted as students move through their education and higher-level math, science, and engineering teachers tend to be male, nor is it confronted in college, where math, science, and engineering professors—especially in advanced technical electives—are nearly all male. Depressingly, every single one of my women engineering faculty friends can count on <u>one hand</u> the number of women professors they've had—in over a *decade* of formal training (college, graduate school, and any postdoctoral study). Most of us have never taken an engineering course from a professor of color. Diversity, equity, and inclusion initiatives at all levels of STEM promote the persistence and retention of women and minoritized racial and ethnic groups for several reasons: first, because we are compelled to confront injustice and underrepresentation from a moral and ethical standpoint, and also because new and different voices and experiences can help the discipline of engineering confront problems from different points of view [2].

Growing up, I equally loved English, history, Spanish, and science. My rural high school didn't have advanced math or engineering classes, but I did take

C. G. P. Berdanier (✉)
Department of Mechanical Engineering, Penn State University, State College, PA, USA
e-mail: cgb9@psu.edu

M. Bailey, L. Shackelford (eds.), *Women in Mechanical Engineering*, Women in Engineering and Science, https://doi.org/10.1007/978-3-030-91546-9_22

chemistry, a subject with which I promptly fell deeply in love. I decided to major in chemistry and Spanish for my bachelor's degree. Engineering wasn't an option: The university I chose didn't have an engineering program. Only through an internship did I realize that I was supposed to be an engineer! I transitioned into aeronautical engineering for my master's degree, which was an uphill battle in more ways than one. First, I had never worked so hard to be *average* at coursework—there were many courses where I had to make up lost time and learn undergraduate material on my own, while simultaneously learning graduate material. Second, I was suddenly plunged into the culture of engineering, which struggles with toxic masculinity [3]. Within engineering subdisciplines, aerospace engineering is among the most male-dominated. If the climate of engineering is generally "chilly" toward women and minoritized populations, then the climate at my time in my laboratory was *arctic*.

The climate in my graduate program directly caused me to question whether I could or should stay in engineering. To exemplify, I'll tell a couple stories. The first is pathetically funny—I have many others that are not. As a new graduate student (freshly twenty-two years old, the only woman in the lab and the youngest student by several years), I was working one morning in the lab and needed to join two pipes together. I didn't know the name of the fitting I was looking for, so I asked a senior colleague for some help, to determine whether we had one in the scrap heap or if I needed to order a new piece. He fumbled for words for several minutes. He and I went around in circles as I kept describing the piece that I needed. Finally, embarrassed, he called the piece what it is: A "nipple"—a segment of pipe that is threaded on both ends to fit two "female" ends of a pipe together. The very definition of equipment as "male" and "female" is deeply engrained into technology, and the fact that my colleague was so embarrassed to tell me the actual name of the piece that he wasted 15 minutes of both of our time demonstrates the gendered nature of STEM. In this case, the very *language* used to name parts served as a gatekeeper to prevent me from being able to do my job. As I tell this story, I'm rolling my eyes. Another time, I was introduced to an important industry sponsor touring the lab as "the token girl" (apparently, I had no name). At the time, I wore that tokenization as a badge of honor, a sign that I was the only girl who could 'hack it' in the lab. Now I'm horrified that this statement—a patronizing verbal pat on the head—was apparently meant to be a compliment.

Handling the gendered and hostile language prevalent in the lab was emotionally taxing. The favored insults for infractions both big and small were misogynist slurs for women, and these verbal assaults were often launched in quick succession, ricocheting around as if in a pinball machine. At first, attempting to fit in, I laughed along and probably used that language myself…but with each passing month, I realized I could no longer be complicit. I began gently questioning those words, stopped laughing at "jokes," and immediately became extremely unpopular, to the detriment of my work. I began getting deliberately left out of work meetings about resource allocation and timelines that affected my research productivity.

Over time, the severity of offences seemed to ratchet up, or maybe I was just more aware: I vividly recall a new (woman) student coming to tour the lab. The instant she left, the guys greedily embarked on a crude review of her physique.

Another time, a woman faculty candidate was touring the lab, obviously dressed formally for her interview. One of her heels got caught for a split second on a grated section of flooring, and the male PhD students scoffed for days how "women couldn't be experimentalists if they didn't know how to dress for the lab" (despite the fact that this woman was and is well-known for her expertise). These types of comments were always followed by chest-thumping commentary about how the percentage of women faculty in each department should be capped at the percentage of women students—a disastrous proposition. As the youngest lab member with already-dismal levels of social capital, my arguments were dismissed.

People ask me if I reported these incidents. After I earned my master's, I did talk to administrators at the request of a few other women who had their own experiences to report. To my knowledge, no penalties resulted. In the thick of things, I didn't have the self-worth to fight the long battle with administration out of fears that if I were to have brought them to the attention of administrators as they occurred, I would seem petty, or worse: "sensitive." I also anticipated it would have wasted a significant amount of my research time to file and contest all the complaints. Events like these often fall into legal "gray" areas and underreported, because both ambient and direct gender harassment events are difficult to prove. However, it is optimistic that in the past few years these issues have drawn the attention from the National Academies [4], which synthesized decades of research in academic settings, documented the lasting harm of climate and harassment issues, and legitimized these often-invisible and stigmatized issues to administrators.

In retrospect, I was in a very dark place, though I didn't identify it at the time. I had a persistent pit in my stomach, developed an eating disorder, and struggled to convince myself to go to the lab. I should have been referred to counseling services. I credit the graduate Women in Engineering Program, my roommate Lucie, and my now-husband, Reid, for helping me gather scraps of emotional well-being and providing me with community in those lonely years. In a moment of clarity, I realized that no matter how much I loved my technical research, for my own well-being, I couldn't spend another four years in that lab to earn a PhD.

I'm going to admit something here that puts me in a vulnerable position in terms of academic merit (which is based on boastful and deceptively performative "perfection.")

Can you keep a secret?

The secret is that I left my plan of study. I got a degree on the way, yes, but I left my plan, which was originally to continue through to a PhD. I earned a master's degree, and that's what counted to the university… no one cared how I got there, or how far I could have gone if things had been different. But I, Catherine Berdanier, am an attrition statistic.

Cutting the Links of My Gordian Knot

As I was writing my master's thesis, I looked for alternate paths to a PhD. Earning a terminal degree was important to me for several reasons. First, I thrive on autonomy, so holding a PhD increased the chance that I'd be able to oversee my own directives in any future career. Second, I really loved research and the creation of new knowledge but wished for an academia where all students felt they could excel. To break the system, I had to work in the system! Lastly, and perhaps most importantly, I never wanted some guy to be "more qualified" for a position than me because of some letters at the end of his name.

I fatefully stumbled into the research field of engineering education and found my calling and life's work. In stark contrast to the climate in my master's program, I was welcomed with open arms. My first interaction with an engineering education professor was profound: I needed special permission to enroll in a class before officially being accepted to that program. In response to my request, Dr. Brent Jesiek wrote back one fateful line, something to the effect of: "Catherine, you may enroll. Welcome to Engineering Education! I look forward to having you in class."

I was floored: I had been socialized into the false connection between rigor and exclusion for so long that the feeling of *in*clusion broke over me like a wave! That little tiny email was no longer than a terse, impersonal email would have been, and "cost" him nothing at all, not even extra typing time…but its warmth changed my trajectory.

His email welcoming me to the discipline was prophetic, as, indeed, I found "my people"—engineers who love both the discipline and the humans who make the discipline. Engineering education researchers love engineering so much they devote their careers to studying and improving the educational system. In engineering education, I took classes on the sociological and psychological aspects of education, exploring how the United States' struggles with race, class, and gender manifest within the culture of engineering. My eyes were opened as I studied rigorous research, methods, and theories that reveal underlying issues in engineering education and how to improve them. As an example, even if our studies are not explicitly about gender or race, the understanding that engineering is a heavily gendered and raced discipline is often interwoven in the literature review and discussion sections of our publications. I loved this true systems-level perspective, and I committed to develop an expertise in studying the systems of graduate-level engineering. Under the mentorship of my esteemed advisor, Dr. Monica Cox, I developed a new sense of agency, purpose, expertise, and the moral compass and strength to work for change. I developed the methodological toolsets to funnel my past experiences with injustice into a research agenda that confronts serious issues in graduate engineering education.

In retrospect, it was fitting that in the process of forming my ideas about the inextricable links between graduate education and workforce sustainability (though I didn't identify these processes as such at the time), I also had severed the links of my own Gordian Knot and escaped a dysfunctional trajectory.

I had a radically different experience in my PhD than my master's degree, with a cohort of supportive colleagues who are now still my closest confidantes. (We have vowed to always share hotel rooms at conferences, even when we're as old as the hills!) Our professors demonstrated and demanded inclusivity in the classroom and in research, and even though I worked harder than I ever had in my life, I had a scholarly mission that supported my God-given calling.

My experiences in my PhD program—starting with that fateful email—reframed my standards of excellence for myself and my colleagues: My professors, who were (and still are!) stellar researchers, also demonstrated how they cared deeply about their students … and their colleagues' students! How transformative! In my mind, these professors were just that much more worthy of awe. I wanted to work even harder for them! This was the first time I had witnessed such care, given that I did not have exposure to many other professors at that time for comparison. Still to this day, I reserve my highest professional respect for academic colleagues across all disciplines and domains who demonstrate technical genius alongside kindness, inclusion, respect, and warmth.

Because the false equivalency between academic merit and absence of emotion had been effectively decoupled for me, I felt "safe" enough to start a family while finishing a PhD. Historically, in academia, women who want families or have babies are often seen as uncommitted to their research. (Across sectors, this "motherhood penalty" is well established in terms of earnings, which correlate with promotions. Conversely, men statistically have a marriage and fatherhood *premium*, with married men and married fathers earning more annually than single men [5].) My husband defended his PhD one week before our first son was born, and in the months following, I finished my dissertation; we went "on the market" for academic jobs, bought a house, and moved across the country to start our current positions at Penn State. When interviewing, we paid close attention to how our potential future colleagues treated one another and used interpersonal climate as a primary criterion to decide which jobs to accept.

I am now an assistant professor of mechanical engineering at Penn State, and happy to report that I am thriving! I teach thermodynamics, the exchange of work and heat in energy systems, at the undergraduate and graduate levels. Unapologetically non-traditional, my research expertise lies in engineering education research, specializing in attrition, persistence, and career trajectories of graduate engineering students. Although graduate school is perceived as a privileged endeavor, graduate students are a highly disenfranchised group in higher education, often stuck in a liminal status between being a student and a university employee. As graduate education systems in the United States are often still based upon antiquated visions of apprenticeship, many aspects of doctoral education outside of coursework are not formalized, and therefore invisible, varied, and highly dependent on timing and access to mentorship. As change agents, my research team and I use my research platform to advocate for disenfranchised groups in engineering, giving voice to those who have left academia or the profession altogether, identifying areas of inequity, and promoting mechanisms for change.

So, sustainability! How does this relate to sustainability?!

Severing the Gordian Knot of Sustainability Through the Lens of Engineering Doctoral Education and Workforce Sustainability

Because of my journey to, through, and nearly out of engineering, I have become convinced that a human-centric education system that sustains a diverse and thriving engineering workforce is the key to accomplishing environmental sustainability.

Related to technical definitions of sustainability, my calling as a professor is to help undergraduate and graduate students in my classes discover their own callings in engineering to solve messy sociotechnical engineering problems, like those related to energy and the environment. In a looser interpretation of "sustainability," I am compelled to help develop and sustain a passionate and inclusive engineering workforce, for populations have been historically underrepresented in engineering.

The two missions are not mutually exclusive: Should our students, especially those with traditionally minoritized voices, be pushed out of engineering research careers either through political or psychosocial pressures, we lose thought leaders in both industry and academia who could drastically influence scientific research agendas, commercial opportunities, and technological innovations that could be leveraged to solve energy and environmental issues. These people also represent thought leaders who could help disrupt disciplinary climate issues in engineering by being advocates, allies, champions, and mentors, or those who might one day rise into administrative roles to help reform policy—perhaps rectifying unhealthy workplace climates and enhancing accountability.

It may seem elitist for my research to focus on the persistence of graduate students—those pursuing master's and PhD degrees in engineering—rather than focusing on elementary school populations or undergraduate students. (Never fear, there are many researchers committed to these populations!) However, I argue that until there is diversity at every tier of engineering—in academia or industry, in administration and leadership—we will not have the chance to hear different and perhaps more creative perspectives.

Without advanced degrees, individuals with good ideas may not be awarded the legitimacy to enact their vision or influence technological progress. This silencing is especially true for women and minoritized groups in STEM, whose accomplishments are often ignored or discounted, seen as "lucky," or resulting only from being from a minoritized population [6]. Attempts to self-promote are perceived as "unfeminine" or "aggressive." As Foor and Walden note, *"[w]omen who wish to answer the call for increased participation in engineering experience a cultural space enmeshed in a web of conflicting threads of possibility and frustration"* [7 , p. 42]. As a complicating factor, in engineering most graduate students are adults in their 20s and early 30s. Therefore, grad school and junior faculty life overlaps with women's healthy reproductive years, which causes some women to question or avoid academic careers [8, 9].

We see the ramifications of a systemic "weeding out" of underrepresented voices in doctoral engineering in the macro-level statistics. True attrition is difficult to capture because it is poorly defined, with many students choosing to leave with a master's degree instead of a PhD, rather than leaving with no degree, such that universities may count these as degree conferrals rather than attrition events. Although the median time-to-degree in the engineering PhD is five years, the ten-year completion rates (so, the numbers of students who have completed their degree in *double* the median time) are estimated to be only 56% for women (compared with 65% for men) and 47% for Black and African American engineering doctoral students [10]. These percentages are large bites out of the already microscopic population pools of women and minoritized students that attempt a bachelor's degree in engineering and then pursue doctoral-level study (estimated to be only 16% women and 13% from all racially minoritized groups across disciplines at the doctoral level together).

In my NSF-funded research, I specialize in using quantitative and qualitative methods to tease apart the factors that can contribute to the attrition of graduate students. While my work studies all doctoral students in engineering, as attrition disproportionately affects women and students from marginalized racial and ethnic groups, I do explicitly keep issues of race and gender in mind.

There are myriad anecdotal myths held by faculty that my research team and I systematically work to disentangle and disprove. The most pervasive of these myths is that any students that leave are not academically prepared or cannot pass doctoral qualifying exams. My team's research has proven that academic coursework is rarely the reason that students choose to leave their programs: In fact, even students with prestigious, nationally recognized, merit-based fellowships consider departing their PhDs. My team's recent research indicates that around 75% of students consider leaving their PhDs at some point in their PhDs. Given this number, if we as faculty trust our admissions standards, then there intuitively must be much more to the attrition factor than academic preparation. Indeed, my group's research reveals deep interconnections between factors related to attrition or persistence, including advisor relationships, a strong support network, development of academic identity, changing motivations and goals, and issues of well-being and mental health. The relative importance of these factors in the attrition calculus is also impacted by a student's prior educational experiences, concurrent experiences, and other social identities such as gender and race.

The second-most pervasive opinion is that students that leave just don't have "it": An amorphous *je ne sais quoi* for success in grad school. Some researchers try to (incompletely) articulate this "it" factor as the ability to fit into the expectations and norms of a department or research group. My research participants overwhelmingly tell a different story: With unclear and unstated expectations or guidance on how to thrive in graduate school, students who have had guidance, mentorship, or role models who (often unknowingly) teach and model hidden competencies seem to have "it." Conversely, students who haven't had sound formative experiences (often those who also come from traditionally minoritized groups in engineering) are playing a game where the rules seem capricious, dependent on the mood of the advisor, and invisible. In a revision of the catchphrase from the old gameshow "Whose Line is it Anyway?" graduate students feel that the rules are made up and the points…well, the points matter a *lot*.

Graduate research is difficult and working doggedly through difficulties toward research breakthroughs is important to the development of independence and expertise. But I argue the hard parts of a PhD should be pushing on the boundaries of science, not working to solve a mystery regarding invisible competencies that are required but not taught. Skills like writing research papers quickly and accurately while anticipating the needs of various academic audiences are infinitely important to graduate students as they work to support their research advisor. But academic engineering writing is rarely taught. (It's not necessary to love writing, just to be effective. But my team's research shows most graduate engineering students avoid writing at all costs and are paralyzed with perfectionism and procrastination when it comes time to write.) Progress in teaching these invisible competencies could greatly improve education for all graduate students but would likely be especially impactful for women and students with marginalized racial and ethnic identities, who have much higher rates of attrition.

Other invisible competencies include self-regulation, discipline, and time management. Few incoming graduate students fresh out of undergraduate programs know how to distribute their effort between multiple projects, including coursework and research obligations (which have few real deadlines, but are infinitely more important than coursework). While an individual's strategies for time management can only be forged in the fire, cluing students into a variety of strategies would be easy. Anticipating an advisor's needs is a related invisible competency, but graduate students are allowed to witness so little of the research cycle that their advisor's demands seemingly come out of nowhere, as previously long-term projects come due and become urgent projects requiring late nights and stress.

My favorite thing about my work is how immediately I can put it into practice. When I give workshops and research talks, I immediately see students calmed when they realize they're not alone, empowered through the "naming" of some of their struggles, and given agency when I articulate practical suggestions to make change. The most validating experience I had was being approached after a workshop by a student who told me that the interconnected model of factors related to graduate student attrition exactly described him, and even articulated some of the issues he hadn't been able to find the words to describe. I also get to be a liaison between graduate students and faculty as I lead faculty workshops to help advisors become better mentors by accurately assessing needs and facilitating the development of invisible competencies.

I anticipate that some of my colleagues might scoff and say, "But this isn't engineering!" I vehemently disagree. Nor are these skills "soft." (I hate the term "soft skills.") These skills, instead, are foundational to the practice of research and the discipline of engineering so students can harness their technical potential. It's true that there is an "it" factor for the ability to excel in graduate school, but we as faculty can't be lazy and make each new generation of students figure "it" out for themselves. The "it" is a combination of these invisible skills, combined with confidence and the sense that they can belong (and want to belong) to the academic community. By articulating and deliberately teaching these invisible competencies, we can provide an even playing field for graduate students who haven't yet been blessed with effective mentors.

My research team also studies the important, but rarely discussed, distinction between graduate students "surviving" a PhD versus "thriving" in their PhD. When universities and departments report degree conferral statistics, they are interested in the number of students who make it to graduation, with no nuance on how emotionally intact and healthy the individuals are that comprise those numbers. If we want to encourage students to consider joining the professorate, they need to want to be in academia! Graduate students are deeply affected by mental health issues stemming from unhealthy work environments that cause isolation, pressure, and overwork [11–14]. In engineering, we generally don't talk about these unsavory emotional things, although they have a direct effect on metrics we do love to talk about, like research productivity [15, 16].

I would also be remiss to not discuss how the "real world" affects students: We don't yet know the long-term impact of COVID-induced isolation and research delays on graduate engineering student populations. It is plausible that the concurrent flares in racism, white nationalism, and anti-immigrant policies and sentiment that affect both our domestic and international engineering students will compound the already fragile issues of isolation and mental health. Only time and research will tell the extent to which these factors affect our future thought leaders.

The disturbing (and often invisible) attrition statistics influence the United States' ability to develop a sustainable workforce of diverse PhD-holding engineers and impair pathways to the professorate that serve as technical thought leaders and role models. Indeed, only about 14.1% of mechanical engineering faculty identify as women, and only 2.6% and 3.3% of mechanical engineering faculty identify as Black/African American and Hispanic, respectively. Only 0.3% (yes, that's right, a third of a percent!) of *all* tenure-line engineering faculty in *all* engineering disciplines identify as Native American [17]. (As a note, the disaggregated demographic data by gender, race, and discipline *cannot be reported* because the numbers are so tragically low they would be immediately identifiable.) The feedback loop of underrepresentation between faculty and students is self-sustaining.

I yearn for the day where all engineering students—both undergraduates and graduate students—would have professors who are women with kids, women without kids, women who perform femininity in all types of ways, women who enjoy being nurturing, women who are blunt, and women from a variety of ethnic backgrounds... rather than having a singular image of a woman faculty member to accept or reject. I yearn for the day when the system of higher education is equitable and overturns the notion that graduate students must be academically hazed to "survive" a PhD. I yearn for the day that all incoming graduate students feel supported, welcomed, and coached in the skills that are critical to thriving in research, and I yearn for the day that all engineers with good ideas to foster our planet's sustainability will be heard, considered, and appreciated.

These visions sustain me, even though my work is hard and oftentimes sad. I carry the stories of my research participants along with me into every committee meeting and university initiative with which I'm involved, letting their experiences guide my decisions. Through support from my friends and mentors who continually champion me (including those who are also telling their stories in this book), a lot of hard work, a bit of good timing, privilege from the family into which I was born, and God's grace, I am positioned now to join the generations of women and scholars before me who have worked to push the boundaries of their respective times and situations.

I acknowledge that I hold a radically different perspective from many other engineers in thinking about graduate-level workforce development and education as a core element of sustainability in mechanical engineering. It's a disruptive thought, but I couldn't care less. Only through an upheaval of old ideas on *what* engineering is and *who* does it can we create a thriving and agile profession that can adapt to the technological needs of the future. We need to upend an ongoing climate of exclusion to keep equity- and justice-minded future thought leaders in the engineering pipeline through to the doctoral degree and beyond, who can be forces for systemic "climate change" to make engineering more inclusive while piloting their visions for science. My scholarly work studying graduate engineering education, attrition, and persistence is personal salve on my own wounds while also being forged armor for those who are fighting against the exclusivity of engineering. I proudly stand on the shoulders of other "disruptors" to make change for future generations of scholars. I maintain that there's no way to disentangle the Gordian Knot of Sustainability—workforce or otherwise—without being delightfully, unapologetically, passionately disruptive!

Would you like to join me?

References

1. B. M. Capobianco, B. M., H. A. Diefes-Dux, I. Mena, and J. Weller, J, "What is an engineer? Implications of elementary school student conceptions for engineering education," *Journal of Engineering Education*, vol. 100, no. 2, pp. 304–328, 2011.
2. K. Beddoes, "Engineering education discourses on underrepresentation: Why problematization matters," *International Journal of Engineering Education*, vol. 27, no. 5, p. 1117–1129, 2011.
3. R. A. Miller, A. Vaccaro, E. W. Kimball, and R. Forester, "'It's dude culture': Students with minoritized identities of sexuality and/or gender navigating STEM majors," *J. Diversity in Higher Educ.*, 2020. doi: http://dx.doi.org/https://doi.org/10.1037/dhe0000171.
4. National Academies of Sciences, Engineering, and Medicine, "Sexual Harassment of Women: Climate, Culture, and Consequences in Academic Sciences, Engineering, and Medicine," Washington, DC: The National Academies Press, 2018 . Accessed: Jun. 8, 2021,[Online]. Available: https://doi.org/10.17226/24994
5. C. Juhn and K. McCue, "Specialization then and now: Marriage, children, and the gender earnings gap across cohorts," *Journal of Economic Perspectives*, vol. 31, no. 1, pp. 183–204, 2017.
6. M. Cabay, B. L. Bernstein, M. Rivers, and N. Fabert, "Chilly climates, balancing acts, and shifting pathways: What happens to women in STEM doctoral programs." *Social Sciences*, vol. 7, no. 2, 2018.

7. C. E. Foor and S. E. Walden, "'Imaginary Engineering' or 'Re-imagined Engineering': Negotiating gendered identities in the borderland of a college of engineering," *NWSA journal*, pp. 41–64, 2009.
8. S. Thébaud and C. J. Taylor, "The specter of motherhood: Culture and the production of gendered career aspirations in science and engineering," *Gender & Society*, 2021, doi: https://doi.org/10.1177/08912432211006037.
9. W. M. Williams and S. J. Ceci, "When scientists choose motherhood: a single factor goes a long way in explaining the dearth of women in math-intensive fields. How can we address it?" *American scientist*, vol. 100, no. 2, pp. 138–145, 2012.
10. R. Sowell, "PhD completion and attrition: Analysis of Baseline Data." *Washington*, DC: Council of Graduate Schools, pp. 1–23, 2008.
11. Council of Graduate Schools and The Jed Foundation, "*Supporting Graduate Student Mental Health and Well-being: Evidence-informed Recommendations for the Graduate Community,*" 2021. Accessed: Jun. 8, 2021,[Online]. Available: https://cgsnet.org/ckfinder/userfiles/files/CGS_JED_Grad%20Student%20Mental%20Health%20Report.pdf
12. National Academies of Sciences, Engineering, and Medicine, "Mental Health, Substance Use, and Wellbeing in Higher Education: Supporting the Whole Student," Washington, DC: The National Academies Press, 2021. Accessed: Jun. 8, 2021,[Online]. Available: https://doi.org/10.17226/26015
13. National Academies of Sciences, Engineering, and Medicine, "Graduate STEM Education for the 21st Century," Washington, DC: The National Academies Press, 2018. Accessed: Jun. 8, 2021,[Online]. Available: doi: https://doi.org/10.17226/25038
14. T. M. Evans, L. Bira, J. B. Gastelum, L. T. Weiss, and N. L. Vanderford, "Evidence for a mental health crisis in graduate education." *Nature Biotechnology*, vol. 36, no. 3, pp. 282–284, 2018.
15. M. Lohela-Karlsson, L. Nybergh, and I. Jensen. "Perceived health and work-environment related problems and associated subjective production loss in an academic population," *BMC Public Health*, vol. 18, no. 1, pp. 1–10, 2018.
16. J. Sheridan, J. N. Savoy, A. Kaatz, Y. G. Lee, A. Filut, and M. Carnes, "Write more articles, get more grants: the impact of department climate on faculty research productivity." *J. Women's Health*, vol. 26, no. 5, pp. 587–596, 2017.
17. American Society for Engineering Education, "*Engineering by the Numbers*," 2019. Accessed: Jun. 8, 2021,[Online]. Available: https://ira.asee.org/by-the-numbers/engineering-faculty/

Catherine G.P. Berdanier is an Assistant Professor and Clyde W. Shuman Jr. and Nancy Shuman Early Career Professor of Mechanical Engineering at the Pennsylvania State University and is the Director of the Online MSME Program. She earned her B.S. in Chemistry from the University of South Dakota, her M.S. in Aeronautical and Astronautical Engineering and Ph.D. in Engineering Education from Purdue University. She directs the Engineering Cognitive Research Laboratory (E-CRL), which focuses attention on graduate-level engineering education and methodological development within a disciplinary setting. Her research interests include graduate-level engineering education including doctoral student attrition and persistence, postdoctoral engineering education, and academic engineering writing. Her research has been published in *Journal of Engineering Education*, *International Journal of Engineering Education*, *IEEE Transactions on Professional Communication*, and many other journal and conference venues. She is the recipient of several National Science Foundation grants in engineering education research and is the winner of National Science Foundation CAREER Award studying master's-level departure from the engineering doctorate.

Appendix

The AdvanceRIT (NSF ADVANCE 1209115) program's **goal** is to increase the representation, retention, and career advancement of women faculty in STEM. The research team examines the unique challenges experienced by women faculty of color and deaf and hard of hearing faculty and refines interventions to address the needs of these key subpopulations. The AdvanceRIT program influences long-term changes that transform RIT's culture, promote inclusion, and expand the representation of women on campus and among leadership.

Program objectives are to:

- Refine and strengthen institutional structures
- Improve the quality of women faculty's work-life including reward structures
- Align systems of power and resources to support and sustain progress
- Enhance the working environment using symbolic measures that emphasize issues of meaning and significance within the organization

Prior to Advance, the capacity to discuss work environment matters, specific but not limited to, inclusion, diversity, conflict resolution, behavior issues, and open dialogue, was very limited. In leading the AdvanceRIT effort, Professor Margaret Bailey, P.E. and the research team facilitated both cultural and environmental transformation initiatives and interventions. The group collaborated effectively with the faculty and all levels of the administration using a multifaceted approach addressing political, structural, symbolic, and human resource aspects of the organization.

In 2018, RIT established a permanent ADVANCE office in Academic Affairs with a director and staff. The project influenced long-term change that transformed RIT's culture, promoted inclusion, and expanded the representation of women on campus and among leadership. Key achievements are below.

M. Bailey, L. Shackelford (eds.), *Women in Mechanical Engineering*, Women in Engineering and Science, https://doi.org/10.1007/978-3-030-91546-9

Key Project Achievements

1. ***Promote Faculty Advancement and Research Success***

(a) Created ***Connect* Grants** program to support leadership and career development for all tenured and pre-tenured faculty at RIT by funding faculty projects consistent with the goals of AdvanceRIT. Seven rounds of grants have been awarded ($330K overall with average award of $5.7K) since 2014 with 58 awards **supporting 111 unique faculty members**. Dozens of scholarly products resulted including funding proposals, publications, short animations, and films. All of the RIT colleges have had *Connect* Grant awardees; 35 grants went to STEM fields, 11 went to SBS fields, and 12 went to other fields.

The *Connect* Grant program has resulted in strong impacts for the grantees of the programs as well as the broader RIT community, based on an evaluation of the program. Individual impacts are related to improving women's work-life, professional development, career advancement, and working environment. In particular, *Connect* grantees strengthened internal and external networks which led to career advancement, benefitted from mentoring relationships as mentors and mentees, built skills (including leadership skills) to support their professional development, increased their research autonomy and project visibility, increased their confidence in disseminating and being experts of their work, increased their value and influence, and supported the advancement of women and other underrepresented groups in their field, through their grant projects. Organizational impacts of the *Connect* Grants at RIT are less common than individual impacts but are related to refining institutional structures and installing practices to promote representation and advancement of women faculty. Some grantees worked to transform their institutional environment by strengthening community and raising internal awareness of their work and created new learning and leadership opportunities that helped to modify institutional policies, practices, and procedures. Examples of past funded efforts include projects focused within the following four areas:

Networking

- Strengthen faculty's professional visibility in national/international networks.
- Utilize social resources to support the creation/expansion of networks among RIT faculty and within professional disciplines.

Research

- Support research proposal development or re-submission efforts for external funding.
- Expand research and/or writing capabilities to enhance competitive proposal development and submissions.
- Promote interdisciplinary collaborations among faculty internal and external to RIT, which could include formal peer-to-peer reciprocal visits to mutual institutions.

Professional Development

- Sponsor and coordinate a visiting scholar's colloquium series.
- Present research at national/regional conferences.
- Specialized training to support scholarly collaborations.
- Engage the assistance of an external career/leadership coach.

Community Building and Engagement

- Unit-level efforts focused on recruiting and retaining exceptional faculty from broadly diverse backgrounds.
- Efforts designed to support organizational development.

(b) Created the ***Connectivity* Series program with 189 professional development events** offered since the program's inception in 2013. Offerings include interactive sessions focused on unconscious bias education, creating an inclusive academic environment, bystander awareness and action, and the recruitment, retention, and advancement of a diverse faculty. The *Connectivity* Series has reached 1086 **unique individuals** (628 women and 458 men). The tenured or pre-tenured (TT) faculty attendees represent **83% of women TT faculty** and **47% of men TT faculty**. The *Connectivity* Series is intentionally inclusive. All RIT faculty are invited to sessions as well as staff who work closely with students. On average, 94% reported that attending events was a valuable use of their time.

The Women of Color (WoC) and Deaf and Hard of Hearing (DHH) *Connectivity* Series' events comprised about 34% (64/189) of all mapped *Connectivity* Series events over the grant period. Faculty in each population lead these events, and content for each series was developed based on quantitative and qualitative research conducted by RIT researchers.

(c) Establish mentor access initiatives and the faculty-driven creation of mentoring clusters through **Mentoring Workshops** for over 200 faculty participants. Piloted multi-level **Women Leadership Development Program** utilizing external and internal facilitators and experts as well as best practice benchmarking.

(d) Address length-in-rank issue at the associate professor level through detailed data analysis. Developed a white paper and a supporting faculty-driven program called **Promotion Package Preparation (P^3) Group**, which is a series of workshops created to address this long-standing issue. Over 100 faculty have participated in a P^3 Workshop since 2016.

2. ***Recruit and Retain Top-Tier Faculty – Focus on Culture Change Efforts –* Following Programs Institutionalized**

To support faculty retention, we focus on **culture change initiatives**. Through carefully crafted workshops, AdvanceRIT has led efforts that promote culture change toward a more inclusive and vibrant learning and work environment. This culture aligns with our university's core values and strategic plan and will enable RIT to attract, retain, and advance top-tier faculty. To date, **AdvanceRIT has hosted 50 workshops on enhancing and improving campus culture** including unconscious bias education workshops, bystander awareness workshops using interactive theater, and ground rule discussions. **Audiences have reached 850 unique participants (59% women)** and have included campus leaders, members of RIT's

Academic Senate, promotion/tenure committees, staff, entire departments and colleges, senior design coaches/sponsors, undergraduate and graduate students, department chairs, and faculty.

Faculty retention efforts have resulted in the development of *Connectivity* Series for Women of Color, the successful **P&T SMARTS** program designed for faculty of color with broad offerings to all faculty, the recently launched **allofus@RIT** organization which hosts events that engage thoughtful campus change agents, and the successful ***Connectivity* Series for Deaf and Hard of Hearing Women Faculty** which has been expanded to include offerings for all members of the National Technical Institute for the Deaf.

3. ***Policies, Practices, Structures, and Supporting Research***

The project team worked closely with key administrative partners on faculty evaluation, compensation, and data analysis. The team developed an important knowledge base of topics through the creation of **tip sheets** and **white papers** reflecting relevant research, benchmarking, and faculty data.

Notable policy changes include the complete rewrite of RIT's tenure policy E05.0 that incorporates an automatic tenure-extension provision; conducting benchmarking research that informed expansion in 2015 of the parental leave benefits to all university members; conducting research to inform campus dialogue concerning key topics while considering work-life integration.

4. ***Institutional Collaboration***

The university's newest Strategic Plan integrates several elements related to AdvanceRIT. The team engaged in productive collaborations resulting in new institutional practices, such as **faculty exit interviews**; **COACHE climate survey**; **dual career program**, **gender-equity salary study**; and **NSF Indicator data**. The project has consistently worked toward increased transparency regarding resource allocation through the creation of the **Resource Allocation Committee**.

Project Evaluation

Achievements based on <u>external evaluation</u> *(2018, Laura Kramer and Alice Hogan, Independent Consultants who served as project external evaluators)*

- AdvanceRIT has been a model of the organizational agility that is part of the university's strategic plan as evidenced by the project's nimble adaptation to changing circumstances.
- Greater integration of faculty/administrators on cross-functional teams, building a more transparent culture.
- Further development of unconscious bias education for RIT faculty focuses on intentional cultural shift.
- Significant grassroots engagement in AdvanceRIT's work over the past years has added positive and amplifying energy to the program, helping to demonstrate the

ongoing necessity and value of the work, while engaging a wider group of RIT faculty.

- Significant policy and practice changes in support of managing work-life integration.
- In terms of the institutionalization of project initiatives, many have been absorbed within the function of Human Resources and Academic Affairs such as faculty exit interviews, salary equity studies and dissemination, and dual career assistance program.
- Based on recipient feedback, the *Connect* Grant initiative is a valuable addition to the array of internal resources for faculty development.
- The AdvanceRIT programming has had a positive effect on senior women in STEM, who appreciate the attention to networking and the changes in awareness of issues among the faculty and administration.
- Important faculty development and networking opportunities have grown from the project's social science research strand. Efforts focused on deaf and hard of hearing women faculty and faculty women of color have each produced important insights about the particular needs of women faculty in these subgroups.

Achievements based on internal evaluation *(2018, Elizabeth Litzler, CERSE Director from University of WA who served as project internal evaluator)*

- **Increased Awareness** was an impact of all of the AdvanceRIT offerings. *Connectivity* Series events often built awareness about bias and how to be proactive at intervening against discriminatory acts. In addition, *Connect* Grants helped to build people's self-awareness of their skills and improved self-confidence that they could achieve their goals. The RIT Women Leaders programs also increased awareness, but in different ways. The RIT Women Leaders program had significant and meaningful impacts on the self-awareness of participating individuals.
- **Increased Action** occurred because of some of the AdvanceRIT grant programs. In particular, the unconscious bias-related workshops often increase attendees' intention to change their behaviors or practices on campus and in the classroom. *Connect* grantees themselves often took additional actions on their research or for the good of the community as part of their grants; sometimes this resulted in broader community action as well.
- **Sense of Community** grew during the AdvanceRIT grant. *Connectivity* Series attendees were able to connect with other women to form new networks. These networks provided mentoring, learning, and sometimes collaboration. *Connect* Grants helped to improve working environments and women's work-life and strengthened women's internal and external networks. The visibility of the women of color and deaf and hard of hearing faculty events had an institutional impact by raising the internal visibility of the groups and strengthening the community. The RIT Women Leaders program had positive impacts on the sense of community for most of the participants.
- **Career Progression and Development** was a key impact of the *Connect* Grants and the *Connectivity* Series. *Connectivity* attendees reported progress toward

promotion and tenure, and finding other researchers with similar interests, which improved productivity. *Connect* grantees saw career-related benefits from mentoring relationships, increased leadership skills, research autonomy, and visibility. Many *Connect* grantees also described a sense that they now had more value and influence with the university.

- **Organizational Impact** – There is an untold story of the work that was done to make these AdvanceRIT events and programs happen. Many of the interviewees described that they were thankful to RIT broadly for providing these programs and events; AdvanceRIT was not always named by the interviewees or directly cited for this work. This suggests that AdvanceRIT has helped to create an environment in which people feel "cared for" by RIT at large. The many hours that went into planning and administering both the *Connect* Grants and *Connectivity* Series were repaid in terms of gratitude to RIT broadly. AdvanceRIT should be given kudos for its work in improving the university climate.

A list of project related publications at https://www.rit.edu/nsfadvance/.

This material is based upon work supported by the National Science Foundation under Grant No. 1209115. Any opinions, findings, and conclusions or recommendations expressed in this material are those of the author(s) and do not necessarily reflect the views of the National Science Foundation.

Index

M. Bailey, L. Shackelford (eds.), *Women in Mechanical Engineering*, Women in Engineering and Science, https://doi.org/10.1007/978-3-030-91546-9

D

MIX
Papier aus verantwortungsvollen Quellen
Paper from responsible sources
FSC® C105338

If you have any concerns about our products,
you can contact us on
ProductSafety@springernature.com

In case Publisher is established outside the EU,
the EU authorized representative is:
Springer Nature Customer Service Center GmbH
Europaplatz 3, 69115 Heidelberg, Germany

Printed by Libri Plureos GmbH
in Hamburg, Germany